Great Mathematicians

Great Mathematicians

SALEM PRESS
A Division of EBSCO Publishing
Ipswich, Massachusetts

Cover Photo: René Descartes © Leonard de Selva/CORBIS

Copyright © 2013, by Salem Press, A Division of EBSCO Publishing, Inc.
All rights reserved. No part of this work may be used or reproduced in any manner whatsoever or transmitted in any form or by any means, electronic or mechanical, including photocopy, recording, or any information storage and retrieval system, without written permission from the copyright owner. For permissions requests, contact proprietary-publishing@ebscohost.com.

ISBN 978-1-4298-3789-7

Printed in the United States of America

Contents

List of Contributors	vii
Niels Henrik Abel	1
Maria Gaetana Agnesi	4
Alhazen	6
Apollonius of Perga	10
Archimedes	13
Āryabhaṭa the Elder	17
Friedrich Wilhelm Bessel	20
Max Born	23
Brahmagupta	25
Gerolamo Cardano	28
Lewis Carroll	30
Jean le Rond d'Alembert	34
Richard Dedekind	37
René Descartes	40
Diophantus	44
Euclid	47
Eudoxus of Cnidus	50
Leonhard Euler	53
Pierre de Fermat	56
Joseph Fourier	59
Gottlob Frege	62
Évariste Galois	65
Carl Friedrich Gauss	68
Sophie Germain	71
Kurt Gödel	73

James Gregory	76
Sir William Rowan Hamilton	78
David Hilbert	81
Hipparchus	84
Sofya Kovalevskaya	86
Joseph-Louis Lagrange	89
Pierre-Simon Laplace	93
Gottfried Wilhelm Leibniz	96
Leonardo of Pisa	101
Nikolay Ivanovich Lobachevsky	103
Colin Maclaurin	106
Benoit B. Mandelbrot	109
Gaspard Monge	112
John Napier	116
John von Neumann	119
Simon Newcomb	123
Omar Khayyám	125
Pappus	128
Blaise Pascal	131
Émile Picard	134
Henri Poincaré	136
Pythagoras	139
Bertrand Russell	143
Charlotte Angas Scott	146
Seki Kōwa	148
Jakob Steiner	151
Simon Stevin	153
Alan Mathison Turing	155
John Wallis	158
Alfred North Whitehead	160
Norbert Wiener	164
Resource Guide	**169**
Index	**173**

LIST OF CONTRIBUTORS

William Aspray
Charles Babbage Institute

Scott Bouvier
California State University, Los Angeles

Jeanie R. Brink
Huntington Library

Celeste Williams Brockington
Independent Scholar

Judit Brody
Independent Scholar

Victor W. Chen
Chabot College

Patricia Cook
Emory University

Thomas Drucker
University of Wisconsin—Whitewater

Steven I. Dutch
University of Wisconsin—Green Bay

Donald R. Franceschetti
University of Memphis

Michael Craig Hillmann
Independent Scholar

James R. Hofmann
California State University, Fullerton

J. Donald Hughes
University of Denver

Wilbur R. Knorr
Stanford University

James Livingston
Northern Michigan University

Paolo Mancuso
Independent Scholar

Lyndon Marshall
College of Great Falls

Michael McCaskey
Georgetown University

D. Gosselin Nakeeb
Pace University

Robert J. Paradowski
Rochester Institute of Technology

Mark Pestana
Grand Valley State University

George R. Plitnik
Frostburg State University

Thomas Rankin
Concord, California

Joseph Rosenblum
Independent Scholar

Roger Sensenbaugh
Indiana University

R. Baird Shuman
University of Illinois at Urbana-Champaign

Genevieve Slomski
Independent Scholar

Roger Smith
Independent Scholar

Joseph L. Spradley
Wheaton College, Illinois

Paul Stuewe
Green Mountain College

Charles R. Sullivan
University of Dallas

Daniel Taylor
Bethel College

Frank Wu
Independent Scholar

Clifton K. Yearley
State University of New York, Buffalo

Ronald Edward Zupko
Marquette University

Niels Henrik Abel

Norwegian mathematician

Abel was a central figure in the evolution of modern mathematics, especially in the field of algebra. Regarded as one of the foremost analysts of his time, he insisted that a rigorous approach to mathematical proof was critical for the further development of abstract mathematics.

Born: August 5, 1802; Finnøy, Norway
Died: April 6, 1829; Froland, Norway

Early Life

Niels Henrik Abel (AH-behl) was the second child of Søren Georg Abel, a second-generation Lutheran minister, and Anne Marie Simonsen, a daughter of a successful merchant and shipowner. Soon after his birth, his father was transferred to the parish of Gjerstad, in southeastern Norway, about 150 miles from Oslo, where Abel spent his childhood with his five brothers and sisters. Abel was an attractive youth, with light ash-brown hair and blue eyes.

Although his father's earnings were never adequate to provide for the large family, the emphasis on educational stimulation in the Abel household was an important formative influence on the young boy. Although his early education was conducted at home, it was sufficient to allow him to attend the Cathedral School at Oslo when he was thirteen years old. It was there that his talent in mathematics was discovered, although his initial efforts were somewhat unpromising.

The Cathedral School had once been quite good, but many positions had been filled by inexperienced or inadequate teachers because their predecessors had been recruited to join the faculty of the newly formed University of Oslo. Indeed, Abel's first mathematics instructor was dismissed abruptly after beating a student to death. The replacement in that position was Bernt Michael Holmboe, who was the first to recognize Abel's talent and who later edited the first edition of his work. Holmboe also assisted Christopher Hansteen, a professor at the university; this connection would prove valuable to Abel.

When Holmboe first arrived at the school, he noticed Abel's ability in mathematics and suggested that the two of them study some of the contemporary mathematics works together. Abel soon outpaced Holmboe and began developing a general solution for the quintic equation, that is, an equation of the fifth degree ($ax^5 + bx^4 + cx^3 + dx^2 + ex + f = 0$). When Abel believed that the work was complete, Holmboe and Hansteen sensed that no one in Norway, including themselves, could review the work capably. They forwarded the paper to Ferdinand Degen of the Danish Academy, who carefully reviewed the work. Before publication, Degen helped Abel discover that his solution was flawed, but

Niels Henrik Abel. (Library of Congress)

he steered Abel into the field of elliptic functions, which Degen believed would be more fruitful.

At about that same time, Abel discovered that several of his predecessors, particularly Leonhard Euler and Joseph-Louis Lagrange, had not completed the reasoning required to prove some of their work. Abel diligently supplied rigorous proofs where they were missing; a noted case is his proof of the general binomial theorem, which had been stated previously in part by Sir Isaac Newton and Euler. The mathematics community later was to find his meticulous treatment of the works he studied invaluable. Unfortunately for his personal life and his financial situation, Abel's father, who had served two terms in the *Storting* (congress), was impeached and disgraced. His father died in 1829, leaving his family in even more desperate financial straits than ever before.

Life's Work

The nineteen-year-old Abel entered the University of Oslo in 1821. Although this entering age would not normally denote a prodigy, the fact that the university granted him a free room and that several professors donated funds for his support does. Abel completed the preliminary requirements for a degree in a single year. He was then free to study mathematics on his own, as he had no peers among the faculty. He developed a love for the theater at that time that lasted throughout his short life. A modest person, he made many lasting friendships.

In addition to studying all available work, he began writing papers, the first of which were published in the journal *Magazin for Naturvidenskaberne* begun by Hansteen. In 1823, Abel's first important paper, "Opläsning afet Par Opgaver ved bjoelp af bestemte Integraler" ("Solution of Some Problems by Means of Definite Integrals"), was published, containing the first published solutions of integral equations. During 1822 and 1823, he also developed a longer paper discussing the integration of functions. This work is recognized as very significant in the evolution of that field of study.

At that time, Abel's work was largely ignored by the international mathematics community because Abel was from Norway and wrote in Norwegian, and the focal point of the mathematics community of the day was Paris, with the language of the learned being French. By applying himself diligently, Abel learned French and began to publish work in that language. The quintic equation still held his attention, and, as he thought of possibilities for its solution, he also considered that there might be no solution that could be found for all such equations. In time, he was able to prove this result. Nevertheless, the mathematicians whose approval he desired so fervently, those in Paris, ignored his work.

Abel began to press for the opportunity to go to Paris, but penniless as he was he was forced to rely on grants. After his first application, it was decided that he needed to study more foreign languages before going abroad. Although it meant delaying his dream for nearly two years, Abel applied himself to learning various languages. Meanwhile, he became engaged to Christine (Krelly) Kemp before he finally received a royal grant to travel abroad in 1825.

This trip was unsuccessful in many ways. When he arrived at Copenhagen, he discovered that Degen had died. Instead of going on to Paris, Abel decided to go to Berlin because several of his friends were there. The time in Berlin was invaluable, for he met and befriended August Leopold Crelle, who became his strongest supporter and mentor. When Abel met him, Crelle was preparing to begin publication of a new journal, *Journal für die reine und angewandte Mathematik* (journal for pure and applied mathematics). Crelle was so taken by Abel's ability that much of the first few issues was devoted to Abel's work in an attempt to win recognition for the young mathematician.

For a variety of reasons, Abel did not proceed to Paris until the spring of 1826. By this time, he had spent most of his grant and was physically tired, and the Parisian mathematicians he had hoped to convince were nearly all on holiday. However, his masterwork, *Mémoire sur une propriété générale d'une classe très-étendue de fonctions* transcendantes (memoir on a general property of a very extensive class of transcendental functions), was presented to the Academy of Sciences on October 30, 1826. The paper was left in the keeping of Augustin-Louis Cauchy, a prominent mathematician, and Cauchy and Adrien-Marie Legendre were to be the referees. Whether the paper was illegible, as Cauchy claimed, or was misplaced, as most historians believe, no judgment was issued until after Abel's death.

Abel felt a great sense of failure, for many young mathematicians had been established by recognition from the academy. He returned first to Berlin and finally to Oslo in May, 1827. His prospects were bleak: He had contracted tuberculosis, there was no prospect for a mathematical position in Norway, and he was in debt. Abel began tutoring and lecturing at the university on a substitute basis in order to support himself.

Another young mathematician, Carl Gustav Jacob Jacobi, soon began publishing work in Abel's foremost field, the theory of elliptic functions and integrals. The rivalry created between them dominated the rest of Abel's life. He worked furiously to prove his ideas, and his efforts were spurred by his correspondence with Legendre. As he finally began to be recognized in Europe, many mathematicians, led by Crelle, attempted to secure a patronage for him. However, he succumbed to tuberculosis on April 6, 1829, two days before Crelle wrote to inform him that such financial support had been found. In June, 1830, he and Jacobi were awarded the Grand Prix of the French Academy of Sciences for their work in elliptic integrals. Abel's original manuscript was found and finally published in 1841.

Significance

Although Niels Henrik Abel's life was short and his work was unrecognized for most of his life, he has exercised a great influence on modern mathematics. His primary work with elliptic functions and integrals led to interest in what became one of the great research topics of his century. Without his preliminary findings, many of the developments in mathematics and, consequently, science, may not have been made. One example of this is his theory of elliptic functions, much of which was developed very quickly during his race with Jacobi. In addition, his proof that there is no general solution to the quintic equation is quite important, as are his other findings in equation theory.

Abel's theory of solutions using definite integrals, including what is now called Abel's theorem, is also widely used in engineering and the physical sciences and provided a foundation for the later work of others. Abelian (commutative) groups, Abelian functions, and Abelian equations are but three of the ideas that commonly carry his name. Given Abel's short life span and his living in Norway, a definite academic backwater at the time, his prolific achievements are amazing.

Abel is also significant because his writing and mathematical styles, which were easily comprehended, made his discoveries available to his contemporaries and successors. Abel's insistence that ideas should be demonstrated in such a way that the conclusions would be supported by clear and easily comprehended arguments, that is, proved rigorously, is the cornerstone of modern mathematics. It is in this regard that Abel is most often remembered.

Celeste Williams Brockington

Further Reading

Abel, Niels Henrik. "From a Memoir on Algebraic Equations, Proving the Impossibility of a Solution of the General Equation of the Fifth Degree." In *Classics of Mathematics*, edited by Ronald Calinger. Oak Park, Ill.: Moore, 1982. This extract of Abel's paper on the general quintic equation demonstrates Abel's style. Although it is too technical for the casual reader, it is of interest to mathematicians and demonstrates Abel's place in the development of mathematics. Also includes a brief biography.

Bell, Eric T. "Genius and Poverty: Abel." In *Men of Mathematics*. New York: Simon & Schuster, 1937. This compilation of brief biographies of famous mathematicians includes a chapter on Abel, focusing more on the subject's life than on his mathematical achievements.

Boyer, Carl B. *A History of Mathematics*. New York: John Wiley & Sons, 1968. This general history of mathematics will help the reader place Abel within the general development of mathematics.

Kline, Morris. *Mathematical Thought from Ancient to Modern Times*. New York: Oxford University Press, 1972. Kline includes both a brief biography of Abel and discussions of his most important work in this history of mathematics.

Ore, Øystein. *Niels Henrik Abel: Mathematician Extraordinary*. Minneapolis: University of Minnesota Press, 1957. This English-language biography gives a detailed account of Abel's life without requiring a specialized knowledge of mathematics.

Pesic, Peter. *Abel's Proof: An Essay on the Sources and Meaning of Mathematical Unsolvability*. Cambridge, Mass.: MIT Press, 2003. Describes Abel's life, focusing on his proof of the quintic equation. Pesic discusses why and how the proof changed the perception of mathematics. Includes a new annotated translation of Abel's original proof.

Stubhaug, Arild. *Niels Henrik Abel and His Times: Called* Too Soon by Flames Afar. Translated by Richard H. Daly. Berlin: Springer, 2000. Stubhaug, a Norwegian mathematician and historian, recounts Abel's brief life and achievements in this readable biography intended for a general audience. Includes illustrations.

Maria Gaetana Agnesi
Italian mathematician and charity worker

In her youth, Maria Agnesi advocated education for young women, in part by demonstrating her own impressive intellectual abilities. Her two-volume textbook on the calculus provided a complete synthesis of the mathematical methods developed during the Scientific Revolution. In her later years, she devoted herself to charitable work for the sick, poor, and aged.

Born: May 16, 1718; Milan (now in Italy)
Died: January 9, 1799; Milan

Early Life

Maria Gaetana Agnesi (mah-REE-ah gah-ay-TAHN-ah ahn-YAY-zee) was born to a wealthy and literate Italian family, the oldest of twenty-one children. According to most accounts, her father, Pietro Agnesi Mariami, became a mathematics professor at the University of Bologna, but there is no record of this. He and her mother, Anna Brivia, recognized Maria's abilities early and encouraged her to develop her intellect. Her special gift for languages was evident, as she could speak French fluently by the age of five. Impressively, at age nine she wrote a lengthy discourse in Latin on the importance of higher education for women entitled *Oratio qua ostenditur artium liberalium studia femineo sexu neutiquam abhorreri* (1727; an oration by which it is shown that the study of the liberal arts should not at all be abhorred by the female sex), which was printed at Milan. By the age of eleven, Maria was known as the Walking Polyglot and the Seven-Tongued Orator for her competence in Italian, French, German, Spanish, Latin, Greek, and Hebrew.

Maria's father engaged leading university professors to be her tutors. Their home became a gathering place for the most distinguished intellectuals of the time, both Italian and foreign. Maria participated in most of these meetings, engaging the guests in abstract academic discussions. Her younger sister, Maria Teresa (b. 1720), was a composer, singer, and harpsichordist who often performed her music at these meetings while Maria Gaetana presented theses in Latin on a wide variety of scientific and philosophical topics and defended them in the native language of her questioners. By the age of fourteen, Agnesi was solving difficult problems in ballistics and geometry. At seventeen, she circulated a critical commentary on the *Traité analytique des sections coniques* (1696; analytical treatise on conic sections) by the French mathematician Guillaume de L'Hôpital.

Life's Work

Although Agnesi had an attractive and agreeable manner, she was not eager to continue the public displays of her extraordinary learning. At the age of twenty, she expressed a desire to enter a convent. However, the death of Pietro Agnesi's second wife gave her another opportunity to retire from public life by assuming responsibility for her father's household and care for her twenty younger siblings. For the next two decades, she devoted herself to these duties along with the study of mathematics and the education of her younger brothers.

After her father's meetings were discontinued, Maria published a collection in Latin of 190 of the theses she had defended as a girl in a book called *Propositiones philosophicae* (pb. 1738; philosophical propositions). The topics covered in this book indicate the breadth of her knowledge in the sciences and philosophy of her day, including mechanics, hydromechanics, elasticity, celestial mechanics, universal gravitation, chemistry, mineralogy, botany, zoology, logic, and ontology. The theses on these topics appear together with a plea for the education of women.

For the next decade, Maria concentrated on mathematics, culminating in the publication of her most important mathematical work as a text for her younger brothers. *Instituzioni analitiche ad uso della gioventù italiana* (1748; *Analytical Institutions*, 1801) consisted of two volumes of 1,020 pages plus 59 pages of figures engraved by Marc' Antonio Dal Rè. Agnesi wrote the book in Italian rather than Latin to make it more accessible to young students. Her facility with languages enabled her to draw from a wide range of authors, producing the most complete synthesis of eighteenth century mathematics for at least the next fifty years. The second volume was later translated into French by Pierre d'Antelmy as *Traités élémentaires de calculus* (1775; elementary treatise on calculus), and the entire book was translated into English by the Cambridge mathematics professor John Colson and published in London in 1801 as *Analytical Institutions*.

The first volume of *Analytical Institutions* deals with the analysis of finite quantities, using algebra and geometry to construct and analyze geometric curves. The second volume develops differential and integral calculus and gives an introduction to the emerging topic of differential equations. It was the first text to give a systematic presentation of both differential and integral calculus, establishing the modern differential notation of Gottfried Wilhelm Leibniz over the more archaic "fluxions" of Isaac Newton. The English translation of 1801 introduced England to modern calculus notation.

One geometric curve discussed by Agnesi in *Analytical Institutions* has come to be uniquely associated with her name in an unusual but confusing way. She called this bell-shaped curve a versed sine or "versiera" (from the Latin *vertere*, meaning "to turn"). John Colson in his 1801 English translation confused this term with the Italian word *avversiera*, which means "witch." Hence, this curve came to be known as the "Witch of Agnesi" in the English-speaking world.

Agnesi dedicated *Analytical Institutions* to Empress Maria Theresa of Austria, who rewarded her with a diamond ring and a letter in a diamond-encrusted crystal case. Pope Benedict XIV sent her a congratulatory letter with a gold medal and a wreath made of gold and precious stones. She was elected to the Bologna Academy of Sciences and commended by the French Academy of Sciences for producing the best book of its kind. Pope Benedict named Agnesi an honorary professor of mathematics at the University of Bologna in 1750, although there is no record of her teaching there.

After her father's death in 1752, Agnesi turned increasingly to religious studies and charitable work. For the rest of her life, she devoted herself to the needs of the poor, starting with a small hospital in her home. During nearly forty years of charitable work, Agnesi gave away her inheritance to the poor, beginning with her diamond ring and crystal box from Maria Theresa, and even begged for money from others to provide for those in need. In 1771, she founded a charitable home for the aged with the Blue Nuns in Milan called Pio Albergo Trivulzio, acting as director of this hospice for several years. According to some sources, she took the blue habit of the Augustinian nuns before she died at the age of eighty.

Significance

Maria Gaetana Agnesi was the first woman to publish a book on mathematics in modern Europe. Her most important contribution was the publication of *Analytical Institutions* in 1748. This was the first comprehensive and systematic textbook covering both differential and integral calculus with a unified notation. It is the first surviving mathematical work written by a woman

Maria Gaetana Agnesi. (Library of Congress)

and the most valuable work in establishing the calculus for at least the next fifty years. It unified the ideas and methods of the greatest mathematicians of the Scientific Revolution, including the analytic geometry of René Descartes and the newly developed calculus of Newton and Leibniz, establishing the superior notation of Leibniz. She also led the way in applying the calculus to many original problems in geometry and physics.

In addition to her mathematical work, Agnesi lived an exemplary life of service. Although science had faltered to some extent after the condemnation of Galileo a century earlier, Agnesi contributed to the revival of scientific work in Italy and the associated Catholic Enlightenment. She was an early advocate of education for women and a champion of this cause in several publications. Her devotion to duty and her lifelong service to the poor and needy set a worthy example of Christian charity.

— Joseph L. Spradley

Further Reading

Alic, Margaret. *Hypatia's Heritage*. Boston: Beacon Press, 1986. Focuses on the historical contributions of women in science, including a discussion of the work of Maria Agnesi.

Gray, S. I. B., and Tagui Malakyan. "The Witch of Agnesi: A Lasting Contribution from the First Surviving Mathematical Work Written by a Woman." *College Mathematical Journal* 30, no. 4 (September, 1999): 258-268. This commemorative article on the two hundredth anniversary of Agnesi's death discusses her life and the geometric curve named for her.

Mazzotti, Massimo. "Maria Gaetana Agnesi: Mathematics and the Making of the Catholic Enlightenment." *Isis* 92, no. 4 (December, 2001): 657-683. Places Agnesi's work in the context of the Italian enlightenment of the mid-eighteenth century.

Osen, Lynn M. *Women in Mathematics*. Cambridge, Mass.: MIT Press, 1974, 1995. Includes a chapter on Agnesi and her mathematical contributions.

Smith, Sanderson. *Agnesi to Zeno: Over One Hundred Vignettes from the History of Math*. Emeryville, Calif.: Key Curriculum Press, 1996. This book on historical developments in mathematics includes a one-page introduction to the life and work of Agnesi followed by questions and projects for math students.

Alhazen

Arab physicist, astronomer, and mathematician

Alhazen, Islam's greatest scientist, devoted his life to physics, astronomy, mathematics, and medicine. His treatise Optics, *in which he deftly used experiments and advanced mathematics to understand the action of light, exerted a profound influence on many European natural philosophers.*

Born: 965; Basra (now in Iraq)
Died: 1039; Cairo, Egypt
Also known as: Ibn al-Haytham; Abū ʿAlī al-Ḥasan ibn al-Haytham (full name)

Early Life

Alhazen (ahl-HAH-zehn) was born in Basra. He was given a traditional Muslim education, but at an early age he became perplexed by the variety of religious beliefs and sects, because he was convinced of the unity of truth. When he was older, he concluded that truth could be attained only in doctrines whose matter was sensible and whose form was rational. He found such doctrines in the writings of Aristotle and in natural philosophy and mathematics.

By devoting himself completely to learning, Alhazen achieved fame as a scholar and was given a political post at Basra. In an attempt to obtain a better position, he claimed that he could construct a machine to regulate the flooding of the Nile. The Fāṭimid caliph al-Ḥākim (r. 996-1021?), wishing to use this sage's expertise, persuaded him to move to Cairo. Alhazen, to fulfill his boast, was trapped into heading an engineering mission to Egypt's southern border. On his way to Aswān, he began to have doubts about his plan, for he observed excellently designed and perfectly constructed buildings along the Nile, and he realized that his scheme, if it were possible, would have already been carried out by the creators of these impressive structures. His misgivings were confirmed when he discovered that the cataracts south of Aswān made flood control impossible. Convinced of the impracticability of his plan, and fearing the wrath of the eccentric and volatile caliph, Alhazen pretended to be mentally deranged; on his return to Cairo, he was confined to his house until al-Ḥākim's death around 1021.

Alhazen then took up residence in a small domed shrine near the al-Azhar Mosque. Having been given back his previously sequestered property, he resumed his activities as a writer and teacher. He may have earned his living by copying mathematical works, including Euclid's *Stoicheia* (c. fourth century B.C.E.; *Elements*, 1570) and Ptolemy's *Mathāmatikā suntaxis* (c. 150; *Almagest*, 1952), and may also have traveled and had contact with other scholars.

Life's Work

The scope of Alhazen's work is impressive. He wrote studies on mathematics, physics, astronomy, and medicine, as well as commentaries on the writings of Aristotle and Galen. He was an exact observer, a skilled experimenter, and an insightful theoretician, and he put these abilities to excellent use in the field of optics. He has been called the most important figure in optics between antiquity and the seventeenth century. Within optics itself, the range of his interests was wide: He discussed theories of light and vision, the anatomy and diseases of the eye, reflection and refraction, the rainbow, lenses, spherical and parabolic mirrors, and the pinhole camera (camera obscura).

Alhazen's most important work was *Kitāb al-Manāzir*, commonly known as *Optics*. Published first in Latin (*Opticae thesaurus Alhazeni libri vii*, 1572) and partially in English as *The Optics of Ibn al-Haytham: Books I-III, On Direct Vision* (1989), it attempted to clarify the subject by inquiring into its principles. He rejected Euclid's and Ptolemy's doctrine of visual rays (the extramission theory, which regarded vision as analogous to the sense of touch). For example, Ptolemy attributed sight to the action of visual rays issuing cortically from the observer's eye and being reflected from various objects. Alhazen also disagreed with past versions of the intromission theory, which treated the visible object as a source from which forms (simulacra) issued. The atomists, for example, held that objects shed sets of atoms as a snake sheds its skin; when a set enters the eye, vision occurs. In another version of the intromission theory, Aristotle treated the visible object as a modifier of the medium between the object and the eye.

Alhazen found the atomistic theory unconvincing because it could not explain how the image of a large mountain could enter the small pupil of the eye. He did not like the Aristotelian theory because it could not explain how the eye could distinguish individual parts of the seen world, since objects altered the entire intervening medium. Alhazen, in his version of the intromission theory, treated the visible object as a collection of small areas, each of which sends forth its own ray. He believed that vision takes place through light rays reflected from every point on an object's surface converging toward an apex in the eye.

According to Alhazen, light is an essential form in self-luminous bodies, such as the sun, and an accidental form in bodies that derive their luminosity from outside sources. Accidental light, such as the moon, is weaker than essential light, but both forms are emitted by their respective sources in exactly the same way: noninstantaneously, from every point on the source, in all directions, and along straight lines. To establish rectilinear propagation for essential, accidental, reflected, and refracted radiation, Alhazen performed many experiments with dark chambers, pinhole cameras, sighting tubes, and strings.

In the first book of *Optics*, Alhazen describes the anatomy of the eye. His description is not original, being based largely on the work of Galen, but he modifies traditional ocular geometry to suit his own explanation of vision. For example, he claims that sight occurs in the eye by means of the glacial humor (what would be called the crystalline lens), because when this humor is injured, vision is destroyed. He also uses such observations as eye pain while gazing on intense light and afterimages from strongly illuminated objects to argue against the visual-ray theory, because these observations show that light is coming to the eye from the object. With this picture of intromission established, Alhazen faces the problem of explaining how replicas as big as a mountain can pass through the tiny pupil into the eye.

He begins the solution of this problem by recognizing that every point in the eye receives a ray from every point in the visual field. The difficulty with this

This frontispiece to Alhazin's Opticae thesaurus *depicts various uses of optics including Archimedes' defense of Syracuse with mirrors.* (Library of Congress)

punctiform analysis is that, if each point on the object sends light and color in every direction to each point of the eye, then all this radiation would arrive at the eye in total confusion; for example, colors would arrive mixed. Simply put, the problem is a superfluity of rays. To explain vision, each point of the surface of the glacial humor needs to receive a ray from only one point in the visual field. In short, it is necessary to establish a one-to-one correspondence between points in the visual field and points in the eye.

To fulfill this goal, Alhazen notices that only one ray from each point in the visual field falls perpendicularly on the convex surface of the eye. He then proposes that all other rays, those falling at oblique angles to the eye's surface, are refracted and so weakened that they are incapable of affecting visual power. Alhazen even performed an experiment to show that perpendicular rays are strong and oblique rays weak: He shot a metal sphere against a dish both perpendicularly and obliquely. The perpendicular shot fractured the plate, whereas the oblique shot bounced off harmlessly. Thus, in his theory, the cone of perpendicular rays coming into the eye accounts for the perception of the visible object's shape and the laws of perspective.

Book 2 of *Optics* contains Alhazen's theory of cognition based on visual perception, and book 3 deals with binocular vision and visual errors. Catoptrics (the theory of reflected light) is the subject of book 4. Alhazen here formulates the laws of reflection: Incident and reflected rays are in the same plane, and incident and reflected angles are equal. The equality of the angles of incidence and reflection allows Alhazen to explain the formation of an image in a plane mirror. As throughout *Optics*, Alhazen here uses experiments to help establish his contentions. For example, by throwing an iron sphere against a metal mirror at an oblique angle, he found that the incident and reflected movements of the sphere were symmetrical. The reflected movement of the iron sphere, because of its heaviness, did not continue in a straight line, as the light ray does, but Alhazen did not contend that the iron sphere is an exact duplicate of the light ray.

Alhazen's investigation of reflection continues in books 5 and 6 of *Optics*. Book 5 contains the famous "Problem of Alhazen": For any two points opposite a spherical reflecting surface, either convex or concave, find the point or points on the surface at which the light from one of the two points will be reflected to the other. Today it is known that the algebraic solution of this problem leads to an equation of the fourth degree, but Alhazen solved it geometrically by the intersection of a circle and a hyperbola.

Book 7, which concludes *Optics*, is devoted to dioptrics (the theory of refraction). Although Alhazen did not discover the mathematical relationship between the angles of incidence and refraction, his treatment of the phenomenon was the most extensive and enlightening before that of René Descartes. As with reflection, Alhazen explores refraction through a mechanical analogy. Light, he says, moves with great speed in a transparent medium such as air and with slower speed in a dense body such as glass or water. The slower speed of the light ray in the denser medium is the result of the greater resistance it encounters, but this resistance is not strong enough to hinder its movement completely. Because the refracted light ray is not strong enough to maintain its original direction in the denser medium, it moves in another direction along which its passage will be easier (that is, it turns toward the normal). This idea of the easier and quicker path was the basis of Alhazen's explanation of refraction, and it is a forerunner of the principle of least time associated with the name of Pierre de Fermat.

Optics was Alhazen's most significant work and by far his best known, but he also wrote more modest treatises in which he discussed the rainbow, shadows, camera obscura, and Ptolemy's optics as well as spheroidal and paraboloidal burning mirrors. The ancient Greeks had a good understanding of plane mirrors, but Alhazen developed an exhaustive geometrical analysis of the more difficult problem of the formation of images in spheroidal and paraboloidal mirrors.

Although Alhazen's achievements in astronomy do not equal those in optics, his extant works reveal his mastery of the techniques of Ptolemaic astronomy. These works are mostly short tracts on minor problems, for example, sundials, moonlight, eclipses, parallax, and determining the *gibla* (the direction to be faced in prayer). In another treatise, he was able to explain the apparent increase in size of heavenly bodies near the horizon, and he also estimated the thickness of the atmosphere.

His best astronomical work, and the only one known to the medieval West, was *Hay'at al-'alan* (tenth or eleventh century; *Ibn al-Haytham's On the Configuration of the World* 1990). This treatise grew out of Alhazen's desire that the astronomical system should correspond to the true movements of actual heavenly bodies. He therefore attacked Ptolemy's system, in which the motions of heavenly bodies were explained in terms of imaginary points moving on imaginary circles. In his work, Alhazen tried to discover the physical reality underlying Ptolemy's abstract astronomical system. He accomplished this task by viewing the heavens as a series of concentric spherical shells whose rotations were interconnected. Alhazen's system accounted for the apparent motions of the heavenly bodies in a clear and nontechnical way, which accounts for the book's popularity in the Middle Ages.

Alhazen's fame as a mathematician has largely depended on his geometrical solutions of various optical problems, but more than twenty strictly mathematical treatises have survived. Some of these deal with geometrical problems arising from his studies of Euclid's *Elements*, whereas others deal with quadrature problems, that is, constructing squares equal in area to various plane figures. He also wrote a work on lunes (figures contained between the arcs of two circles) and on the properties of conic sections. Although he was not successful with every problem, his performance, which exhibited his masterful command of higher mathematics, has rightly won for him the admiration of later mathematicians.

Significance

Alhazen was undoubtedly the greatest Muslim scientist, and *Optics* was the most important work in the field from Ptolemy's time to the time of Johannes Kepler. He extricated himself from the limitations of such earlier theories as the atomistic, Aristotelian, and Ptolemaic and integrated what he knew about medicine, physics, and mathematics into a single comprehensive theory of light and vision. Although his theory contained ideas from older theories, he combined these ideas with his new insights into a fresh creation, which became the source of a new optical tradition.

His optical theories had some influence on Islamic scientists, but their main impact was on the West. *Optics* was translated from Arabic into Latin at the end of the twelfth century. It was widely studied, and in the thirteenth century, Witelo (also known as Vitellio) made liberal use of Alhazen's text in writing his comprehensive book on optics. Roger Bacon, John Pecham, and Giambattista della Porta are only some of the many thinkers who were influenced by Alhazen's work. Indeed, it was not until Kepler, six centuries later, that work on optics progressed beyond the point to which Alhazen had brought it. Even Kepler, however, used some of Alhazen's ideas, for example, the one-to-one correspondence between points on the object and points in the eye. It would not be going too far to say that Alhazen's optical theories defined the scope and goals of the field from his day to today.

Robert J. Paradowski

Further Reading

Bakar, Osman. *The History and Philosophy of Islamic Science*. Cambridge, England: Islamic Texts Society, 1999. Discusses questions of methodology, doubt, spirituality and scientific knowledge, the philosophy of Islamic medicine, and how Islamic science influenced medieval Christian views of the natural world.

Grant, Edward, ed. *A Source Book in Medieval Science*. Cambridge, Mass.: Harvard University Press, 1974. A compilation of readings from medieval natural philosophers, including several selections in English translation from the works of Alhazen. Bibliography, index.

Hayes, John R., ed. *The Genius of Arab Civilization: Source of Renaissance*. 3d ed. New York: New York University Press, 1992. In this beautifully illustrated book, several international authorities discuss the intellectual achievements of Islamic culture. Sabra's chapter on the exact sciences contains an account of Alhazen's work in the context of Islamic intellectual history. Bibliography, index.

Hogendijk, Jan P., and Abdelhamid I. Sabra, eds. *The Enterprise of Science in Islam: New Perspectives*. Cambridge, Mass.: MIT Press, 2003. A collection surveying the history of science, including mathematics, optics, and astronomy, in Islam, with a chapter on Alhazen's work. Illustrations, bibliography, index.

Huff, Toby E. *The Rise of Early Modern Science: Islam, China, and the West*. 2d ed. New York: Cambridge University Press, 2003. Provides a strong cross-cultural background for the rise of science and medicine in the Muslim world. Illustrations, bibliography, index.

Lindberg, David C. *Theories of Vision from al-Kindi to Kepler*. Chicago: University of Chicago Press, 1981. Surveys visual theory against the background of ancient accomplishments. The chapter on Alhazen's intromission theory is excellent. Bibliography, index.

_____, ed. *Science in the Middle Ages*. Chicago: University of Chicago Press, 1978. Through the expertise of several historians of medieval science, this book examines in depth all major aspects of natural philosophy in the Middle Ages. The approach is not encyclopedic but interpretative. Includes a chapter on optics, in which Alhazen's work is clearly explained.

Nasr, Seyyed Hossein. *Islamic Science: An Illustrated Study*. Westerham, England: World of Islam Festival, 1976. The first illustrated study ever undertaken of the whole of Islamic science. Using traditional Islamic concepts, the author discusses various branches of science, including optics.

_____. *Science and Civilization in Islam*. Cambridge, Mass.: Harvard University Press, 1968. This book is the first one-volume work in English to deal with Islamic science from the Muslim rather than the Western viewpoint. Its approach is encyclopedic rather than analytic, but it does contain a discussion of Alhazen's work in its Muslim context.

Sabra, Abdelhamid I. *Optics, Astronomy, and Logic: Studies in Arabic Science and Philosophy*. Brookfield, Vt.: Variorum, 1994. A history of science, specifically optics and astronomy, in the Muslim world of Alhazen's time. Includes discussion of Alhazen and his work. Illustrations, index.

_____. *Theories of Light from Descartes to Newton*. New York: Cambridge University Press, 1981. Though this book is mainly centered on seventeenth century theories of light, the author discusses in detail the impact of Alhazen's ideas on the optical discoveries of such thinkers as Descartes and Christiaan Huygens. Bibliography, index.

Apollonius of Perga

Pergan geometer

One of the ablest geometers in antiquity, Apollonius systematized the theory of conic sections. His study of circular motion established the foundation for Greek geometric astronomy.

Born: c. 262 B.C.E.; Perga, Pamphylia, Asia Minor (now Murtana, Turkey)
Died: c. 190 B.C.E.; Alexandria, Egypt
Also known as: The Great Geometer

Early Life
Information on the life of Apollonius (ap-uh-LOH-nee-uhs) is meager. Born at Perga around the middle of the third century B.C.E., he studied mathematics with the successors of Euclid at Alexandria. His activity falls near the time of Archimedes (c. 287-212 B.C.E.), but links between their work are indirect. In his surviving work, Apollonius once mentions the Alexandria-based geometry Conon of Samos, but his principal correspondents and colleagues (Eudemus, Philonides, Dionysodorus, Attalus I) were active at Pergamum and other centers in Asia Minor. It appears that this circle benefited from the cultural ambitions of the new Attalid Dynasty during the late third and the second centuries B.C.E.

Life's Work
Apollonius's main achievement lies in his study of the conic sections. Two properties of these curves can be distinguished as basic for their conception: First, they are specified as the locus of points whose distances x, y from given lines satisfy certain second-order relations: When $x^2 = ay$ (for a constant line segment a) the curve of the locus is a parabola, when $x^2 = ay - ay^{2/b}$ the curve is an ellipse (it becomes a circle when $b = a$), and when $x^2 = ay + ay^{2/b}$ it is a hyperbola. The same curves can be produced when a plane intersects the surface of a cone: When the plane is parallel to the side of the cone, there results a parabola (a single open, or infinitely extending, curve); when the plane is not parallel to the side of the cone but cuts through only one of its two sheets, there results an ellipse (a single closed curve); and when it cuts through both sheets of the cone, there results a hyperbola (a curve consisting of two separate branches, each extending indefinitely).

The curves were already known in the fourth century B.C.E., for the geometer Menaechmus introduced the locus forms of two parabolas and a hyperbola in order to solve the problem of doubling the cube. By the time of Euclid (c. 300 B.C.E.), the formation of the curves as solid sections was well understood. Euclid himself produced a major treatise on the conics, as had a geometer named Aristaeus somewhat earlier. As Archimedes often assumed theorems on conics, one supposes that his basic reference source (which he sometimes cited as the "Conic Elements") was the Euclidean or Aristaean textbook. Also in the third century, Eratosthenes of Cyrene and Conon pursued studies in the conics (these works no longer survive), as did Diocles in his writing on burning mirrors (extant in an Arabic translation).

Apollonius thus drew from more than a century of research on conics. In the eight books of his treatise, *Cōnica* (*Treatise on Conic Sections*, 1896; best known as *Conics*), he systematized the elements of this field and contributed many new findings of his own. Only the first four books survive in Greek, in the edition prepared by Eutocius of Ascalon (active at Alexandria in the early sixth century C.E.), but all of its books except for the eighth exist in an Arabic translation from the ninth century C.E.

Among the topics that Apollonius covers are these: book 1, the principal constructions and properties of the three types of conics, their tangents, conjugate diameters, and transformation of axes; book 2, properties of hyperbolas, such as their relation to their asymptotes (the straight lines they infinitely approach but never meet); book 3, properties of intersecting chords and secants drawn to conics; book 4, how conics intersect one another; book 5, on the drawing of normal lines to conics; book 6, on similar conics; book 7, properties of the conjugate diameters and principal axes of conics; book 8 (lost), problems solved via the theorems of book 7.

As Apollonius states in the prefaces to the books of his treatise, the chief application of conics is to geometric problems—that is, propositions seeking the construction of a figure satisfying specified conditions. Apollonius includes only a few examples in the *Conics*: for example, to find a cone whose section produces a conic curve of specified parameters (1: 52-56), or to draw tangents and normals to given conics (2: 49-53 and 5: 55-63). Much of the content of the *Conics*, however, deals not with problems but with theorems auxiliary to problems. This is the case with book 3, for example, which Apollonius says is especially useful for problem solving but which actually contains no problems. In his preface, he explicitly mentions the problem of the "locus relative to three (or four) lines," all cases of which, Apollonius proudly asserts, can be worked out by means of his book 3, whereas Euclid's earlier effort was incomplete.

The significance of problem solving for the Greek geometric tradition is evident in works such as Euclid's *Stoicheia* (*Elements*) and *Ta dedomena* (*Data*). In more advanced fields such as conic theory, however, the surviving evidence is only barely representative of the richness of this ancient activity. A notable exception is the *Synagogē* (*Collection*), a massive anthology of geometry by Pappus of Alexandria (fourth century C.E.), which preserves many examples of problems. Indeed, the whole of its book 7 amounts to an extended commentary on the problem-solving tradition—what Pappus calls the "analytic corpus" (*topos analyomenos*), a group of twelve treatises by Euclid, Apollonius, and others. Of the works taken from Apollonius, two are extant—*Conics* and *Logou apotomē* (*On Cutting Off a Ratio*, 1987)—while another five are lost—*Chōriou apotomē* (cutting off an area), *Diōrismenē tomē* (determinate section), *Epaphai* (tangencies), *Neyseis* (vergings), and *Topoi epipedoi* (plane loci). Pappus's summaries and technical notes preserve the best evidence available regarding the content of these lost works. Thus it is known that in *Epaphai*, for example, Apollonius covered all possible ways of constructing a circle so as to touch any combination of three given elements (points, lines, or circles). In *Neyseis* he sought the position of a line verging toward a given point and such that a marked segment of it lies exactly between given lines or circles. In *Topoi epipedoi* circles were produced as loci satisfying stated conditions, several of these being equivalent to expressions now familiar in analytic geometry.

It is significant that these last three works were restricted to planar constructions—that is, ones requiring only circles and straight lines. Pappus classifies problems in three categories: In addition to the planar,

Apollonius of Perga. (Library of Congress)

he names the solid (solvable by conics) and the linear (solvable by special curves, such as certain curves of third order, or others, such as spirals, now termed "transcendental," composed of coordinated circular and rectilinear motions). For Pappus, this scheme is normative; a planar solution, if known, is preferable to a solid one, and, similarly, a solid solution to a linear. For example, the problems of circle quadrature, cube duplication, and angle trisection can be solved by linear curves, but the last two can also be solved by conics and so are classed as solid.

Historians often misinterpret this classification as a restriction on solutions, as if the ancients accepted only the planar constructions. To the contrary, geometers throughout antiquity so fully explored all forms of construction as to belie any such restriction. Presumably, in his three books on planar constructions, Apollonius sought to specify as completely as possible the domain of such constructions rather than to eliminate those of the solid or linear type. In any event, from works before Apollonius there is no evidence at all of a normative conception of problem-solving methods.

There survive isolated reports of Apollonian studies bearing on the regular solids, the cylindrical spiral, irrationals, circle measurement, the arithmetic of large numbers, and other topics. For the most part, little is known of these efforts, and their significance was slight in comparison with his treatises on geometric constructions.

Ptolemy reports in *Mathēmatikē suntaxis* (c. 150 C.E.; *Almagest*) that Apollonius made a significant contribution to astronomical theory by establishing the geometric condition for a planet to appear stationary relative to the fixed stars. Since, according to Ptolemy, he proved this condition for both the epicyclic and the eccentric models of planetary motion, Apollonius seems to have had some major responsibility for the introduction of these basic models. Apollonius studied only the geometric properties of these models, however, for the project of adapting them to actual planetary data became a concern only for astronomers such as Hipparchus a few decades later in the second century B.C.E.

Significance
If Apollonius of Perga did indeed institute the eccentric and epicyclic models for planetary motion, as seems likely, he merits the appellation assigned to him by historian Otto Neugebauer: "the founder of Greek mathematical astronomy. These geometric devices, when adjusted to observational data and made suitable for numerical computation, became the basis of the sophisticated Greek system of astronomy. Through its codification by Ptolemy in the *Almagest*, this system flourished among Arabic and Hindu astronomers in the Middle Ages and Latin astronomers in the West through the sixteenth century. Although Nicolaus Copernicus (1473-1543) made the significant change of replacing Ptolemy's geocentric arrangement with a heliocentric one, even he retained the basic geometric methods of the older system. Only with Johannes Kepler (1571-1630), who was first to substitute elliptical orbits for the configurations of circles in the Ptolemaic-Copernican scheme, can one speak of a clear break with the mathematical methods of ancient astronomy.

Apollonius's work in geometry fared quite differently. The fields of conics and advanced geometric constructions he so fully explored came to a virtual dead end soon after his time. The complexity of this subject, proliferating in special cases and lacking convenient notations (such as the algebraic forms, for example, of modern analytic geometry that first appeared only with François Viète, René Descartes, and Pierre de Fermat in the late sixteenth and the seventeenth centuries), must have discouraged further research among geometers in the second century B.C.E.

In later antiquity, interest in Apollonius's work revived: Pappus and Hypatia of Alexandria (fourth to early fifth century C.E.) and Eutocius (sixth century) produced commentaries on the *Conics*. Their work did not extend the field in any significant way beyond what Apollonius had done, but it proved critical for the later history of conic theory, by ensuring the survival of Apollonius's writing. When the *Conics* was translated into Arabic in the ninth century, Arabic geometers entered this field; they approached the study of Apollonius with considerable inventiveness, often devising new forms of proofs, or contributing new results where the texts at their disposal were incomplete. Alhazen (early eleventh century), for example, attempted a restoration of Apollonius's lost book 8.

In the early modern period, after the publication of the translations of Apollonius and Pappus by Federigo Commandino in 1588-1589, the study of advanced geometry received new impetus in the West. Several distinguished mathematicians in this period (François Viète, Willebrord Snel, Pierre de Fermat, Edmond Halley, and others) tried their hand at restoring lost analytic works of Apollonius. The entirely new field of projective geometry emerged from the conic researches of Gérard Desargues and Blaise Pascal in the seventeenth century.

Thus, the creation of the modern field of geometry owes much to the stimulus of the *Conics* and the associated treatises of Apollonius.

Wilbur R. Knorr

FURTHER READING

Apollonius. *On Cutting Off a Ratio*. Translated by Edward Macierowski. Fairfield, Conn.: Golden Hind Press, 1987. This translation is literal and provisional; a full critical edition is being prepared by Macierowski.

_____. *Treatise on Conic Sections*. Translated and edited by Thomas Little Heath. Cambridge, England: Cambridge University Press, 1896. Translation in modern notation, with extensive commentary. Heath surveys the older history of conics, including efforts by Euclid and Archimedes, and then summarizes the characteristic terminology and methods used by Apollonius. A synopsis appears in Heath's *History of Greek Mathematics* (Oxford, England: Clarendon Press, 1921), together with ample discussions of the lost Apollonian treatises described by Pappus.

Fried, Michael N. *Apollonius of Perga's "Conica": Text, Context, Subtext*. Boston: E. J. Brill, 2001. A scholarly analysis. Bibliographic references, index.

Hogemdijk, J. P. *Ibn al-Haytham's Completion of the "Conics."* New York: Springer-Verlag, 1984. This edition of the Arabic text of Alhazen's restoration of the lost book 8 of the *Conics* is accompanied by a literal English translation, a mathematical summary in modern notation, and discussions of the Greek and Arabic traditions of Apollonius's work.

Knorr, W. R. *Ancient Tradition of Geometric Problems*. Cambridge, Mass.: Birkhauser Boston, 1986. A survey of Greek geometric methods from the pre-Euclidean period to late antiquity. Chapter 7 is devoted to the work of Apollonius, including his *Conics* and lost analytic writings.

Neugebauer, Otto. *A History of Ancient Mathematical Astronomy*. New York: Springer-Verlag, 1975. The section on Apollonius in this work provides a detailed technical account of his contributions to ancient astronomy.

Pappus of Alexandria. *Book 7 of the "Collection."* Translated by A. Jones. New York: Springer-Verlag, 1986. A critical edition of Pappus's Greek text (collated with the former edition of F. Hultsch in volume 2 of *Pappi Collectionis Quae Supersunt*, 1877), with English translation and commentary. Pappus's book preserves highly valuable information on Apollonius's lost works on geometric construction. Jones surveys in detail Pappus's evidence of the lost works and modern efforts to reconstruct them.

_____, ed. *Apollonius: Conics Books V to VII: The Arabic Translation of the Lost Greek Original in the Version of the Banu Musa*. New York: Springer-Verlag, 1990. This is the first literal English translation of this work ever to be published. Based on all known manuscripts, it includes the Arabic text with a full critical apparatus, an accurate English translation, and a commentary to elucidate both mathematical and historical difficulties.

Waerden, Bartel Leendert van der. *Science Awakening*. Translated by Arnold Dresden. 4th ed. Princeton Junction, N.J.: Scholar's Bookshelf, 1988. In this highly readable survey of ancient mathematics, Waerden includes a useful synopsis of the geometric work of Apollonius.

ARCHIMEDES

Greek mathematician and engineer

The greatest mathematician of antiquity, Archimedes did his best work in geometry and also founded the disciplines of statics and hydrostatics.

Born: c. 287 B.C.E.; Syracuse, Sicily (now in Italy)
Died: 212 B.C.E.; Syracuse, Sicily

EARLY LIFE

Few details are certain about the life of Archimedes (ar-kuh-MEED-eez). The birth date of 287 B.C.E. was established from a report, about fourteen hundred years after the fact, that he was seventy-five years old at his death in 212 B.C.E. Ancient writers agree in calling him a Syracusan by birth, and he himself provides the information that his father was the astronomer Phidias, the author of a treatise on the diameters of the sun and moon. His father's profession suggests an explanation for the son's early interest in astronomy and mathematics. Some scholars have characterized Archimedes as an aristocrat who actively participated in the Syracusan court and who may have been related to King Hiero II, the ruler of Syracuse. He certainly was friendly with Hiero and Hiero's son Gelon, to whom he dedicated one of his works. (Original titles of Archimedes' works are not known, but most of his books were first translated into English by Thomas L. Heath in 1897 in the volume *The Works of Archimedes*.)

Archimedes traveled to Egypt to study in Alexandria, then the center of the scientific world. Some of his teachers had, in their youth, been students of Euclid. He made two close friends in Alexandria: Conon of Samos, a gifted mathematician, and Eratosthenes of Cyrene, also a good mathematician. From the prefaces to his works, it is clear that Archimedes maintained friendly relations with several Alexandrian scholars, and he played an active role in developing the mathematical traditions of this intellectual center. It is possible that he visited Spain before returning to Syracuse, and a return trip to Egypt is also a possibility. This second visit would have been the occasion for his construction of dikes and bridges reported in some Arabian sources.

In Syracuse, Archimedes spent his time working on mathematical and mechanical problems. Although he was a remarkably ingenious inventor, his inventions were, according to Plutarch, merely diversions, the work of a geometer at play. He possessed such a lofty intellect that he considered these inventions of much less worth than his mathematical creations. Plutarch may have exaggerated Archimedes' distaste for engineering, because there is evidence that he was fascinated by mechanical problems from a practical as well as a theoretical point of view.

In the stories that multiplied about him, Archimedes became a symbol of the learned man—absentminded and unconcerned with food, clothing, and the other necessities of life. In images created long after his death, he is depicted as the quintessential sage, with a heavily bearded face, massive forehead, and contemplative mien. He had a good sense of humor. For example, he often sent his theorems to Alexandria, but to play a trick on some conceited mathematicians there, he once slipped in a few false propositions, so that these individuals, who pretended to have discovered everything by themselves, would fall into the trap of proposing theorems that were impossible.

LIFE'S WORK

The range of Archimedes' interest was wide, encompassing statics, hydrostatics, optics, astronomy, and engineering, in addition to geometry and arithmetic. It is natural that stories should tell more about his engineering inventiveness than his mathematical ability, for clever machines appealed to the average mind more than abstract mathematical theorems. Unfortunately, many of these stories are doubtful. For example, Archimedes is supposed to have invented a hollow, helical cylinder that, when rotated, could serve as a water pump, but this device, now called the Archimedean screw, antedates its supposed inventor.

In another well-known story, Archimedes boasted to King Hiero that, if he had a place on which to stand, he could move the earth. Hiero urged him to make good this boast by hauling ashore a fully loaded, three-masted merchantman of the royal fleet. Using a compound pulley, Archimedes, with modest effort, pulled the ship out of the harbor and onto the shore. The compound pulley may have been Archimedes' invention, but the story, told by Plutarch, is probably a legend.

The most famous story about Archimedes is attributed to Vitruvius, a Roman architect under Emperor Augustus. King Hieron, grateful for the success of one of his ventures, wanted to thank the gods by consecrating a golden wreath. On delivery, the wreath had the weight of the gold supplied for it, but Hiero suspected that it had been adulterated with silver. Unable to make the goldsmith confess, Hiero asked Archimedes to devise some way of testing the wreath. Because it was a consecrated object, Archimedes could not subject it to chemical analysis. He pondered the problem without success until one day, when he entered a full bath, he noticed that the deeper he descended into the tub, the more water flowed over the edge. This suggested to him that the amount of overflowed water was equal in volume to the portion of his body submerged in the bath. This observation gave him a way of solving the problem, and he was so overjoyed that he leapt out of the tub and ran home naked through the streets, shouting: "Eureka! Eureka!" Vitruvius then goes on to explain how Archimedes made use of his newly gained insight. By putting the wreath into water, he could tell by the rise in water level the volume of the wreath. He also dipped into water lumps of gold and silver, each having the same weight as the wreath. He found that the wreath caused more water to overflow than the gold and less than the silver. From this experiment, he determined the amount of silver admixed with the gold in the wreath.

As amusing and instructive as these legends are, much more reliable and interesting to modern historians of science are Archimedes' mathematical works. These treatises can be divided into three groups: studies of figures bounded by curved lines and surfaces, works on the geometrical analysis of statical and hydrostatical problems, and arithmetical works. The form in which these treatises have survived is not the form in which they left Archimedes' hand: They have all undergone transformations and emendations. Nevertheless, one still finds the spirit of Archimedes in the intricacy of the questions and the lucidity of the explanations.

In finding the areas of plane figures bounded by curved lines and the volumes of solid figures bounded by curved surfaces, Archimedes used a method originated by Eudoxus of Cnidus, unhappily called the "method of exhaustion." This indirect proof involves inscribing and circumscribing polygons to approach a length, area, or volume. The name "exhaustion" is based on the idea that, for example, a circle would finally be exhausted by inscribed polygons with a growing number of sides. In *Peri sphairas kai kylindron* (c. 240 B.C.E.; *On the Sphere and the Cylinder*, 1897), Archimedes compares perimeters of inscribed and circumscribed polygons to prove that the volume of a sphere is two-thirds the volume of its circumscribed cylinder. He also proves that the surface of any sphere is four times the area of its greatest circle.

Having successfully applied this method to the sphere and cylinder, Archimedes went on to use the technique for many other figures, including spheroids, spirals, and parabolas. *Peri konoeideon kai sphaireodeon* (c. 240 B.C.E.; *On Conoids and Spheroids*, 1897) treats the figures of revolution generated by conics. His spheroids are what are now called oblate and prolate spheroids, which are figures of revolution generated by ellipses. Archimedes' object in this work was the determination of volumes of segments cut off by planes from these conoidal and spheroidal solids. In *Peri helikon* (c. 240 B.C.E.; *On Spirals*, 1897), Archimedes studies the area enclosed between successive whorls of a spiral. He also defines a figure, now called Archimedes' spiral: If a ray from a central point rotates uniformly about this point, like the hand of a clock, and if another point moves uniformly along this line (marked by the clock hand), starting at the central point, then this linearly moving and rotating point will trace Archimedes' spiral.

Tetragonismos ten tou orthogonion konoy tomes (c. 250 B.C.E.; *On the Quadrature of the Parabola*, 1897), when translated, is not Archimedes' original title for the treatise, as "parabola" was not used in the sense of a conic section in the third century B.C.E. On the other hand, quadrature is an ancient term: It denotes the process of constructing a square equal in area to a given

Archimedes. (Library of Congress)

surface, in this case a parabolic segment. Archimedes, in this treatise, proves the theorem that the area of a parabolic segment is four-thirds the area of its greatest inscribed triangle. He was so fond of this theorem that he gave different proofs for it. One proof uses a method of exhaustion in which the parabolic segment is exhausted by a series of triangles. The other consists of establishing the quadrature of the parabola by mechanically balancing elements of the unknown area against elements of a known area. This latter method gives an insight into how Archimedes discovered theorems to be proved. His most recently discovered work, *Peri tfn mechanikon theorematon* (c. 250 B.C.E.; *On the Method of Mechanical Theorems*, 1912), provides other examples of how Archimedes mathematically balanced geometrical figures as if they were on a weighing balance. He did not consider that this mechanical method constituted a demonstration, but it allowed him to find interesting theorems, which he then proved by more rigorous geometrical methods.

15

Archimedes also applied geometry to statics and hydrostatics successfully. In his *Epipledon isorropion* (c. 250 B.C.E.; *On the Equilibrium of Planes*, 1897), he proves the law of the lever geometrically and then puts it to use in finding the centers of gravity of several thin sheets of different shapes. By center of gravity, Archimedes meant the point at which the object can be supported so as to be in equilibrium under the pull of gravity. Earlier Greek mathematicians had made use of the principle of the lever in showing that a small weight at a large distance from a fulcrum would balance a large weight near the fulcrum, but Archimedes worked this principle out in mathematical detail. In his proof, the weights become geometrical magnitudes acting perpendicularly to the balance beam, which itself is conceived as a weightless geometrical line. In this way, he reduced statics to a rigorous discipline comparable to what Euclid had done for geometry.

Archimedes once more emphasizes geometrical analysis in *Peri ochoymenon* (c. 230 B.C.E.; *On Floating Bodies*, 1897). The cool logic of this treatise contrasts with his emotional discovery of the buoyancy principle. In this work, he proves that solids lighter than a fluid will, when placed in the fluid, sink to the depth where the weight of the solid will be equal to the weight of the fluid displaced. Solids heavier than the fluid will, when placed in the fluid, sink to the bottom, and they will be lighter by the weight of the displaced fluid.

Although Archimedes' investigations were primarily in geometry and mechanics, he did perform some interesting studies in numerical calculation. For example, in *Kykloy metresis* (c. 230 B.C.E.; *On the Measurement of the Circle*, 1897) he calculated, based on mathematical principles rather than direct measurement, a value for the ratio of the circumference of a circle to its diameter (this ratio was not called pi until much later). By inscribing and circumscribing regular polygons of more and more sides within and around a circle, Archimedes found that the ratio was between 223/71 and 220/71, the best value for π (pi) ever obtained in the classical world.

In *Psiammites* (c. 230 B.C.E.; *The Sand-Reckoner*, 1897), Archimedes devises a notation suitable for writing very large numbers. To put this new notation to a test, he sets down a number equal to the number of grains of sand it would take to fill the entire universe. Large numbers are also involved in his treatise concerned with the famous "Cattle Problem." White, black, yellow, and dappled cows and bulls are grazing on the island of Sicily. The numbers of these cows and bulls have to satisfy several conditions. The problem is to find the number of bulls and cows of each of the four colors. It is unlikely that Archimedes ever completely solved this problem in indeterminate analysis.

Toward the end of his life, Archimedes became part of a worsening political situation. His friend Hiero II had a treaty of alliance with Rome and remained faithful to it, even after the Second Punic War began. After his death, however, his grandson Hieronymus, who became king, was so impressed by Hannibal's victories in Italy that he switched sides to Carthage. Hieronymus was then assassinated, but Sicily remained allied with Carthage. Consequently, the Romans sent a fleet under the command of Marcellus to capture Syracuse. According to traditional stories, Archimedes invented devices for warding off the Roman enemy. He is supposed to have constructed large lenses to set the fleet on fire and mechanical cranes to turn ships upside down. He devised so many ingenious war machines that the Romans would flee if so much as a piece of rope appeared above a wall. These stories are grossly exaggerated, if not totally fabricated, but Archimedes may have helped in the defense of his city, and he certainly provided the Romans with a face-saving explanation for their frustratingly long siege of Syracuse.

Because of treachery by a cabal of nobles, among other things, Syracuse eventually fell. Marcellus ordered that the city be sacked, but he made it clear that his soldiers were to spare the house and person of Archimedes. Amid the confusion of the sack, however, Archimedes, while puzzling over a geometrical diagram drawn on sand in a tray, was killed by a Roman soldier. During his lifetime he had expressed the wish that on his tomb should be placed a cylinder circumscribing a sphere, together with an inscription giving the ratio between the volumes of these two bodies, a discovery of which he was especially proud. Marcellus, who was distressed by the great mathematician's death, had Archimedes' wish carried out. More than a century later, when Cicero was in Sicily, he found this tomb, overgrown with brush but with the figure of the sphere and cylinder still visible.

SIGNIFICANCE

Some scholars rank Archimedes with Sir Isaac Newton (1642-1727) and Carl Friedrich Gauss (1777-1855) as one of the three greatest mathematicians who ever lived, and historians of mathematics agree that the theorems Archimedes discovered raised Greek mathematics to a new level of understanding. He tackled very difficult and original problems and solved them through boldness and vision. His skill in using mechanical ideas in mathematics was paralleled by his ingenious use of mathematics in mechanics.

The Latin West received its knowledge of Archimedes from two sources: Byzantium and Islam. His works were translated from the Greek and Arabic into Latin in the twelfth century and played an important role in stimulating the work of medieval natural philosophers. Knowledge of Archimedes' ideas multiplied during the Renaissance, and by the seventeenth century his insights had been almost completely absorbed into European thought and had deeply influenced the birth of modern science. For example, Galileo was inspired by Archimedes and tried to do for dynamics what Archimedes had done for statics. More than any other ancient scientist, Archimedes observed the world in a way that modern scientists from Galileo to Albert Einstein admired and sought to emulate.

Robert J. Paradowski

FURTHER READING

Aaboe, Asger. *Episodes from the Early History of Mathematics*. New York: Random House, 1964. After a brief account of Archimedes' life and a survey of his works, the third chapter of this book presents three samples of Archimedean mathematics: the trisection of an angle, the construction of a regular heptagon, and the determination of a sphere's volume and surface area.

Bell, E. T. *Men of Mathematics*. New York: Simon and Schuster, 1965. A collection of biographical essays on the world's greatest mathematicians. Bell discusses Archimedes, along with Zeno of Elea and Eudoxus, in an early chapter on "Modern Minds in Ancient Bodies."

Dijksterhuis, E. J. *Archimedes*. Princeton, N.J.: Princeton University Press, 1987. This edition of the best survey in English of Archimedes' life and work also contains a valuable bibliographical essay by Wilbur R. Knorr.

Finley, Moses I. *Ancient Sicily*. Vol. 1 in *A History of Sicily*. New York: Viking Press, 1968. Finley's account of the history of Sicily from antiquity to the Arab conquest has a section explaining how the politics of the Second Punic War led to Archimedes' death.

Heath, T. L. *A History of Greek Mathematics*. 2 vols. 1921. Reprint. New York: Dover, 1981. A good general survey of ancient Greek mathematics that contains, in volume 2, a detailed account of the works of Archimedes.

Kline, Morris. *Mathematical Thought from Ancient to Modern Times*. New York: Oxford University Press, 1990. Kline's treatment of Archimedes emphasizes the themes of his work rather than the events of his life.

Stein, Sherman K. *Archimedes: What Did He Do Besides Cry Eureka?* Washington, D.C.: Mathematical Association of America, 1999. An accessible account of Archimedes' accomplishments, as well as Archimedes' life, the 1906 discovery of his manuscript, and his methods. Includes bibliography and index.

Van der Waerden, B. L. *Science Awakening*. 4th ed. Princeton Junction, N.J.: Scholar's Bookshelf, 1988. A survey of ancient Egyptian, Babylonian, and Greek mathematics. The chapter on the Alexandrian era (330-220 B.C.E.) contains a detailed account of Archimedes' life, legends, and mathematical accomplishments.

ĀRYABHAṬA THE ELDER

Indian mathematician and astronomer

Āryabhaṭa the Elder's treatise, The Aryabhatiya, *is the first work of Indian mathematics that has a definite author and date. It indicates what Indian mathematicians had accomplished by the end of the fifth century.*

Born: c. 476; possibly Ashmaka or Kusumapura (now in India)
Died: c. 550; place unknown
Also known as: Āryabhaṭa I; Āryabhaṭṭa

EARLY LIFE

Āryabhaṭa (ahr-yah-BAH-tuh) the Elder is a figure of whom almost nothing is known other than his work. His only surviving book, the *Aryabhatiya* (499; *The Aryabhatiya*, 1927), indicates that it was written in 499 and that he was twenty-three years old at the time, which provides the traditional date for his birth. There is a good deal of disagreement about the exact date, as well as about the place of his birth. One of his successors in the field of Indian mathematics refers to Āryabhaṭa as being from Ashmaka, a region in the northwestern part of modern India. There was a migration from that region to a more southerly part of India, and Āryabhaṭa is likely to have been born there, as it is close to the only geographic region in which he is known to have worked. He is called Āryabhaṭa the Elder to distinguish him from a later mathematician who wrote under the same name. The later Āryabhaṭa may have chosen to use that as the signature for his works as a tribute to his predecessor.

In the course of the mathematical section of *The Aryabhatiya*, the author refers to the town of Kusumapura. Most scholars believe that this refers to the town where he was doing his academic work rather than his birthplace, but others argue that this town was where the mathematician was born. Kusumapura is frequently identified as the modern-day Patna, the capital of the region of Bihar. Because his work was to prove influential in the Kerala school of astronomy, practiced at the very south tip of India, some scholars have claimed that he spent his career there, but evidence to support that claim is no easier to come by than evidence regarding any other feature of Āryabhaṭa's life.

The nature of the education that Āryabhaṭa received is unknown, but from the text of his work it is clear that he studied Greek mathematics and astronomy as well as the Indian traditions in both areas. Because Āryabhaṭa's work is the first systematic treatment of these subjects that has survived from that part of the world, it may be that he was dependent on odd scraps of knowledge from his predecessors. He would have had plenty of examples of particular sorts of calculation from which to learn. Because he wrote in Sanskrit verse, his training would have been literary as well as scientific. From the religious tone of his invocations, it seems safe to conclude that he was not a Buddhist, as he pays reverence to Brahman and the astronomical bodies.

LIFE'S WORK

From later references, it appears that Āryabhaṭa wrote a book on astronomy as well as *The Aryabhatiya*. There were periods of time when neither work was available to students, but *The Aryabhatiya* was subsequently made generally available. The work on astronomy is not known to have survived.

The Aryabhatiya is a Sanskrit poem divided into four sections: an introduction, a mathematical section, a section on time measurement, and a section on celestial spheres. The opening section is written in a different verse form than the others, as though it may have been more like an invocation. There is an invocation at the start of the section on mathematics as well. The entire mathematical section is only thirty-three verses long, which attests to the brevity of his style. As a result, there is no illustration of the techniques and ideas he describes by numerical examples.

By Āryabhaṭa's time, the use of numerals that were the ancestors of the Hindu-Arabic system was already common. Because Āryabhaṭa was writing a poem, however, that form of number was not convenient for fitting into the meter. As a result, he introduced a system of words in the first section of the poem to correspond to numbers. His numerical vocabulary therefore is not as wide as the range of numbers already in use in Indian computation but sufficed for the writing of his poem.

After the opening invocation in the section devoted to mathematics (*ganita*, in Sanskrit), Āryabhaṭa lists in the second stanza the names for classes of numbers. The third stanza is on the subject of geometry; it points out that the product of two equal numbers was called by the same name as both the square of a figure and its area. He points that the same consideration applies in three dimensions to the cube and the product of three equal quantities.

The following two stanzas (four and five) present methods for calculating the square root and the cube root of a number. From the form of the instruction, it is clear that Āryabhaṭa recognized that the square root and cube root were not generally going to terminate in a finite number of places. The geometrical points made in the previous stanzas could have come out of the Greek tradition of theoretical mathematics, but these techniques for calculation could be part of a native Indian tradition. In general, the issue of how far Āryabhaṭa was indebted to Greek sources is hard to resolve, and the discussion often revolves about political issues as well as mathematical or scholarly ones.

The next four stanzas all take up the area and volume of various two-dimensional and three-dimensional figures. The formulae for the area of a triangle and a circle are correct, as is that for the volume of a pyramid, but the formula for the volume of a sphere is incorrect. If Āryabhaṭa were drawing on the Greek geometrical tradition throughout, it is hard to see how he could have made this error, which seems to be based on an analogy with the area of a circle. In addition, he gives the area for a trapezoid and plane figures more generally. The more general formulae may be suggested by the kind of averaging involved in handling the trapezoid. It is worth remembering that all the so-called formulae are really expressed in the language of poetry rather than written out as equations.

The tenth stanza includes one of the most interesting features of *The Aryabhatiya*, an approximation for π (the ratio of the circumference of a circle and its diameter). The specific numerical values given by Āryabhaṭa lead to a decimal approximation good to four places, and there is plenty of discussion about how this compares with the best possible values that the Greeks obtained. In principle, for example, the method of exhaustion of Archimedes (c. 287-212 B.C.E.) could produce approximations to an arbitrary degree of exactness. What is clear is that Āryabhaṭa recognized that his value remained only an approximation.

The following two stanzas deal with the calculation of quantities related to the trignometric sine function. The centrality of the sine function in the mathematics of Āryabhaṭa came from its appearance in the tradition of calculating orbits of planets that goes back to Greek explanations of observations, records of which survive from Babylonian times. The mathematical astronomy of Ptolemy (c. 100-178 C.E.) was probably brought to India in the course of the centuries before Āryabhaṭa's work, as there were Greek and Egyptian traders in India and Indian traders in the centers of culture of Europe and the Mediterranean. However, there was also a native tradition of astronomy going back many centuries, although it is not clear how mathematical a form it had taken. Āryabhaṭa's interests seemed to merge those of Greek mathematics and Hindu astronomy.

Greek mathematics took issues of construction as seriously as proofs, and Āryabhaṭa includes a stanza on the construction of various figures. The next three stanzas take up matters relating to the construction of sundials, not surprising in view of the section following that on mathematics being devoted to time measurement. The seventeenth stanza then turns to the Pythagorean theorem about the relationship between the sides of a right triangle. Historians have found references to the theorem in many mathematical cultures other than the Greek, and it could easily have been part of an Indian tradition as well. Because there is no reference to a proof, it is hard to see any connection with the version given in Euclid (c. 330-c. 270 B.C.E.).

After a stanza on the lengths of chords in a circle, Āryabhaṭa explains how to calculate the sum of the terms in an arithmetic series (where the difference between successive terms is constant). He also provides the way to calculate the number of terms involved if one has the sum, and this presupposes a knowledge of the quadratic formula, the way to derive the roots of a quadratic (second-degree) equation. The quadratic formula had been used by Babylonian mathematicians much earlier, and it is hard to tell how it fit into the rest of Āryabhaṭa's algebraic knowledge. The formulae he then proceeds to give for the sum of the squares and cubes of arithmetic series also come without derivation.

A couple of stanzas are devoted to getting the values of individual terms from various ways in which they can be combined. After a stanza (number 25) on interest (in which compounding takes place), Āryabhaṭa introduces the rule of three, the standard method of dealing with proportions for algebraists for millennia. In subsequent stanzas, he provides a guide to calculating common denominators for fractions and resolving problems by using the inverse of the operations in which the problems are posed. The range of subjects continues to widen, as he gives a way of getting a sum from differences, of calculating the value of an object with an agreed monetary standard, and of calculating the distances between planets.

The height of Āryabhaṭa's mathematical originality comes in the final two stanzas, in which he presents the general techinique for solving Diophantine equations of the first degree. Diophantus (fl. c. 250 C.E.) had presented solutions of number of special cases, but Āryabhaṭa clearly developed more general forms. The additional detail given to the method (by taking an extra stanza) and its position at the end of the section suggests that Āryabhaṭa recognized the extent of what he had accomplished.

The third section of *The Aryabhatiya* looks at issues relating to time but largely through the medium of astronomy. The final section takes up the celestial sphere, and both sections had an influence on subsequent Indian astronomy as well as being brought subsequently to the West. The work as a whole ends with a couple of stanzas celebrating his own accomplishments and criticizing those who disparage the field of astronomy. It is safe to say that this was in response to critics, but it is not clear what the basis was for their objections.

After Āryabhaṭa's death, subsequent Indian mathematicians, especially Brahmagupta (c. 598-c. 660), disparaged his accomplishments for not having been presented systematically. On the other hand, his work remained the starting point for Indian mathematics and astronomy for many years. Commentaries were published on *The Aryabhatiya* for more than a thousand years, and it has also drawn the attention of modern scholars. As historians of mathematics look for sources outside the Greek tradition, Āryabhaṭa's work is both attractive for its style and worthy of investigation.

SIGNIFICANCE

Āryabhaṭa stands at the head of the Indian tradition of mathematics. There was clearly a stream of predecessors on which he drew, but their anonymity makes their importance a trifle elusive. The work of Āryabhaṭa provided a spur for a thousand years of mathematical research. The form in which he presented his work also helped make sure that it could be taken as a classic contribution to Indian scholarship, not merely a mathematical thesis.

When India launched its first satellite in 1976, it was the 1500th anniversary of Āryabhaṭa's birth, and the satellite was named after him. At present, there is an Āryabhaṭa Forum centered in the United Kingdom working on issues in the history of mathematics. The legacy of Āryabhaṭa continues to serve as a reminder that not all mathematics can be assumed to have come through the Greek pipeline.

Thomas Drucker

FURTHER READING

Menon, K. N. *Aryabhata: Astronomer, Mathematician.* New Delhi, India: Ministry of Information, 1977. Part of the spate of literature produced by the 1500th anniversary of Āryabhaṭa's birth, this work is typical in that it claims all sorts of achievements for Āryabhaṭa at the expense of both other mathematical cultures (like the Greek) and other Indian mathematicians.

Proceedings of the International Seminar and Colloquium on Fifteen Hundred Years of Aryabhateeyam. Kochi, India: Kerala Sastra Sahitya Parishad, 2002. This was sparked by the 1500th anniversary of Āryabhaṭa's book rather than his birth. The collection of articles covers a good deal not connected with Āryabhaṭa, but it does provide evidence for continued disagreements over some of the basic facts of his life.

Velukutty, K. K. *Heritages to and from Aryabhatta.* Elipara, India: Sahithi, 1997. Another assault on the standard accounts of the details of Āryabhaṭa's life. It follows up some other Indian mathematical traditions with attention to details of practice.

FRIEDRICH WILHELM BESSEL
German astronomer

Bessels contribution to astronomy was to increase greatly the accuracy of measuring the positions of stars by using more advanced instruments and developing methods to account for instrument and observer error. The most famous discovery resulting from these observations was the first accurate determination of the distance to a star.

Born: July 22, 1784; Minden, Brandenburg (now in Germany)
Died: March 17, 1846; Königsberg, Prussia (now Kaliningrad, Russia)

EARLY LIFE
Friedrich Wilhelm Bessel (BEHS-ehl) was born to a civil servant and a minister's daughter. One of nine children, he went to the local gymnasium but left after only four years to become a merchant's apprentice. He showed no particular talent at school. At the age of fifteen, he began his unpaid seven-year apprenticeship to a merchant firm in Bremen. He excelled at his accounting job and received a small salary after one year. He spent his spare time teaching himself geography and languages because of his interest in foreign trade. He also learned about ships and practical navigation through self-study. Determining the position of a ship at sea—a long-standing problem in navigation—intrigued him. He therefore began to study astronomy and mathematics to understand the theory behind the existing methods.

Bessel began to make observations of stars on his own, which he was able to compare to observations reported in numerous professional journals of astronomy. One of his earliest tasks was to determine the orbit of Halley's Comet based on several observations of its position. He studied existing methods to determine the easiest way to do this, and he used observations made in 1607 to supplement his own. The precision of his observations and the scrupulous care given to minimizing or correcting for observational errors would characterize his professional work throughout his life. He made the observations, adjusted (or reduced) the 1607 data to make them directly comparable to his own, and submitted the results to Wilhelm Olbers in 1804.

Olbers, a physician and highly esteemed amateur astronomer, was impressed with the agreement between Bessel's observations and Edmond Halley's calculation of the orbit. He urged Bessel to improve it further with more observations. Olbers was impressed enough to recommend Bessel in 1806 for a position as assistant at a private observatory near Bremen. Bessel made further observations of the comet and published the results in 1807 to wide acclaim.

LIFE'S WORK

In 1809, Bessel was appointed director of the new observatory at Königsberg, where he remained for the rest of his life. His early fame came from his reduction of the earlier observations of James Bradley. Bradley's measurement of the apparent position of stars had to be corrected for the motion of Earth, the bending of starlight as it passes through the air, and instrument errors. With sufficient care, any observation can be reduced to a universal coordinate system.

While waiting for the construction of the observatory, Bessel worked on reducing Bradley's observations of more than thirty-two hundred stars with the goal of producing a reference system for measuring the positions of other stars. Bessel received the Lalande Prize of the Institute of France for his production of tables of refraction based on these observations. In 1818, he completed thereduction and published the results in his work *Fundamenta astronomiae* (1818; fundamental astronomy). This work provided the most accurate positions of a chosen set of stars. Accurate positions of a few stars are required to form the basis for extremely accurate measurements of positions for all other stars. This work has been said to mark the birth of modern astrometry. Bessel also provided accurate proper motions of many stars. (The so-called fixed stars actually move a very small amount over the centuries. When all perturbing effects are removed, the motion that is left to the star is called its proper motion.)

Bessel's next important contribution was to increase the accuracy of the measurement of stellar positions and motions. In 1820, he determined the position of the vernal equinox with great accuracy. The equinox is employed as the origin of the coordinate system used to record a star's position. He further improved accuracy in his work *Tabulae regiomontanae* (1830; *Refraction Tables*, 1855), in which he published the mean positions for thirty-eight stars for the period 1750-1850. In 1821, he noticed a systematic error in observation that was peculiar to each observer and called it the personal equation. This systematic error was reduced as each observer became more experienced, but it never disappeared. Bessel devised a method to remove the error.

Identifying and measuring the proper motion of starswas crucial in producing an important contribution to astronomy. The slight but periodic variation in proper motion of a few stars was not accountable by considering the motion of Earth or instrumental factors. In making the observations that later appeared in the *Refraction Tables*, Bessel suggested that the variations in proper motion of the stars Sirius and Procyon could be explained by the existence of an as-yet-unseen companion star. More than a century later, the companions to these stars were observed, as well as companions to many others.

Bessel also made important contributions to mathematics. Prior to Bessel, it was common for observers of the heavens to record their data and only later, if ever, reduce that data. Bradley, whose observations Bessel used extensively, carefully noted any possible perturbing effects in his observations. Nevertheless, reduction of the data for the positions of the stars was put aside in favor of recording lunar data. Bessel emphasized the need for the data reduction to be done immediately by the observer. Such reduction required extensive manipulation of complicated equations. In the process of developing ways to remove errors, Bessel noticed that he could use a class of functions that solved problems involving the perturbing influence of one planet on the orbit of another. He systematically investigated and described this class of functions in 1824. These functions, which bear Bessel's name, are not restricted to astronomy: They are used in the solution of a wide variety of problems in physics, mathematics, and engineering.

Another direct benefit of increased accuracy in stellar positions was Bessel's determination of the distance to a star, which is his most important contribution to astronomy. Although many astronomers had earlier claimed to have measured stellar distance using methods based on questionable assumptions, and although two of Bessel's contemporaries also correctly determined such distances, Bessels comprehensive treatment of the data and his high accuracy of observations were convincing to his contemporaries.

As Earth moves in its annual orbit around the sun, the stars appear to move across the dome of the sky. By determining the location of a star at opposite ends of

Earth's orbit and using some simple trigonometry, the distance to a star can be measured. However, this so-called parallax (the apparent motion of an object caused by the motion of the observer) was small because the stars are very far away. The parallax had therefore never been measured. Indeed, some opponents of the heliocentric theory, according to which Earth revolves around the sun, used this failure to measure parallax as an argument against the theory. Astronomers had a rough figure for the radius of Earth's orbit and some idea of the extent of the solar system, but there did not seem to be a way to determine stellar distances that did not require the assumption that all stars had the same intrinsic brightness.

Earlier attempts at measuring parallax involved circumventing the problem of the immeasurably small parallax by looking at two stars that appeared to be very near to each other but were of different brightness. It was thought that the dimmer star would be farther away and that observing the relative change of position could lead to a determination of stellar distances. This method did not work, because not all stars are of the same intrinsic brightness. Most stars that appear near to each other are in fact binary stars and really are near each other.

Bessel used a different approach: He assumed that stars with large proper motions are closer than stars with small proper motions. He chose the star known as 61 Cygni, because it had the largest proper motion known. He used a new measuring device called the Fraunhofer heliometer (after Joseph von Fraunhofer, a nineteenth century optician), which was designed to measure the angular diameter of the sun and the planets. Its manner of comparing the images from two objects to determine angular diameter was more accurate than earlier instruments. Using nearby stars for comparison and observing for eighteen months, Bessel was able to measure a parallax of slightly under one-third of a second of arc, which is equal to the width of a dime viewed from twenty miles. From this amount of parallax, Bessel calculated that 61 Cygni was 10.9 light years or seventy trillion miles away. He completed his calculations and published the results in 1838.

The last six years of Bessel's life were marked by deteriorating health, but he managed to complete a number of works before his death from cancer in March, 1846.

Significance

Friedrich Wilhelm Bessel made important contributions to astronomy, mathematics, and geodesy. His work marks the turning point from a concern with planetary, solar, and lunar observations to investigations of the stars. His measurement of the distance to a star is noteworthy because it settled the centuries-old question of whether stars exhibited parallax. The care he took in his observations set much higher standards for the science of astronomy. Bessel's goal was to observe the stars accurately enough to predict their motion and to establish a reference system for their positions. As part of that plan, he developed methods for the careful determination of instrument and observer error, conducted years of observations himself, and developed the mathematical techniques to reduce the data. Bessel's lasting achievement was to raise the science of observing, reducing, and correcting astronomical data to an art.

Roger Sensenbaugh

Further Reading

Clerke, Agnes. *A Popular History of Astronomy During the Nineteenth Century*. New York: Macmillan, 1887. Although written during the period it was supposed to cover, this work has several redeeming qualities. Expressly written for a general audience, the book's language is clear and precise. A valuable record of what near-contemporaries thought of Bessel.

"The Deep Sky." *Astronomy* 31, no. 8 (August, 2003): 65. Describes the special place the star system Cygnus holds in the history of astronomy because of Bessel's measurements of 61 Cygni.

Herrmann, Dieter B. *The History of Astronomy from Herschel to Hertzsprung*. Translated by Kevin Krisciunos. Rev. ed. New York: Cambridge University Press, 1984. Traces the history of astronomy from 1780 to 1930. Written from a Marxist perspective.

Hirshfeld, Alan. *Parallax: The Race to Measure the Cosmos*. New York: W. H. Freeman, 2001. Bessel's measurement of 61 Cygni is recounted in this history of astronomers' attempts to measure the distances to stars and plants by parallax observation.

Hoskin, Michael A. *Stellar Astronomy: Historical Studies*. Chalfont, England: Science History, 1982. A collection of material published in *Journal for the History of Astronomy*, with the addition of some new material. Attempts a synthesis of existing scholarship on the history of stellar (as opposed to planetary) astronomy as of the early 1980s.

Pannekoek, Anton. *A History of Astronomy*. New York: Interscience, 1961. Traces the history of astronomy from antiquity to the present. Part 3, "Astronomy Surveying the Universe," contains information on

Bessel and places him in the historical context of nineteenth century astronomy.

Williams, Henry Smith. *The Great Astronomers*. Westport, Conn.: Greenwood Press, 1930. Reprint. New York: Newton, 1932. Book 5 deals with Bessel, among other subjects. Concerns parallax and Bessels contributions to measuring the distance to 61 Cygni. Describes the work of Bessel's contemporaries who measured the parallax of other stars at about the same time.

Max Born
German physicist

Born made fundamental contributions to the emerging theory of quantum mechanics. He also wrote extensively on relativity and atomic physics for popular readers. As professor of physics at Göttingen, he served as mentor to emerging physicists, including J. Robert Oppenheimer and Werner Heisenberg. He was honored with a Nobel Prize in Physics in 1954.

Born: December 11, 1882; Breslau, Germany (now Wrocław, Poland)
Died: January 5, 1970; Göttingen, West Germany (now in Germany)

Early Life

Max Born was the son of Gustav Born, a physician and embryologist, and Margarethe Kauffmann, daughter of a wealthy industrialist. Born's mother gave birth to a sister, Käthe, about eighteen months later but died during her third pregnancy, when Born was only four years old. The Borns were nonreligious Jews but nonetheless experienced anti-Semitism. Born's father remarried in 1892, providing him with a stepmother, but his father died in 1900, before Born's eighteenth birthday.

Born began his education at the Kaiser Wilhelm Gymnasium, a school that stressed Latin and Greek but did offer solid mathematics education. He was a satisfactory but not exceptional student. His family was relieved when he passed the exam that permitted his entry to a university. Initially, Born had difficulty deciding what and where to study. He first attended the University of Breslau as a student in the philosophy faculty, which included the sciences and mathematics. In 1902 he moved on to Heidelberg and returned to Breslau. In the summer of 1904, he entered the University of Göttingen.

Göttingen at the time was a center for mathematics study, featuring three world-class intellects: department chair Felix Klein, David Hilbert, and the young Hermann Minkowski, brought in by Hilbert and one of only two professors of Jewish background. Born at first sought to concentrate in pure mathematics but incurred the displeasure of Klein. Born then chose Carl Runge, an applied mathematician, to be his dissertation adviser. He completed a thesis on the mathematical theory of elasticity and passed his Ph.D. oral examination on July 11, 1906.

Before Born could begin a university teaching career, he had to habilitate, that is, present a research paper beyond his Ph.D. work and pass a further oral examination. Unwilling to present himself as a candidate in pure mathematics, he decided to qualify as a physicist, traveling first to Cambridge University in England and then back to the University of Breslau. Ultimately, he returned to Göttingen, where he was able to habilitate in 1909. Habilitation made Born a Privatdocent, one who could lecture to paying students but did not receive a salary from the university.

On August 2, 1913, Born and Hedwig Ehrenberg were married in the Berlin suburb of Granau. Hedwig's father had converted from Judaism to the Lutheran Church in order to marry her mother; he expected Born to do likewise. At first Born resisted, although he agreed to complete some religious instruction. Months after the ceremony, Born relented and allowed himself to be baptized, although he never attempted to distance himself from his Jewish origins.

A daughter, Irene, was born the following year, and was christened in the Lutheran Church. The Borns would have two other children, daughter Margarethe and a son, Gustav.

Life's Work

In 1912, Born and Hungarian physicist Theodore von Kármán, who had been a student at Göttingen, published an analysis of lattice vibrations in crystalline solids in an attempt to explain anomalies in the heat capacities of solids. A simpler treatment by Dutch physical chemist Peter Debye appeared in the same year, diverting attention from Born and Kármán's work. The dynamics of crystals would remain an interest for Born throughout his career.

In 1915, the young Born family moved to Berlin, where Born had been named an *extraordinarius*, or associate professor. Berlin was then the center of German physics, with both Max Planck and Albert Einstein in residence. However, World War I had begun in August of 1914, and the conflict rapidly devolved into prolonged trench warfare. Born had to face the possibility of doing war research as well as being called to military service. Fritz Haber, an eminent German physical chemist, was in charge of the physical chemistry institute in Berlin and an advocate of gas warfare. He sought Born's assistance in the development of protective masks, but Born resisted, believing the use of poison gas to be immoral. Instead, Born was attached as a sergeant to the army unit charged with artillery research. This provided an environment in which he could devote some time to fundamental physical issues, including the nature of ionic crystals. Born and Haber would together develop the Born-Haber cycle, by which chemically important quantities could be calculated from physical data.

In 1919, Born was called to Frankfurt as *ordinarius*, or full professor. Postwar conditions in Frankfurt were rather dismal, however, and the Borns accepted an invitation to return to Göttingen in 1921.

As head of the physical institute at Göttingen, Born had the opportunity to hire some of the best physicists of his day as assistants. These included Wolfgang Pauli and Werner Heisenberg, both of whom would win Nobel Prizes. With Heisenberg, Born worked out the matrix formulation of quantum mechanics and, later, the famous uncertainty principle, which asserts that it is not possible to simultaneously determine the position and velocity of a subatomic particle. Born's strong command of mathematics supplemented Heisenberg's more intuitive insight.

It was Born and Heisenberg's uncertainty principle that Einstein objected to in his famous statement that "God does not play dice." At scientific conferences and through numerous letters and publications, Einstein maintained that the quantum mechanics being developed by Born, Heisenberg, and others could not be an adequate description of nature. Einstein would propose experiments to circumvent the work of Born and Heisenberg. Danish physicist Niels Bohr would, in turn, point out the problems in Einstein's proposals. Austrian physicist Erwin Schrödinger complicated matters by introducing a wave formulation of quantum mechanics, which gave the same predictions as the matrix mechanics of Heisenberg. Born first suggested the generally accepted interpretation of Schrödinger's wave function as specifying the probability for finding the particle of interest at a given point in space.

Born made contributions to many areas of physics and served as doctoral mentor to a number of students who would gain eminence in their own right. In 1933, with the rise of the political power of the Nazis in Germany, Born and other professors of Jewish ancestry were suspended from their teaching positions and then dismissed. The Born family made its way to Cambridge University in England, where Born was named Stokes lecturer. In 1936 he assumed the Tait Chair in Natural Philosophy at the University of Edinburgh, becoming a British subject in 1939. Long after the war, in 1953, he retired to Göttingen, where he could enjoy the salary that was due him as a professor under the German restitution laws. In 1954 he shared the Nobel Prize in Physics with Walther Bothe of Germany.

Significance

Born's contributions to physics began with the emergence of Einstein's relativity theory and continued through the development of quantum mechanics in its early twenty-first century form. Born's effectiveness as a collaborator and mentor may have overshadowed his own contributions. His work was not recognized with a Nobel Prize until 1954. His many contributions include his work on the dynamics of crystal lattices, relativity theory, and the uncertainty principle. He served as doctoral adviser to a number of eminent physicists, including J. Robert Oppenheimer, who became prominent in the American physics community and became director of the Manhattan Project. The Born-Oppenheimer approximation remains the starting point for most molecular quantum mechanical calculations. Perhaps ironically, Heisenberg, generally credited as the discoverer of the uncertainty principle and head of the German atomic bomb program, was Born's doctoral student as well.

Born wrote extensively, for both general and professional audiences. He is the author of a monograph on the dynamics of crystal lattices, updated with Chinese physicist Huang Kun, that is still considered a standard work in the field. Likewise, a monograph on optics, coauthored with Czech-born American physicist Emil Wolf, is a standard reference in that field. Born also wrote on relativity and atomic physics for general audiences.

Donald R. Franceschetti

Further Reading

Born, Max. *My Life: Recollections of a Nobel Laureate.* New York: Charles Scribner's Sons, 1975. Born began this autobiography in 1940 and completed it near the end of his life. A richly detailed recounting of his early life. His treatment of his later years is somewhat abbreviated.

_____, ed. *The Born-Einstein Letters, 1916-1955: Friendship, Politics, and Physics in Uncertain Times.* New York: Macmillan, 2005. Born and Einstein maintained an extensive correspondence over a period of forty years on topics ranging from politics to the quantum theory, a portion of which is presented here and edited by Born.

Greenspan, Nancy Thorndike. *The End of the Certain World: The Life and Science of Max Born, the Nobel Physicist Who Ignited the Quantum Revolution.* New York: Basic Books, 2005. A full-length biography written by a friend of the Born family.

Brahmagupta

Indian mathematician

Brahmagupta wrote Brahmasphuzasiddhānta, *a book in verse expounding a complex system of astronomy and containing two chapters on arithmetic, algebra, and geometry. His work on indeterminate equations and the introduction of negative numbers greatly influenced the development of science in both India and Arabia.*

Born: c. 598; Bhillamala, Rajputana (now Bhinmal, India)
Died: c. 660; possibly Ujjain, Kingdom of Magadha (now in India)

Early Life

The Hindu astronomer and mathematician Brahmagupta (brah-mah-GEWP-tah) was born to a man named Jishnugupta from the town of Bhillamala. The suffix *-gupta* may indicate that the family belonged to the Vaiśya caste (composed mostly of farmers and merchants).

In contrast to his predecessor, Āryabhaṭa the Elder (c. 476-c. 550), who lived in relative obscurity at Kusumapura (modern-day Patna, Bihar), Brahmagupta had the opportunity to live, study, and teach in Ujjain, a town in the state of Gwalior, Central India. Ujjain was then the center of Hindu mathematics and astronomy and had the best observatory in India. At Ujjain, Brahmagupta also had access to the writings of many great scientists who came before him, including Hero of Alexandria (fl. 62-late first century C.E.), Ptolemy (c. 100-178 C.E.), Diophantus (fl. c. 250 C.E.), and Āryabhaṭa the Elder. Brahmagupta later drew heavily on these sources in his own writings, often correcting their errors. For example, he corrected Āryabhaṭa's mistake regarding the formulas for the surface areas and volumes of the pyramid and cone. Brahmagupta even borrowed mathematical problems, including one calling for the calculation of the position of a break in a bamboo pole. This problem had first appeared in the Chinese text *Jiuzhang shuanshu* (c. 50 B.C.E.-100 C.E.; arithmetic in nine sections), the authorship and date of which are uncertain.

Another influence of Ujjain was on Brahmagupta's style of writing. Like other Hindu scientists, including Āryabhaṭa, he wrote his mathematical texts as poetry. The Indian practice was to clothe all arithmetical problems, especially those in schoolbooks, in poetic garb, fashioning them into puzzles that served as a popular amusement. Brahmagupta wrote that his mathematical problems were undertaken only for pleasure and that a wise man could invent a thousand more or solve those presented by others, thereby eclipsing their brilliance, just as the sun eclipses the other stars in the sky.

At Ujjain, the thirty-year-old Brahmagupta completed his masterwork, the *Brahmasphuzasiddhānta* (c. 628; the improved astronomical system of Brahma). The date of this work has been determined by consulting both commentary from later Hindu scholars and, appropriately, astronomical data.

Life's Work

The first ten chapters of Brahmagupta's *Brahmasphuzasiddhānta* deal with various astronomical issues, including the mean and true longitudes of the planets, diurnal motion, lunar and solar eclipses, heliacal risings and settings, the lunar crescent and "shadow," conjunctions of the planets, and their conjunctions with the stars. The following thirteen chapters take up an examination of previous work on astronomy (including Āryabhaṭa's), additions and problems (and their solutions) supplementing six of the earlier chapters, mathematics, the gnomon, meters, the sphere, instruments, and measurements. The work's twenty-fourth and final chapter summarizes the principles of Brahmagupta's astronomical system in a compendious treatise on astronomical spheres. (Some manuscripts include an additional chapter containing tables.) All but two of the chapters deal with astronomy, but scholars have chosen those two chapters, 12 and 18, which deal with algebra and mathematics, to study most intensely.

Although Brahmagupta studied mathematics only for its applicability to astronomy and considered knowledge of the rules of arithmetic a prerequisite to be a *ganaca* (a student of astronomy), most scholars in the ages since he lived have studied his mathematics more closely than his astronomy. Of particular interest is his work on indeterminate equations, building on the work of both Diophantus and Āryabhaṭa. Brahmagupta's work, along with that of Bhāskara II (1114-c. 1185), solved the so-called Pell equation, $y^2 = ax^2 + 1$, where a is a nonsquare integer. Brahmagupta showed that from one solution where x, y, and xy do not equal zero, a general formula indicating an infinite number of solutions could be derived. Brahmagupta also stated that the equation $y^2 = ax^2 - 1$ could not be solved with integral values of x and y unless a was equal to the sum of the squares of any two integers. Brahmagupta's work on these equations, with additions by Bhāskara, is highly regarded because it was not for several centuries that another mathematician, namely Joseph-Louis Lagrange (1736-1813), could completely work out the Pell equation.

Brahmagupta also studied indeterminate equations of the first order, such as this one: Two ascetics live on top of a hill of h units of height, whose base is mh units away from a nearby town. One ascetic descends the hill and walks directly to the town. The other, being a wizard, flies straight up a certain distance, x, then proceeds in a straight line toward the town. If the distance traveled by each ascetic is the same, and h is 12 and mh is 48, find x. The solution comes from the formula $x = mh/(m + 2)$, or in this case, $x = 8$.

Brahmagupta's work on the geometry of quadrilaterals, which was probably inspired by his studies of Ptolemy and Hero, is also a landmark in the history of Hindu mathematics. Brahmagupta found the formulas, for the first time, for the diagonals (defined as m and n) of a quadrilateral having sides of length a, b, c, and d

and opposite angles of A and B, and C and D. He calculated the diagonals thus:

$$m^2 = (ab + cd)(ac + bd)/(ad + bc) \text{ and}$$
$$n^2 = (ac + bd)(ad + bc)/(ab + cd).$$

These formulas were later studied by Bhāskara, who, failing to understand that they applied only to quadrilaterals inscribed in a circle, incorrectly pronounced them unsound. Brahmagupta also figured that, if a, b, c, A, B, and C are positive integers such that $a^2 + b^2 = c^2$ and $A^2 + B^2 = C^2$, then the cyclic quadrilateral having consecutive sides aC, cB, bC, and cA (which came to be called a Brahmagupta trapezium) has rational area and diagonals, and the diagonals are perpendicular to each other. These formulas are most remarkable; nothing like them had previously appeared in Hindu geometry.

Brahmagupta borrowed from Hero of Alexandria the formula for the triangular area, but he brilliantly extended Hero's formula to work with quadrilaterals that can be inscribed within circles. This idea was later built on by the ninth century Hindu mathematician Mahāvīra and was much admired by later commentators. Brahmagupta's other advances in mathematics included proving the Pythagorean theory of the right triangle, deriving formulas for the areas of a square and a triangle inscribed in a circle, and showing that a rectangle whose sides were the radius and semiperimeter of a circle had the same area as that circle.

Although he is now remembered mostly for his advances in mathematics and his influence on the mathematical work of later Hindus such as Mahāvīra and Bhāskara, Brahmagupta considered himself primarily an astronomer. Almost every Hindu commentator on astronomy discusses his work. Indeed, some of his work in astronomy is quite admirable. He provided fairly accurate figures for the circumference of Earth and the length of the calendar year. Brahmagupta gives a figure different from Āryabhaṭa's for the circumference of Earth: 5,000 *yojanas*. Assuming that Brahmagupta's *yojana* was a short league of about 4.5 miles, that would convert his figure to 22,500 miles, which is not too far off the mark. He also tried to correct Āryabhaṭa's computation for the length of the year, which was 365 days, 15 *ghati*, 31 *pala*, and 15 *vipala*, or 365 days, 6 hours, 12 minutes, and 30 seconds. His own figure was slightly more accurate: 365 days, 15 *ghati*, 30 *pala*, 22 *vipala*, and 30 *pratipala* (365 days, 6 hours, 12 minutes, and 9.0 seconds).

Much of his astronomy, however, is quite erroneous. Like many Hindu scientists of the time, Brahmagupta was vehemently opposed to Āryabhaṭa's ideas that Earth revolved around the Sun and spun on its axis. Why then, Brahmagupta asked, do not the lofty bodies fall down to Earth? He also questioned Āryabhaṭa's theory of an aerial fluid that causes Earth to rotate.

Significance

Although Brahmagupta greatly extended the work of many preceding mathematicians and presented numerous valid theories of his own, it must be acknowledged that he did make some serious scientific errors. In addition to denying Āryabhaṭa's theories of the place of Earth in the solar system, he gave a faulty formula for the area of an equilateral triangle. In his studies on the circle, he alternately used 3 and the square root of 10 as values for π.

Yet Brahmagupta's importance as a scientist must have been recognized during his lifetime, because he was accused of propagating scientific falsehoods to please the priests and the ignorant commonfolk. The priests were particularly opposed to the ideas that Earth was round and that it rotated around the Sun. Perhaps Brahmagupta had lied to avoid the fate of Socrates (c. 470-399 B.C.E.).

Despite these accusations, at least two of Brahmagupta's algebraic formulations, although originally devised for use in astronomy, became widely used by Hindu traders. Of particular practical use was his rule of three, in which the Argument, the Fruit, and the Requisition are the names of the terms. The first and last terms have to be similar. The Requisition multiplied by the Fruit and divided by the Argument yielded the Produce.

Brahmagupta also introduced the use of negative numbers, which he used to unify three of Diophantus's quadratic equations under a general equation. These negative numbers were especially useful to merchants in representing debts, along with positive numbers, which represented assets. Another advance in mathematics that the merchants must have found helpful was Brahmagupta's work on interest rates.

By 700, Hindu merchants had introduced Brahmagupta's mathematics to the Arabs, with whom they carried on a high volume of trade. In 772, a table of sines from Brahmagupta—which, incidentally, was probably based on work by Āryabhaṭa—reached the ʿAbbāsid caliph al-Manṣūr, and it was ordered to be translated into Arabic. The entirety of the *Brahmasphuzasiddhānta* was translated into Arabic by 775, around the time works by other Greek and Hindu mathematicians were being translated by Arab scholars. Together, these works would greatly influence the nascent Arabic mathematics, with Brahmagupta's greatest contributions coming in the study of negative numbers and indeterminate equations.

Frank Wu

Further Reading

Ball, W. W. Rouse. *A Short Account of the History of Mathematics*. 4th ed. London: Macmillan, 1908. A thorough overview of the history of mathematics, with a section on Brahmagupta and his work on quadratic equations, right triangles, and algebra, plus scattered information on his later influence on Hindu and Arab mathematicians.

Cajori, Florian. *A History of Mathematics*. 5th ed. Providence, R.I.: AMS Chelsea, 2000. Gives the solution to Brahmagupta's broken bamboo problem, plus formulas for Brahmagupta's work on triangles and quadrilaterals.

Eves, Howard. *An Introduction to the History of Mathematics*. 6th ed. Philadelphia: Saunders College, 1990. Includes information on Brahmagupta's studies on indeterminate equations, the Pell equation, cyclic quadrilaterals, and the rule of three, along with a discussion of his place in the history of mathematics. Some problems (with solutions) based on his formula for the cyclic quadrilateral are included.

Joseph, George Gheverghese. *The Crest of the Peacock: The Non-European Roots of Mathematics*. Rev. ed. Princeton, N.J.: Princeton Unversity Press, 2000. Joseph examines the history of mathematics in cultures throughout the world, including India. Its wide coverage places India within the greater scope of mathematical development. Bibliography and indexes.

Lakshmikantham, V., and S. Leela. *The Origin of Mathematics*. Lanham, Md.: University Press of America, 2000. Lakshmikantham argues that the importance of the early Indian mathematicians has been underestimated. Bibliography and index.

Prakash, Satya. *A Critical Study of Brahmagupta and His Works*. New Delhi, India: Indian Institute of Astronomical and Sanskrit Research, 1968. A comprehensive study of Brahmagupta, his works, his sources, and the influence of his work on later writers. Contains an extensive bibliography.

Puttaswamy, T. K. "The Mathematical Accomplishment of Ancient Indian Mathematicians." In *Mathematics Across Cultures: The History of Non-Western Mathematics*, edited by Helaine Selin. Boston: Kluwer Academic, 2000. Examines the early Indian mathematicians and their importance.

Gerolamo Cardano

Italian mathematician

Cardano, best known for a quarrel over intellectual property with fellow mathematician Niccolò Fontana Tartaglia, also helped to transmit the results of a flurry of sixteenth century work in algebra. Cardano also initiated studies in the field of probability theory, which evolved into games of chance in the seventeenth century.

Born: September 24, 1501; Pavia, duchy of Milan (now in Italy)
Died: September 21, 1576; Rome, Papal States (now in Italy)
Also known as: Jerome Cardan; Girolamo Cardano

Early Life

Gerolamo Cardano (jay-RAW-lah-moh kahr-DAH-noh) wrote an autobiography describing the activities of his early life, but much subsequent scholarship has clarified his recollections. Cardano denied having been born illegitimately, but his parents, Fazio Cardano and Chiara Michena, apparently were not married at the time of his birth. They did marry, however, in 1524.

Cardano's father was a distinguished scholar as well as a lawyer, and he encouraged Cardano in his intellectual pursuits. At the time of his death, he left his son a small inheritance to help support him in his studies. In general, Cardano seems not to have enjoyed his childhood, and he accused those who were looking after him of neglect, even with regard to food.

Cardano began his university studies at Pavia in 1520 but proceeded to a medical degree in Padua in 1526. While his medical degree may have been intended to provide a livelihood for Cardano, his attention was not restricted to medical issues. He was perhaps fated to be known more for his mathematical work (which he presented to the public) than for his medicine.

It was his medical income, though, that enabled him to marry in 1531, and he had two sons and a daughter. It can be said that his family life was not a happy one, as one of his sons poisoned his wife and was executed. By 1534, Cardano had become an instructor of mathematics in Milan, where he also practiced medicine, earning a reputation in the process.

Life's Work

Algebra—from the Arabic word *al-jabr*—and Hindu-Arabic numerals were introduced to the West through the work of the Arabic mathematician and astronomer al-Khwarizmi (c. 780-c. 850). His work *Kitāb al-jabr wa al-muqābalah* (c. 820), which gave the word *al-jabr*, "algebra," to the West, means "the book of integration and equation." In the Western world, even into the last years of the Middle Ages, the influence of Euclid's *Elements* guaranteed that mathematics would be approached as a branch of geometry. While this was not much of a handicap when it came to solving algebraic equations in which the highest power of the variable is a second power (called a quadratic equation), it was not helpful in tackling equations in which there was a third power of the variable (called a cubic equation). With al-Khwarizmi's work, there had been progress in dealing with cubic equations, but this was not immediately known to Western Europe. The end of the Eastern Roman Empire with the fall of Constantinople in 1453 helped to direct a flow of scholarly material from the Middle East to Europe.

Another source of difficulty among Europeans in solving the cubic equation was the lack of a suitable notation. The Roman numeration system was scarcely designed for mathematical work, but it was still in common use in the sixteenth century. Also, the description of mathematical problems and their solutions was usually carried out with words rather than with the symbols later used to put together algebraic equations. The more complicated the problem, the more the lack of a helpful notation was felt, and the leap in level of difficulty between quadratic and cubic equations was substantial.

The first large step toward solving the cubic equation was taken by Scipione del Ferro. While he developed a method for solving a whole class of cubic equations, he did not reveal the method publicly because the method's secrecy was worth something as a weapon in public disputations. Such disputations helped one build a reputation in the intellectual circles of Italy at that time. Del Ferro passed along the secret to his student Antonio Fiore, who tried using it as a tactic in a public disputation with the mathematician Niccolò Fontana Tartaglia. Tartaglia, however, more so than del Ferro, had managed to solve an even broader class of cubic equations and was able to emerge triumphant from his dispute with Fiore.

Cardano learned of Tartaglia's success and wanted to profit from his discovery. Tartaglia followed del Ferro in refusing to bring his technique before the public, but he did disclose the technique to Cardano under condition of confidentiality. Cardano's subsequent actions have been the subject of detailed scrutiny, but he seems to have felt absolved from his vow to Tartaglia for two reasons.

First, he discovered that del Ferro had a version of the formula for solving cubic equations before Tartaglia, even if it was not so general. Then Cardano worked with his own son-in-law, Ludovico Ferrari, who pushed the ideas of Tartaglia even further. Once Ferrari came up with a method for solving an equation with a fourth power of the variable (called a quartic equation), Cardano felt that he owed it to the world to reveal these discoveries.

As a result, Cardano published a volume called *Artis magnae, sive de regulis algebraicis* (1545; *The Great Art: Or, The Rules of Algebra*, 1968). In it he detailed the various contributions of his predecessors and the work of Ferrari. Since the solution of the quartic equation depended in part on the solution of the cubic equation, he discussed the cubic equation and gave credit to Tartaglia, but he assumed credit for the quartic equation. Tartaglia was outraged, and his subsequent denunciations of Cardano's infidelity did a great deal to blacken Cardano's reputation in the scholarly world. A scholarly community that regards publication as an important part of the process of scientific discovery views Cardano with less distrust than does a community that believes discoveries are the property of their initial discoverer.

The other branch of mathematics to which Cardano made his most notable contributions was the field of probability. There had been a certain amount of discussion of counting cases (the number of possible outcomes for experiments) through the Middle Ages, but there was no basic mathematical formula connecting the ideas of probability from philosophy and religion with the calculation of possible outcomes, as in the work of Raymond Lull (c. 1235-1316). Cardano was the first to offer a definition of the probability of an event: the number of outcomes where that event occurs to the total number of possible outcomes. From this definition he went on to state a form of the law of large numbers. Cardano's important contributions, however, remained unpublished until after his death and the start of the work of the French mathematicians Pierre de Fermat (1601-1665) and Blaise Pascal (1623-1662) in the seventeenth century, to which are traced most subsequent developments in the field of probability.

Cardano obtained the chair of medicine at the University of Pavia in 1542 and remained there for almost

Gerolamo Cardano. (National Library of Medicine)

twenty years. His autobiography bears witness to the envy of his contemporaries and also the extent to which his life was embittered by their comments. He proceeded to the chair of medicine at Bologna in 1562 and was involved in public disputations on Galen (the Greek physician) as part of the intellectual life of the city and the university.

In 1570, however, Cardano was arrested by the Inquisition and imprisoned for casting the horoscope of Jesus. Theological objections to Cardano's act suggested that the events of the life of Jesus were the result of the influence of the stars rather than direct divine intention. That Cardano had worked on various ways of concealing texts of messages probably helped make him an object of suspicion.

After a time in prison, Cardano was forced to recant and abandon teaching. Cardano managed to outwait the ban and proceed to Rome by the time that a new pope had been elected. In this new environment Cardano was able to secure an annuity. Perhaps more important to Cardano was that he was able to use the more tolerant reception for his writings to write his autobiography *De propria vita liber* (1576; *The Book of My Life*, 1930).

Significance

Cardano contributed to many areas of scholarship, such as geology, hydrodynamics, and mechanics, and he argued against the continued influence of Aristotle in the physical sciences. His most enduring legacy, however, remains the creation of a discipline of algebra based on the researches of the Italian school to which he belonged. After generations of secrecy, Cardano brought recent advances in mathematics to the scholarly community at large.

Cardano introduced variations on the methods he had learned from others and took seriously the possibility of solutions that involved imaginary numbers. Even though he may have suffered abuse from contemporaries, posterity benefited as much from his arrangement of solution methods as from his own particular discoveries.

Thomas Drucker

Further Reading

Cardan, Jerome. *The Book of My Life*. Translated by Jean Stoner. New York: E. P. Dutton, 1930. An English translation of Cardano's autobiography. Includes a brief bibliography.

Cardano, Girolamo. *The Great Art: Or, The Rules of Algebra*. Translated and edited by T. Richard Witmer. Cambridge, Mass.: M.I.T. Press, 1968. A translation of Cardano's *Artis magnae*. Illustrations, bibliographical footnotes.

Eckman, James. *Jerome Cardan*. Baltimore: Johns Hopkins University Press, 1946. A supplement to a bulletin of the history of medicine, but giving an overall view of Cardano's life written in English, with Latin chapter titles.

Mankiewicz, Richard. *The Story of Mathematics*. Princeton, N.J.: Princeton University Press, 2000. Captures the difficulty of trying to do algebra in the absence of suitable notation and vocabulary.

Ore, Oystein. *Cardano: The Gambling Scholar*. Princeton, N.J.: Princeton University Press, 1953. By a distinguished mathematician, especially devoted to Cardano's work on probability. Includes a translation of Cardano's book on games of chance.

Wrixon, Fred B. *Codes, Ciphers, and Other Cryptic and Clandestine Communication*. New York: Black Dog and Leventhal, 1998. Description of some of Cardano's contributions to the field that may have led to his facing the Inquisition.

Lewis Carroll

English writer and mathematician

The creator of the immortal fairy tale Alice in Wonderland, *Lewis Carroll wrote stories and poems that fundamentally changed and enlivened childrens literature. He also pioneered childrens photography and published books that advanced the fields of logic and mathematics.*

Born: January 27, 1832; Daresbury, Cheshire, England
Died: January 14, 1898; Guildford, Surrey, England
Also known as: Charles Lutwidge Dodgson (birth name)

Early Life

Lewis Carroll was the pen name of Charles Lutwidge Dodgson, who retained his real name throughout his life. His father, Charles Dodgson, had given up his fellowship and lectureship in mathematics at Christ Church, Oxford University, to marry Frances Jane Lutwidge in 1827. Carroll was the first son of their eleven children. The family lived in Daresbury, where Carroll's father was parish curate, until 1843. Carroll showed an early talent for mathematics, cultivated by his father, and comic verse. Growing up in a close-knit upper-middle-class family, living in a secluded village, and deeply influenced by a stern but doting father, Carroll found childhood a time of innocent exploration and wonder, a view that colored his later literary works.

In 1844 Carroll attended a grammar school in Yorkshire, and in 1846 he went on to Rugby, one of England's leading private schools. Instructors at both schools helped him develop his mathematical and literary talents, but he disliked boarding away from his family. At Rugby the often harsh discipline administered by older students especially repelled him. For the rest of his life he disliked boys. At home on vacations, he helped his father teach in the local school, was a leader in games (many of which he invented), and wrote poetry and stories for magazines that he issued to amuse family and friends. These early poems were usually parodies of moralistic verse common during the early nineteenth century, and he explored several themes that appeared in his mature writing: violence, dreams and nightmares, family relationships, and the childs view of a bewildering adult world.

Carroll was a brilliant student. He won a scholarship to Christ Church, his father's alma mater. Like his father, he won a first in mathematics, the highest scholastic distinction for an undergraduate, and was awarded a fellowship even before he earned his bachelor of arts degree in 1854. The fellowship provided him a yearly stipend and rooms at Christ Church for life. He was appointed a lecturer in mathematics in 1855 and took vows as an Anglican deacon in 1861, becoming the Reverend C. L. Dodgson. From then on he dressed in black clerical clothes almost exclusively. A stammer made him shy of public speaking, and his clean-shaven boyish features, thick dark hair, and retiring manner made him seem ethereal to some contemporaries.

Using his birth name, Dodgson began to attract attention as a comic poet soon after earning his degree by publishing in newspapers and magazines some of the poems that he later incorporated into his children's books. In 1856 he published his first work under the name Lewis Carroll, an anagram of the latinized form of his first two names. The same year, he met Alice Pleasance Liddell, the daughter of Christ Church's dean, and took up photography. These interests—literature, photography, and Alice—blended to produce the most creative period of his life during the next twenty years.

LIFE'S WORK

Carroll remained a fellow at Christ Church for the rest of his life. He never married; in fact, the terms of his fellowship forbade it. He devoted himself to tutoring, lecturing, and performing religious and administrative duties at the college, but such work could not use up his creative energy, and he also pursued social and cultural interests outside his academic work.

Carroll was fastidious and almost obsessive with details, and he loved gadgets. Photography was ideally suited to him. At the time he took it up in 1856, it was a cumbersome art with bulky equipment. The photographer had to smear a glass plate with a colloid and dip it into silver nitrate, insert the plate into the camera, expose it for as much as a minute, and then develop the plate in a darkroom. During the exposure the subject had to stay perfectly still. Carroll soon mastered the techniques and tested his skill on architectural subjects and celebrities. Among those he photographed were the poet laureate Alfred, Lord Tennyson; the poet-painter Dante Gabriel Rossetti; and Queen Victoria's youngest son, Prince Leopold.

Photography was a means for Carroll to enter intellectual society and make friends. He soon ingratiated himself with parents by photographing their children. Although he occasionally photographed boys, he preferred girls. He believed that girls of about ten to fourteen years of age epitomized innocent beauty. Carroll told the girls stories or posed riddles of his own invention to put them at ease, dressed them in romantic costumes, and carefully posed them. Sometimes he photographed them nude. In all cases he first obtained the mothers permission and arranged for chaperonage. As well as photographing girls, he regularly sought their company, taking them to plays and museums and entertaining them at dinner parties. Deeply religious, conservative, and rigidly correct in his Victorian-era manners, he behaved with propriety; nonetheless, modern literary critics have speculated that Carrolls interest in girls came from suppressed pedophilia.

Carroll's favorite was Alice Liddell. On July 4, 1862, Carroll and his friend Robinson Duckworth took Alice and two of her sisters on a boating trip up the Thames River for a picnic. To entertain them on the return journey, he told them a tale, making it up as he

Lewis Carroll. (Library of Congress)

rowed the boat. This became the nucleus of *Alice's Adventures in Wonderland* (1865), which became an international best seller. In the story, the fictional Alice falls asleep and dreams of a bewildering array of eccentric characters in the form of animals such as the White Rabbit and March Hare, people such as the Mad Hatter, and playing cards. The blending of fantasy, puns, games, nonsense poetry, and adventure story appealed to both children and adults. Even if comically presented, its themes—violence and punishment, growth, obedience, education and games, correct behavior, and the development of identity—nevertheless addressed the world from a childs point of view.

The sequel, *Through the Looking-Glass and What Alice Found There*, followed in 1871. Again, Alice falls asleep. This time she dreams of crawling through a mirror and becoming a pawn in a mammoth chess game. As she advances across the chessboard countryside to be crowned a queen, she meets some of Carroll's most beloved characters, such as the twins Tweedledee and Tweedledum and the White Knight, and hears more nonsense verse, including his most celebrated poem, Jabberwocky. Many of the same themes appear, and the narrative emphasizes logical absurdities as the basis of humor as well as wordplay and bizarre behavior. Though not as successful as the first book, *Through the Looking-Glass and What Alice Found There* still sold well and brought praise from reviewers.

In 1876 Carroll published *The Hunting of the Snark*, his last masterpiece of nonsense literature. The puns of the subtitle—*An Agony in Eight Fits* (a struggle in eight chapters)—hint that it parodies epic poetry. Indeed, the story's quest for a mythical beast, the snark, by a brave band is a typical epic plot, but little else in the poem makes rational sense. Narrative jumps, eerie illogic, andthe tension of supernatural peril give it a nightmarishquality. The poem attracted dedicated fans, especially among intellectuals, but it was not a popular success.

Carroll wrote other books of poetry, short stories, and games, including *Phantasmagoria* (1869), *Rhyme? and Reason?* (1883), *A Tangled Tale* (1885), and *Three Sunsets and Other Poems* (published posthumously in 1898). His last book-length prose tale for children came out in two parts: *Silvie and Bruno* (1889) and *Silvie and Bruno Concluded* (1893). It contains some of his best comic verse, but the story is convoluted and at times sermonizing. It had few admirers.

Additionally, Carroll wrote articles and pamphlets on social problems and university affairs and left behind more than ninety-two thousand letters.

As Dodgson, Carroll earned a modest reputation during his lifetime as a mathematician. He was best known for clarifying the works of the classical Greek geometer Euclid in *Euclid and His Modern Rivals* (1879) and *Curiosa Mathematica, Part I* (1888), and for expositions of logical analysis such as *The Game of Logic* (1887) and *Symbolic Logic* (1896). He also published writings on number theory, and his work on voting theory was pioneering.

Carroll died of complications from a bronchial infection while staying at his sister's house in Guildford, Surrey, in 1898. Because his health had always been excellent, the sudden death surprised and saddened his readers, colleagues, and friends. Although Carroll could be prickly and prudish, he was famous for his kindness, having supported his sisters and helped friends through financial straits.

Significance

Lewis Carroll's biographers claim that he is second only to William Shakespeare as the most quoted English author. Certainly, his Alice books were widely popular, supporting almost continuous republication in many media. A musical called *Alice in Wonderland*, based on *Alice's Adventures in Wonderland*, was staged in 1886 with Carrolls help, and the first of sixteen motion picture versions appeared in 1903. The story has been told in cartoons, coloring books, pop-up books, audio cassettes, and audio-visual teaching guides. It was translated into at least seventy languages.

Before the Alice books, children's literature aimed to teach correct behavior and practical knowledge rather than entertain. Carroll mocked the moralism of this

Carroll's Works for Children

1865	*Alice's Adventures in Wonderland*
1867	"Bruno's Revenge"
1869	*Phantasmagoria*
1871	*Through the Looking-Glass and What Alice Found There*
1876	*The Hunting of the Snark: An Agony in Eight Fits*
1883	*Rhyme? and Reason?*
1898	*Three Sunsets, and Other Poems*
1977	*The Wasp in a Wig: The "Suppressed" Episode of "Through the Looking-Glass and What Alice Found There"*

style to the delight of his young readers and gave them a character who embodied their point of view. His approach was revolutionary, and it inspired many imitators. Moreover, Carroll was among the first children's authors to make a girl the main character.

Almost immediately, intellectuals and artists began adapting the skewed logic and naïveté in Carrolls stories, poems, and photographs for purposes that spanned the variety of modern culture. For example, scientists employed his logic to explain the theories of relativity and quantum mechanics, surrealist painters borrowed his images, politicians quoted him, songwriters echoed his phrasing, psychoanalysts found archetypes and pathology in his characters, and philosophers pondered his elusive remarks on existence and reality. Many features of the Alice stories entered popular culture as well and became familiar even to people who never read the books, including the image of Alice at the Mad Hatter's tea party, such coinages as "chortle" and "galumphing," and such phrases as "off with his head!" and "curiouser and curiouser."

Roger Smith

FURTHER READING

Carroll, Lewis. *The Annotated Alice*. Introduction and notes by Martin Gardner. New York: Bramhall House, 1960.

———. *The Annotated Snark*. Introduction and notes by Martin Gardner. New York: Simon & Schuster, 1962. These two books have abundant marginal notes that relate references in the Alice tales and *The Hunting of the Snark* to Carroll's life, events, controversies in Victorian England, and mathematics. They also reproduce the original illustrations.

Cohen, Morton N. *Lewis Carroll: A Biography*. New York: Alfred A. Knopf, 1995. Cohen argues that Carroll rigidly conformed to Victorian Christian morality but that beneath his conservatism raged a painful incompleteness in his life, which he palliated through chaste friendships with girls and young women. His attempts to amuse these friends inspired his most beloved books. Carroll's photographs and drawings accompany the text.

Collingwood, Stuart Dodgson. *The Life and Letters of Lewis Carroll*. New York: Century, 1899. As Carroll's nephew, Collingwood had firsthand knowledge of his uncle's life. The biography is accordingly full of anecdotes. The letters quoted in the text often exemplify Carroll's dexterity with humor.

Elwyn Jones, Jo, and J. Francis Gladstone. *The Alice Companion: A Guide to Lewis Carrolls Alice Books*. New York: New York University Press, 1998. Commentary on the people and places that made up Carroll's and Alice Liddell's worlds in mid-nineteenth century Oxford and a source to existing literature on this period of Carroll's life.

———. *The Red Kings Dream: Or, Lewis Carroll in Wonderland*. London: Jonathan Cape, 1995. By the authors of *The Alice Companion*, a study of Carroll's life and times, including his literary milieu, friends, and influences.

Guiliano, Edward. *Lewis Carroll Observed*. New York: Clarkson N. Potter, 1976. This handsome, large-format book contains many drawings and photographs by Carroll, illustrations of his tales, and clips from early films. The text comprises fifteen essays about the childrens books, photography, Carrolls style of humor and reputation, logic, and film versions of the Alice stories.

Leach, Karoline. *In the Shadow of the Dreamchild: A New Understanding of Lewis Carroll*. Chester Springs, Pa.: Peter Owen, 1999. Using original research, Leach refutes the commonly accepted image of Lewis Carroll as a sexual deviant, obsessed with little girls like Alice Liddell. Her book describes Carroll's sexual relations with grown women and other aspects of his life to destroy what she claims are the fictionalized accounts of previous biographers.

Thomas, Donald. *Lewis Carroll: A Portrait with Background*. London: John Murray, 1996. Thomas surmises the formative influences on Carroll's personality and intellect as he describes Victorian England. An invaluable guide for readers who want to understand how manners and ideas changed during Carroll's lifetime.

Jean le Rond d'Alembert
French mathematician and philosopher

A pioneer in the use of differential calculus, d'Alembert applied his mathematical genius to solving problems in mechanics. He provided valuable assistance with Denis Diderot's Encyclopedia *and wrote a number of treatises on musical theory.*

Born: November 17, 1717; Paris, France
Died: October 29, 1783; Paris, France
Also known as: Jean-Baptiste Daremberg

Early Life

On the night of November 17, 1717, Mme Claudine-Alexandrine Guérin, marquise de Tencin, gave birth to a son whom she promptly abandoned on the steps of the Church of Saint-Jean-Le-Rond. There, he was baptized with the name of the church, Jean le Rond d'Alembert (zhah luh-roh dah-lahm-behr); he was then sent to the Maison de la Coucher, from which he went to a foster home in Picardy. When his father, Louis-Camus Destouches, a military officer, returned to Paris, he sought his son and arranged for the child to be cared for by Mme Rousseau, the wife of a glazier. D'Alembert would always regard Mme Rousseau as his real mother and would continue to live with her until 1765, when illness compelled him to seek new quarters in the home of Julie de Lespinasse.

Destouches continued to watch over his illegitimate child, sending him to private schools; when Destouches died in 1726, he left the boy a legacy of twelve hundred livres a year. The sum, though not luxurious, guaranteed him an independence he cherished throughout his life. Through the interest of the Destouches family, the young man entered the Jansenist Collège des Quatre-Nations, where he took the name Jean-Baptiste Daremberg, later changing it, perhaps for euphony, to d'Alembert. Although he, like many other Enlightenment figures, abandoned the religious training he received there, he never shed the Cartesian influence that dominated the school.

After receiving his *baccalauréat* in 1735, he spent two years studying law, receiving a license to practice in 1738. Neither jurisprudence nor medicine, to which he devoted a year, held his interest. He turned to mathematics, for which he had a natural talent. At the age of twenty-two, he submitted his first paper to the Academy of Sciences. In that piece, he corrected a number of errors in Father Charles Reyneau's *Analyse demontrée* (1714). A second paper, on refraction and fluid mechanics, followed the next year, and in May, 1741, he was made an adjunct member of the Academy of Sciences.

Life's Work

Two years later, d'Alembert published a major contribution to mechanics, *Traité de dynamique* (1743), which includes his famous principle stating that the force that acts on a body in a system is the sum of the forces within the system restraining it and the external forces acting on that system. Although Sir Isaac Newton and Johann Bernoulli had already offered similar observations, neither had expressed the matter so simply. The effect of d'Alembert's principle was to convert a problem of dynamics to one of statics, making it easier to solve. The treatise is characteristic of d'Alembert's work in several ways: It illustrates his exceptional facility with mathematics, it reveals a desire to find universal laws in a discipline, and it indicates his ability to reduce complex matters to simple components. Over the next several years, he wrote a number of other innovative works in both mathematics and fluid mechanics.

Jean le Rond d'Alembert. (Library of Congress)

At the same time that d'Alembert was establishing himself as one of Europe's leading mathematicians—in 1752, Frederick the Great offered him the presidency of the Berlin Academy—he emerged as a leading figure of the Parisian salons. In 1743, he was introduced to the influential Mme du Deffand, who would secure his election to the French Academy in 1754. He remained a fixture of her assemblies until Julie de Lespinasse, whom he met there, established her own salon following a quarrel with the older woman. Later in the 1740's, he also joined the gatherings at the homes of Mme Marie-Thérèse Rodet Geoffrin and Anne-Louise Bénédicte de Bourbon, duchesse du Maine. Not striking in appearance—he was short and, according to a contemporary, "of rather undistinguished features, with a fresh complexion that tends to ruddiness," his eyes small and his mouth large—he compensated for his looks with his excellent ability with mimicry and his lively conversation.

While enjoying the female-dominated world of the salons, d'Alembert was also meeting a number of important male intellectuals, with whom he dined weekly at the Hôtel du Panier Fleuri—Denis Diderot, Jean-Jacques Rousseau (no relation to his stepmother), and Étienne Bonnot de Condillac. He probably also knew Gua de Malves, a fellow mathematician and member of the Academy of Sciences, who was chosen as the first editor of the *Encyclopédie: Ou, Dictionnaire raisonné des sciences, des arts, et des métiers* (1751-1772; *Encyclopedia*, 1965), and Malves may have been the one who introduced d'Alembert to the project; after Malves resigned, d'Alembert was named coeditor with Diderot.

D'Alembert did not plan to assume as much responsibility for the work as his coeditor. He wrote to Samuel Formey in September, 1749:

> I never intended to have a hand in [the *Encyclopedia*] except for what has to do with mathematics and physical astronomy. I am in a position to do only that, and besides, I do not intend to condemn myself for ten years to the tedium of seven or eight folios.

It was Diderot who conceived of the work as a summation of human knowledge, but d'Alembert's involvement extended well beyond the mathematical articles that the title page credits to him.

His contributions took many forms. He used his scientific contacts to solicit articles, and his connection with the world of the salons, which Diderot did not frequent, permitted him to enlist support among the aristocracy and upper middle class. Not only was such backing politically important, given the controversial nature of the enterprise, but also the financial assistance d'Alembert secured may well have prevented its collapse. Mme Geoffrin alone is reported to have donated more than 100,000 livres.

Also significant are the fifteen hundred articles that d'Alembert wrote, including the important *Discours préliminaire* (1751; *Preliminary Discourse to the Encyclopedia of Diderot*, 1963). Praised by all the great French intellectuals as well as Frederick the Great, it seeks to explain the purpose and plan of the *Encyclopedia* by showing the links between disciplines and tracing the progress of knowledge from the Renaissance to 1750. In its view of the Enlightenment as the culmination of progress in thought, it reflects the philosophes optimistic, humanistic attitude. D'Alembert's own understanding of the role of the philosopher and the nature of learning also emerges clearly in this essay. For him, "The universe is but a vast ocean, on the surface of which we perceive certain islands more or less large, whose link with the continent is hidden from us." The goal of the scientist is to discover, not invent, these concealed links, and mathematics would provide the means for establishing these connections. Just as physicists of the twenty-first century seek the one force that impels all nature, so d'Alembert sought the single principle that underlies all knowledge.

In 1756, d'Alembert went to Geneva to visit Voltaire, his closest friend among the philosophes, and to gather information for an article on this center of Calvinism. In an earlier work, d'Alembert had antagonized the Church by criticizing ecclesiastical control of education. "Genève," with its intended praise of Protestant ministers, provoked sharp protests from the Catholic establishment in France, and Calvinists were upset as well by d'Alembert's portrait of them as virtual agnostics.

Opposition to the *Encyclopedia* was growing in court circles; in March, 1759, permission to publish would be withdrawn. Never as daring as Voltaire or Diderot, d'Alembert resigned as coeditor in 1758, despite protests from his friends and associates. He did, however, continue to write articles on mathematics and science.

While the controversy surrounding the enterprise, especially "Genève," was the primary reason for d'Alembert's distancing himself from the *Encyclopedia,* another important factor was his growing disagreement with Diderot over the direction the work had been taking. By 1758, Diderot, who had published a treatise on mathematics, *Mémoires sur différens sujets de mathématiques* (1748), had come to believe that no further progress was possible in that field, so he rejected his coeditor's emphasis on mathematics as the key to knowledge, stating that "the reign of mathematics is over." D'Alembert's Cartesian theories also troubled Diderot. Like René Descartes, d'Alembert believed that matter is inert; Diderot disagreed. While d'Alembert maintained that the most precise sciences were those such as geometry that relied on abstract principles derived from reason, Diderot regarded experimentation and observation—empiricism—as the best guarantees of reliability. For d'Alembert, the more abstruse the science the better, for he sought to solve problems. Diderot preferred knowledge that directly affected life. In later years, Diderot continued to praise d'Alembert's mathematical abilities, and d'Alembert unsuccessfully tried to secure Diderot's election to the French Academy, but the two remained only distant friends.

Withdrawing from the *Encyclopedia* did not signal d'Alembert's rejection of the Enlightenment. Instead, he sought to use the French Academy as a forum to promulgate the views of the philosophes. His first speech before the French Academy urged toleration and freedom of expression, and in 1769 he nearly succeeded in having the body offer a prize for the best poem on the subject of "The Progress of Reason Under Louis X," the notion of such progress being a fundamental tenet of the Enlightenment. In 1768, when the king of Denmark, Christian VII, visited the French Academy, and again in March, 1771, when Gustavus III of Sweden attended a session, d'Alembert spoke of the benefits of enlightened policies. Through his influence in the salons, he arranged for the election of nine philosophes to the French Academy between 1760 and 1770, and a number of others sympathetic to their cause also entered because of d'Alembert. Elected permanent secretary of the body in 1772, he threafter used his official eulogies to attack the enemies of the Enlightenment and to encourage advanced ideas.

D'Alembert also continued to publish. The first three volumes of *Opuscules mathématiques* (1761-1780) contain much original work on hydrodynamics, lenses, and astronomy. His anonymous *Sur la destruction des Jésuites en France* (1765; *An Account of the Destruction of the Jesuits in France,* 1766), occasioned by the suppression of the order, discusses the danger of linking civil and ecclesiastical power because theological disputes then disturb domestic peace. In addition to attacking the Jesuits, d'Alembert urged the suppression of their rivals, the Jansenists.

Active as he was in the French Academy, d'Alembert's last years were marked by physical and emotional pain. Devoted to Julie de Lespinasse, he was doubly distressed by her death in 1776 and the discovery of love letters to her from the comte de Guibert and the marquis de Mora. As permanent secretary of the French Academy, d'Alembert was entitled to a small apartment in the Louvre, and there he spent the final seven years of his life, which ended on October 29, 1783. Although he produced little original work of his own during this period, he remained an important correspondent of Voltaire and Frederick the Great, urging the monarch to grant asylum to those persecuted for their views. He also encouraged young mathematicians such as Joseph-Louis Lagrange, Pierre-Simon Laplace, and the marquis de Condorcet.

Significance

Voltaire sometimes doubted Jean le Rond d'Alembert's zeal for the cause of Enlightenment, and d'Alembert's distancing himself from the *encyclopédistes* reveals that he was not one to take great risks. He observed that "honest men can no longer fight except by hiding behind the hedges, but from that position they can fire some good shots at the wild beasts infesting the country." From his post in the salons and the French Academy, he worked, as he told Voltaire, "to gain esteem for the little flock" of philosophes.

If Voltaire could accuse d'Alembert of excessive caution, d'Alembert could in turn charge Voltaire with toadying to the powerful. In his 1753 *Essai sur les gens de lettres*, d'Alembert urged writers to rely solely on their talents, and he reminded the nobility that intellectuals were their equals. "I am determined never to put myself in the service of anyone and to die as free as I have lived," he wrote Voltaire. Neither Frederick the Great's repeated invitations to assume the presidency of the Berlin Academy nor Catherine the Great's offer of 100,000 livres a year to tutor her son Grand Duke Paul could lure him away from France and independence.

In both his life and thought he was loyal to the ideals of the philosophes, so it is fitting that early twentieth century scholar Ernst Cassirer should choose him as the representative of the Enlightenment and call him "one of the most important scholars of the age and one of its intellectual spokesmen." His belief in the ability of reason to solve any problem epitomizes the view of eighteenth century intellectuals, but he also recognized the role of experimentation and imagination. In his *Eléméns de musique théorique et practique suivant les principes de M. Rameau* (1752), d'Alembert dissented from Jean-Philippe Rameau's view that one can devise mathematical rules for composition. As in his article on elocution in the *Encyclopedia,* he argued that rules are necessary, but only genius can elevate a work beyond mediocrity. Excellent scientist though he was, he ranked the artist above the philosopher.

Joseph Rosenblum

FURTHER READING

Cassirer, Ernst. *The Philosophy of the Enlightenment.* Translated by Fritz A. C. Koelln and James P. Pettegrove. Princeton, N.J.: Princeton University Press, 1951. Explores how Enlightenment thinkers looked at nature, psychology, religion, history, society, and aesthetics. Includes a great deal of information about d'Alembert.

Essar, Dennis F. *The Language Theory, Epistemology, and Aesthetics of Jean Lerond d'Alembert.* Oxford, England: Voltaire Foundation at the Taylor Institution, 1976. A study of d'Alembert's philosophy. Argues that d'Alembert's "position in the Enlightenment remains of central, pivotal importance." Also treats d'Alembert's mathematical and scientific contributions.

Grimsley, Ronald. *Jean d'Alembert, 1717-83.* Oxford, England: Clarendon Press, 1963. A topical study of d'Alembert's contributions to the *Encyclopedia,* his relations with other philosophers, and his own views. Largely ignores the scientific and mathematical aspects of d'Alembert's career.

Hankins, Thomas L. *Jean d'Alembert: Science and the Enlightenment.* Oxford, England: Clarendon Press, 1970. An ideal complement to Grimsley's book, for it concentrates on d'Alembert's contributions to science and mathematics. Relates d'Alembert's achievements to those of other scientists and the role of science to that of philosophy in the eighteenth century.

James, Ioan. *Remarkable Mathematicians: From Euler to von Neumann.* Washington, D.C.: Mathematical Association of America, 2002. Includes a chapter on d'Alembert's contributions to mathematics.

Kafker, Frank A. *The Encyclopedists as a Group: A Collective Biography of the Authors of the "Encyclopédie."* Oxford, England: Voltaire Foundation, 1996. Examines the life and thought of d'Alembert and the other authors who created the *Encyclopedia*.

Pappas, John Nicholas. *Voltaire and d'Alembert.* Bloomington: Indiana University Press, 1962. Drawing heavily on the correspondence between the two, this study seeks to rectify the view, fostered in large part by Voltaire, that d'Alembert was a hesitant follower of the older intellectual. Notes that the influence was mutual and shows where the two differed.

Van Treese, Glen Joseph. *D'Alembert and Frederick the Great: A Study of Their Relationship.* New York: Learned Publications, 1974. Treats the origin, nature, and consequences of the friendship between d'Alembert and the Prussian ruler. Offers a portrait of the two men and their age.

RICHARD DEDEKIND

German mathematician

Dedekind gave a new definition to the mathematical concept of irrational numbers, based exclusively on arithmetic principles. He helped clarify the notions of infinity and continuity and contributed to the establishment of rigorous theoretical foundations for mathematics.

Born: October 6, 1831; Brunswick, Duchy of Brunswick (now in Germany)
Died: February 12, 1916; Brunswick, Duchy of Brunswick
Also known as: Julius Wilhelm Richard Dedekind (full name)

EARLY LIFE
Julius Wilhelm Richard Dedekind (DAY-dih-kihnd) was one of four children born to a well-established professional family in the Germanic duchy of Brunswick. His father was a professor of jurisprudence at the local Collegium Carolinum, and his mother was a professor's daughter. In school, Dedekind was primarily interested in physics and chemistry, but when he enrolled in the Collegium Carolinum, it was as a student of mathematics. From a résumé, written somewhat later and in Latin, it is clear that this change was based on his dissatisfaction with the lack of rigor in the natural sciences.

In 1850, Dedekind was matriculated at the University of Göttingen, where he followed various courses in mathematics (studying under Carl Friedrich Gauss), astronomy, and experimental physics. In 1852, Dedekind presented his doctoral dissertation, which, in the opinion of Gauss, showed promise. At that time, the standard of mathematics at Göttingen was not very high, and Dedekind spent the following two years studying privately and preparing himself to become a first-class mathematician. No doubt his friendship with the brilliant Georg Friedrich Bernhard Riemann, at Göttingen at the same time, was also a positive influence. In fact, Dedekind attended Riemann's lectures even after he himself qualified as a university lecturer in 1854. When Gauss died in 1855, Peter Gustav Lejeune Dirichlet,Dirichlet, Peter Gustav Lejeune previously professor in Berlin, succeeded him. Dedekind described Dirichlet's arrival in Göttingen as a life-changing event. Dedekind not only attended Dirichlet's lectures but also became a personal friend of the new professor.

Life's Work
In 1858, the Federal Institute of Technology in Zurich, Switzerland, appointed Dedekind as professor of mathematics on Dirichlet's recommendation. Riemann also applied for the post but his work was considered too abstract. Dedekind stayed in Zurich until 1862 and then accepted an invitation from his old college in Brunswick, which had become a polytechnic by then.

While in Zurich, Dedekind taught differential and integral calculus and was disturbed by having to use concepts that had never been properly defined. In particular, he wrote: "Differential calculus deals with continuous magnitude, and yet an explanation of this continuity is nowhere given." He also deplored accepting without proof the belief that an increasing infinite sequence with an upper bound converges to a limit. He was dissatisfied that the notions of limit and continuity were based solely on geometrical intuition. On November 24, 1858, Dedekind succeeded in securing "a real definition of the essence of continuity." He waited until 1872 to publish this definition in book form, with the title *Stetigkeit und Irrationale Zahlen* ("Continuity and Irrational Numbers," translated in *Essays on the Theory of Numbers*, 1901).

Dedekind's problem was essentially that of irrational numbers, known already to the ancient Greeks. Rational numbers are dense in the sense that between any two rational numbers there is always another rational number, although there are infinitely many gaps between them. These gaps can be thought of as irrational numbers, and, before Dedekind began his work, they were characterized by infinite, nonrecurring decimal fractions. Dedekind devised a method, using "cuts," to define irrational numbers in terms of the rationals. If rational numbers are divided into two sets such that every number in the first set is smaller than every number in the second set, this partition defines one and only one real number. Should there be a largest or smallest number in one of the sets, the Dedekind cut corresponds to that rational number, while an irrational number is defined if neither set has a smallest or largest member.

A Dedekind cut can be imagined as severing a straight line composed of only rational numbers into two parts. Rational and irrational numbers together form the set of real numbers, and this set can now be made to correspond to all the points of a straight line. With this method, Dedekind not only managed to define irrational numbers in terms of rationals without recourse to geometry but also showed that a line, and by implication three-dimensional space, is complete, containing no holes. Furthermore, Dedekind upheld his philosophical principles, according to which numbers do not exist in a Platonic sense but are free creations of the human mind.

Closely connected to this work was the introduction of the concept of "ideals." Dedekind edited and published Dirichlet's lectures on number theory after the death of the latter. Dedekind can, in fact, be considered the author of the book, because Dirichlet left only an outline plan for publication, and that was already based on Dedekind's notes. In the tenth supplement to the second edition of this influential book, Dedekind developed the theory of ideals, following to a certain extent a line Ernst Eduard Kummer had already taken. Dedekind, however, went far beyond Kummer, avoided his mistakes, and made the theory more exact.

Ideals are an extension and generalization of the common number concept. According to the fundamental theory of arithmetic, ordinary integers either are prime numbers or can be uniquely factorized into primes. Unique factorization is a useful feature but does not generally apply to all algebraic integers in a given algebraic number field, algebraic numbers being defined as the roots of polynominal equations with integer coefficients. With the introduction of ideals, unique factorization can be restored. Dedekind subsequently revised and further developed this theory. In an important paper coauthored by Heinrich Weber, the analogy between algebraic numbers and algebraic functions was demonstrated with the help of ideals.

In *Was sind und was sollen die Zahlen* (1888; "The Nature and Meaning of Numbers," translated in *Essays on the Theory of Numbers*, 1901), Dedekind utilized the concept of what he called systems, which later became known as sets, and developed logical theories of original and cardinal numbers and of mathematical induction. In addition to contributing papers to mathematical journals, Dedekind coedited Riemann's collected works and supplied a biography of Riemann.

Dedekind stayed at Brunswick until his death and became a director of the polytechnic between 1872 and 1875. It seems that Dedekind was not offered the posts he would have accepted, while he refused the posts, most notably the one at Halle, that he was offered. Dedekind never married but lived with one of his sisters until her death in 1914. Although he lived in relative isolation, he was never a recluse. He was an excellent musician: He played the cello as a young man and the piano in later life. His portraits show a fine-featured man with thoughtful eyes; his character was described as modest, mild, and somewhat shy.

Significance

Although Richard Dedekind was a corresponding member of several academies and an honorary doctor of several universities, he never received the recognition he so fully deserved. It can be seen that his work was one of the most influential in shaping twentieth century mathematics. He is one of only thirty-one mathematicians meriting an individual entry in *Iwanami Sugaku Ziten* (1954; *Encyclopedic Dictionary of Mathematics*, 1977), in which he is described as a pioneer of abstract algebra. Transcending pure calculation, Dedekind made an attempt to find theoretical foundations to concepts used in algebraic number theory and in infinitesimal calculus. He defined and thereby created new mathematical structures that generalize the notions of number and serve as examples for further generalization.

Dedekind met Georg Cantor on a holiday in Switzerland and became his friend and also, at times, his frequent correspondent. Cantor submitted his theories to Dedekind for comment and criticism, and Dedekind was one of the first to support set theory in the face of hostility by other mathematicians. Independently of Cantor, he also utilized the concept of the actual, or concrete, infinite—a concept that was then regarded as taboo because there existed no theoretical foundation for its existence. Dedekind's work assisted in finding just such a foundation.

Judit Brody

Further Reading

Bashmakova, I. G., and G. S. Smirnova. *The Beginnings and Evolution of Algebra*. Translated from the Russian by Abe Schneitzer. Washington, D.C.: Mathematical Association of America, 2000. Chapter 8 in this history of algebra includes information about Dedekind and the birth of numbers theory.

Bell, Eric T. "Arithmetic the Second." In *Men of Mathematics*. New York: Simon & Schuster, 1937. Reprint. New York: Penguin Books, 1965. This short chapter in a well-known collective biography of mathematicians discusses the life and work of Kummer and Dedekind. Bell makes a good attempt to explain the abstract and often difficult concepts that are necessary for the understanding and appreciation of Dedekind's work.

Corry, Leo. *Modern Algebra and the Rise of Mathematical Structures*. 2d rev. ed. Boston: Birkhäuser, 2004. Traces the development of algebra from the mid-nineteenth century to the present, focusing on ideas concerning algebraic structures. This revised edition includes a revised chapter on Dedekind.

Dauben, J. W. *Georg Cantor: His Mathematics and Philosophy of the Infinite*. Cambridge, Mass.: Harvard University Press, 1979. Not a biography of Cantor, but a study of the emergence of a new mathematical theory. Dedekind's life, work, and influence on Cantor are featured extensively, but these references are dispersed throughout the book. Readers whose main interest is in Dedekind can rely on the well-constructed index and the twenty-four-page bibliography.

Dedekind, Richard. *Theory of Algebraic Integers*. Translated by John Stillwell. New York: Cambridge University Press, 1996. Dedekind's theory, first published in French in 1877, was the genesis of modern algebraic numbers theory. Stillwell provides a detailed introduction offering historical background and outlining the challenges Dedekind faced in devising his theory.

Edwards, Harold M. "Dedekind's Invention of Ideals." *The Bulletin of the London Mathematical Society* 15 (1983): 8-17. Traces the influences on Dedekind's set theoretic approach mainly to Dirichlet but also to Kummer and Riemann. Évariste Galois's influence was limited and resulted in steering Dedekind toward conceptual thinking as opposed to mere calculating. Dedekind went beyond Dirichlet, and against the accepted classical doctrine, by using completed infinites. The author stresses the innovative nature of Dedekind's theories and the analogy between cuts and ideals.

———. "The Genesis of Ideal Theory." *Archive for History of Exact Sciences* 23 (1980): 321-378. Analyzes Kummer's, Leopold Kronecker's, and Dedekind's versions of the theory of ideal factorization of algebraic integers. The author advances the thesis that as Dedekind revised the theory several times to match his philosophical principles, it did not improve from the mathematical point of view, and the first formulation remained the best.

Gillies, D. A. *Frege, Dedekind, and Peano on the Foundations of Arithmetic.* Assen, the Netherlands: Van Gorcum, 1982. A short paperback with an adequate index and a list of references. Investigates the relationship between logic and arithmetic in the work of the three men. Gillies regards Dedekind as fundamentally a logician and compares him to Gottlob Frege, who denied that a set was a logical notion, and to Giuseppe Peano, who thought that arithmetic could not be reduced to logic.

RENÉ DESCARTES
French philosopher and mathematician

Descartes extended mathematical method, the erasure of doubt by reaching certainty, to all fields of knowledge, and argued that "I think, therefore I am" is the only undoubtedly true statement that can be made. His radical distinction between mind and body and his revolutionary method of metaphysical inquiry have had a profound effect on the history of philosophy.

Born: March 31, 1596; La Haye, Touraine, France
Died: February 11, 1650; Stockholm, Sweden

EARLY LIFE
René Descartes (reh-nay day-kahrt) was born to one of the most respected families among the French-speaking nobility in Touraine. His father, Joachim, held the post of counselor to the Parlement de Bordeaux. Descartes's mother died of tuberculosis only a few days after giving birth to her son, leaving a frail child of chronically poor health to the sole care of his father. René's physical condition remained delicate until he was in his twenties.

Joachim Descartes was a devoted and admiring father, determined to obtain the best education for "his philosopher." When Descartes was ten, he was sent to the College of La Flèche, newly established by the Jesuits under the auspices of Henry IV. Descartes was an exemplary student of the humanities and of mathematics. When, at the age of sixteen, he began his study of natural philosophy, he came to the insight that would later give rise to his revolutionary contributions to modern thought. Uncertainty and obscurity, he discovered, were hallmarks of physics and metaphysics. These disciplines seemed to attract a contradictory morass of opinions that yielded nothing uniform or definite. By contrast, Descartes's studies in mathematics showed him something firm, solid, and lasting. He was astonished to find that while mathematical solutions had been applied to scientific problems, the method of mathematics had never been extended to important practical matters. At La Flèche, Descartes concluded that he would have to break with the traditions of the schools if he were to find knowledge of any worth.

Descartes left his college without regret, and his father subsequently sent him to Paris. Social life there failed to amuse him, and he formed his most intimate friendships with some of France's leading scholars and teachers. When he was twenty-one, he joined the army but spent little time campaigning. In his spare time, he wrote a compendium of music and displayed his mathematical genius by instantaneously solving puzzles devised for him by soldiers in his company.

Descartes was housed with a German regiment in winter quarters at Ulm, waiting for active campaign, when the whole core of his subsequent thought suddenly took shape. On the night of November 10, 1619, after a day of intense and agitated reflection, Descartes went to bed and had three dreams. He interpreted these dreams as a divine sign that he was destined to found a unified science based on a new method for the correct management of human reason. Descartes's sudden illumination and resolve on that night to take himself as the judge of all values and the source of all certainty in knowledge was momentous for the world of ideas.

LIFE'S WORK
Descartes spent the next ten years formulating his method while continuing scientific researches and

René Descartes. (Library of Congress)

occupied himself with travel in order to study what he called "the great book of the world." He had come to the view that systems of human thought, especially those of the sciences and philosophy, were better framed by one thinker than by many, so that systematizing a body of thought from the books of others was not the best method. Descartes wanted to be disabused of all the prejudices he had acquired from the books of others; thus, he sought to begin anew with his own clear and firm foundation. This view was codified in his *Regulae ad directionem ingenii* (1701; *Rules for the Direction of the Mind*, 1911). In this work, Descartes set forth the method of rational inquiry he thought requisite for scientific advance, but he advocated its use for the attainment of any sort of knowledge whatever.

Descartes completed a scientific work entitled *Le Monde* (*The World*, 1998) in 1633, the same year that Galileo was condemned by the Inquisition. Upon hearing this news, Descartes immediately had his own book suppressed from publication, for it taught the same Copernican cosmology as did Galileo and made the claim that indicted Galileo's orthodoxy: that human beings could have knowledge as perfect as that of God. A few years later, Descartes published a compendium of treatises on mathematics and physical sciences that were written for the educated but nonacademic French community; this work obliquely recommended his unorthodox views to the common persons of "good sense" from whom Descartes hoped to receive a fair hearing. This work was prefaced by his *Discours de la méthode* (1637; *Discourse on Method*, 1649) and contained the *Geometry*, the *Dioptric*, and the *Meteors*.

Discourse on Method provided the finest articulation of what has come to be known as Descartes's method of doubt. This consisted of the four following logical rules: to admit as true only what was so perfectly clear and distinct that it was indubitable; to divide all difficulties into analyzable elements; to pass synthetically from what is easy to understand to what is difficult; and to make such accurate enumerations of the steps of reasoning so as to be certain of having omitted nothing.

The method is fundamentally of mathematical inspiration, and it is deductive and analytical rather than experimental. It is a heuristic device for solving complex problems that yields explicit innovation and discovery. Descartes employed his method to this end in the tract on geometry when he discovered a way to resolve the geometric curves into Cartesian coordinates. Such an invention could hardly have come from the traditional Euclidean synthetic-deductive method, which starts from assumed axioms and common notions in order to generate and prove logically entailed propositions.

Descartes's new method was akin to those found in the writings of Francis Bacon and Galileo, and it was the architectonic of the new science. "Old" science, leftover from ancient and medieval researches, merely observed and classified, and explained its findings in terms of postulated natural purposes of things. The new science inaugurated in the seventeenth century sought, in Descartes's words, to make humans the "masters and

Descarte's Major Works	
1633	*Le Monde* (*The World*, 1998)
1637	*Discours de la méthode* (*Discourse on Method*, 1649)
1641	*Meditationes de prima philosophia* (*Meditations on First Philosophy*, 1680)
1644	*Principia philosophiae* (*Principles of Philosophy*, 1983)
1649	*Les Passions de l'âme* (*The Passions of the Soul*, 1950)
1701	*Regulae ad directionem ingeni* (*Rules for the Direction of the Mind*, 1911)

possessors of nature." This goal involved invention and discovery, the generation of new and nonspeculative knowledge, to be put in the service of practical ends. For Descartes and the other seventeenth century "new" scientists, human wonder and understanding were without intrinsic value; what was without practical use or application for humankind, Descartes remarked in *Discourse on Method*, was absolutely worthless. The new science aimed to create effects, not merely to understand causes.

Descartes intended his method not for mathematics and science only. He envisioned the unity of all knowledge. He employed his method in a purely metaphysical inquiry in *Meditationes de prima philosophia* (1641; *Meditations on First Philosophy*, 1680) to "establish something firm and lasting in the sciences." He fashioned in this a primary certainty by rejecting at the outset everything about which it was possible to have the least doubt.

He set aside as false everything learned from or through the senses, and the truths of arithmetic and geometry. Only the proposition, "I think, therefore I am," remained an indubitable truth. One cannot doubt one's existence, Descartes reasoned, without existing while one doubts. Thus, *cogito ergo sum* became his first and most certain principle. Further days of meditation on this principle revealed the certitudes that he was a substance whose whole essence it was to think, entirely independent of his body and of all other material things. His primary truth also made him believe he had proven the existence of God.

In this one epochal week of meditations, Descartes made privacy the hallmark of mental activity, moved the locus of certitude to inner mental states, and rejected faith and revelation in favor of clarity and distinctness. Reason itself had previously governed the coherence of what had to be taken as truth; now inner representation, and its correspondence with the external, material world, governed the kingdom of relevant truth. Most philosophers after Descartes have followed his conception of inner representations as the foundation of knowledge of all outer realities. Only in the twentieth century has this position, and its attendant problems, been systematically examined and contested.

The years that followed the publication of *Meditations on First Philosophy* were marked by controversies resulting from attacks by theologians. Descartes's orthodoxy was impugned and his arguments were assailed. In 1647, formal objections to the Cartesian metaphysics, along with the author's replies, were published

---

Descartes Thinks, Therefore He Is

René Descartes doubted the existence of everything that could be doubted, and given the power of the mind, the doubtable could include everything. He famously noted, however, that the only thing that cannot be doubted is that one thinks, that one exists as a thinking thing.

I shall proceed by setting aside all that in which the least doubt could be supposed to exist, just as if I had discovered that it was absolutely false; and I shall ever follow in this road until I have met with something which is certain, or at least, if I can do nothing else, until I have learned for certain that there is nothing in the world that is certain. . . .

What of thinking? I find here that thought is an attribute that belongs to me; it alone cannot be separated from me. I am, I exist, that is certain. But how often [do I exist]? Just when I think; for it might possibly be the case if I ceased entirely to think, that I should likewise cease altogether to exist. I do not now admit anything which is not necessarily true: to speak accurately I am not more than a thing which thinks, that is to say [I am not] a mind or a soul, or an understanding, or a reason, which are terms whose significance was formerly unknown to me. I am however, a real thing and really exist; but what thing? I have answered: a thing that thinks.

Source: Descartes, *Meditations on First Philosophy* (1641), in *Descartes: Selections*, edited by Ralph M. Eaton (New York: Charles Scribner's Sons, 1927), pp. 95-96, 99.

---

as a companion volume to a second edition of the *Meditations on First Philosophy* in French translation.

Descartes's next project was to be his last. *Les Passions de l'âme* (1649; *The Passions of the Soul*, 1950) was a treatise of psychology that explained all mental and physiological phenomena by mechanical processes. This work has striking moral overtones as well. Descartes's implicit prescription for the best human life is reminiscent of that of the ancient Stoics: Humans should strive to conquer their passions in order to attain peace of mind. Descartes maintained in *The Passions of the Soul* that while people who feel deep passions are capable of the most pleasant life, these passions must be controlled with the intervention of rational guidance. In the end, he claimed that teaching one to be the master of one's passions was the chief use of wisdom.

In 1649, Descartes responded to the request of Queen Christina of Sweden to join a distinguished circle of scholars she was assembling in Stockholm to instruct her in philosophy. The cold Swedish climate and the rigorous schedule demanded by the queen took their toll; Descartes caught pneumonia and died the following year.

Significance

Descartes's thought epitomizes the transition from the medieval epoch of the Western world to the modern period, in which personal freedom was deified. This tendency originated with the privatization of consciousness and the drive to overcome the rigors of nature. For Descartes, only absolutely certain knowledge counted as wisdom. Descartes envisaged wisdom as having practical benefits for the many, as opposed to being a mere cerebral exaltation for the educated few. Descartes saw the improvement of the mental and physical health of humankind as being the best of these benefits of wisdom. This prospect was ratified by the enterprises of centuries to come.

Descartes was one of the pioneers of modern mathematics. He conceived the possibility of treating problems of geometry by reducing them to algebraic operations and devised the necessary means for making geometric operations correspond to those of arithmetic. He also introduced the notion of deducing solutions from the assumption of the problem's being solved. This has become such a fundamental technique in algebra and higher mathematics that one can scarcely imagine its having had a genesis.

Patricia Cook

Further Reading

Alanen, Lilli. *Descartes's Concept of Mind*. Cambridge, Mass.: Harvard University Press, 2003. Examines Descartes's influential ideas about the mind and the relation of mind and body.

Balz, Albert G. A. *Descartes and the Modern Mind*. New Haven, Conn.: Yale University Press, 1952. Balz analyzes the pervasive influence of Cartesianism on the last three centuries. The analysis proceeds topically, with exposition of a particular facet of Descartes's thought followed by analysis of its legacy.

Bordo, Susan, ed. *Feminist Interpretations of René Descartes*. University Park: Pennsylvania State University Press, 1999. Collection of essays offering a feminist perspective of Descartes's philosophy. Includes a select bibliography on Descartes, Cartesianism, and gender.

Cottingham, John G. *Descartes*. New York: Basil Blackwell, 1986. Most commentators focus on Descartes's theory of knowledge; Cottingham takes a broader view of Cartesian philosophy and offers a profound Cartesian understanding of human nature. Excellent for beginning students; clear on, and faithful to, Descartes's texts.

Davies, Richard. *Descartes: Belief, Skepticism, and Virtue*. Studies in Seventeenth Century Philosophy 3. New York: Routledge, 2001. Analyzes Descartes's thoughts on credulity, skepticism, and the search for reason and eternal truth.

Gaukroger, Stephen, ed. *Descartes: Philosophy, Mathematics, and Physics*. Totowa, N.J.: Barnes & Noble Books, 1980. Ten authors offer different perspectives on Descartes's interest in providing a philosophical foundation for mathematical physics. Thorough index.

Haldane, Elizabeth S. *Descartes: His Life and Times*. New York: American Scholar Publications, 1966. An artfully crafted and detailed (nearly 400 pages) biography. Haldane is especially good at providing historical notes on circumstances that influenced Descartes's thought and development.

Keeling, S. V. *Descartes*. New York: Oxford University Press, 1968. Still one of the best overviews of Descartes's thought and influence, this book connects Descartes's development to his ideas, gives a systematic reading of his work, and critically analyzes the merits and defects of Cartesianism.

Kenny, Anthony. *Descartes: A Study of His Philosophy*. New York: Random House, 1968. A standard commentary for beginning students of Descartes that emphasizes his epistemology (theory of knowledge). Treats philosophical issues topically in brief, clear chapters.

Moriarty, Michael. *Early Modern French Thought: The Age of Suspicion*. New York: Oxford University Press, 2003. Examines the philosophy of Descartes, Blaise Pascal, and Nicolas Malebranche.

Diophantus

Greek mathematician

Diophantus wrote a treatise on arithmetic that represents the most complete collection of problems dating from Greek times involving solutions of determinate and indeterminate equations. This work was the basis of much medieval Arabic and European Renaissance algebra.

Flourished: c. 250 C.E.; place unknown

Early Life

Almost nothing is known about the life of Diophantus (di-oh-FAHN-tuhs), and there is no mention of him by any of his contemporaries. A reference to the mathematician by Hypsicles (active around 170 B.C.E.) in his tract on polygonal numbers and a mention of him by Theon of Alexandria (fl. 365-390 C.E.) give respectively a lower and an upper bound for the period in which Diophantus lived. There is also evidence that points to the middle of the third century C.E. as the flourishing period of Diophantus. Indeed, the Byzantine Michael Psellus (latter part of the eleventh century) asserts in a letter that Anatolius, bishop of Laodicea around 280 C.E., wrote a brief work on the Diophantine art of reckoning. Psellus's remark seems to fit well with the dedication of Diophantus's masterpiece *Arithmētika* (*Arithmetica*, 1885) to a certain Dionysius, who might possibly be identified with Saint Dionysius, bishop of Alexandria after 247. The only dates known about Diophantus's life are obtained as a solution to an arithmetical riddle contained in the *Greek Anthology*, which gives thirty-three for his wedding age, thirty-eight for when he became a father, and eighty-four for the age of his death. The trustworthiness of the riddle is hard to determine. During his life, Diophantus wrote the *Arithmetica*, the *Porismata*, the *Moriastica*, and the tract on polygonal numbers.

Life's Work

Diophantus's main achievement was the *Arithmetica*, a collection of arithmetical problems involving the solution of determinate and indeterminate equations. A determinate equation is an equation with a fixed number of solutions, such as the equation $x^2 - 2x + 1 = 0$, which admits only 1 as a solution. An indeterminate equation usually contains more than one variable, as for example the equation $x + 2y = 8$. The name indeterminate is motivated by the fact that such equations often admit an infinite number of solutions. The degree of an equation is the degree of its highest degree term; a term in several variables has degree equal to the sum of the exponents of its variables. For example, $x^2 + x = 0$ is of degree two, and $x^3 + x^2 y^4 + 3 = 0$ is of degree six but of degree three in x and degree four in y.

Although Diophantus presents solutions to arithmetic problems employing methods of varying degrees of generality, his work cannot be fairly described as a systematic exposition of the theory of solution of determinate and indeterminate equations. The *Arithmetica* is in fact merely a collection of problems and lacks any deductive structure whatsoever. Moreover, it is extremely hard to pinpoint exactly which general methods may constitute a key for reading the *Arithmetica*. This observation, however, by no means diminishes Diophantus's achievements. The *Arithmetica* represents the first systematic collection of such problems in Greek mathematics and thus by itself must be considered a major step toward recognizing the unity of the field of mathematics dealing with determinate and indeterminate equations and their solutions, in short, the field of Diophantine problems.

The *Arithmetica* was originally divided into thirteen books. Only six of them were known until 1971, when the discovery of four lost books in Arabic translation greatly increased knowledge of the work. The six books that were known before that discovery were transmitted to the West through Greek manuscripts dating from the thirteenth century (these will be referred to as books IG-VIG). The four books in Arabic translation (henceforth IVA-VIIA) represent a translation from the Greek attributed to Qusṭā ibn Lūqā al-Baʾlabakkī (fl. mid-ninth century). The Arabic books present themselves as books 4 through 7 of the *Arithmetica*. Because none of the Greek books overlaps with the Arabic books, a reorganization of the Diophantine corpus is necessary.

Scholars agree that the four Arabic books should probably be spliced between IIIG and IVG on grounds of internal coherence: The techniques used to solve the problems in IVA-VIIA presuppose only the knowledge of IG-IIIG, whereas the techniques used in IVG through VIG are radically different and more complicated than those found in IVA-VIIA. There is also compelling external evidence that this is the right order. The organization of problems in al-Karaji's *al-Fakhri* (c. 1010), an Islamic textbook of algebra heavily dependent on Diophantus, shows that the problems taken from IG-IIIG are immediately followed by problems found in IVA. The most interesting difference between IG-VIG and IVA-VIIA consists in the fact that in the Greek books, after having found the sought solutions (analysis), Diophantus never checks the correctness of the results obtained; in the Arabic books, the analysis is always followed by a computation establishing the correctness of the solution obtained (synthesis).

Before delving into some of the contents of the *Arithmetica*, the reader must remember that in Diophantus's work the term "arithmetic" takes a whole new meaning. The Greek tradition sharply distinguished between arithmetic and logistics. Arithmetic dealt with abstract properties of numbers, whereas logistics meant the computational techniques of reckoning. Diophantus dropped this distinction because he realized that although he was working with numerical examples, the techniques he used were quite general. Diophantus has often been called "the father of algebra," but this is inaccurate: Diophantus merely uses definitional abbreviations and not a system of notation that is completely symbolic. At the outset of the *Arithmetica*, Diophantus gives his notation for powers of the unknown x, called *arithmoi* (and indicated by the symbol σ), and for their reciprocals. (For example, x^2 is denoted by Δ^v and x^3 by K^v.) Diophantus has no signs for addition and multiplication, although he has a special sign for minus and a special word for "divided by."

It is impossible to summarize the rich content of the 290 problems of the *Arithmetica* (189 in the Greek and 101 in the Arabic books), but from the technical point of view a very rough description of the books can be given as follows: IG deals mainly with determinate equations of the first and second degree; IIG and IIIG address many problems that involve determinate and indeterminate equations of degree no higher than two; IVA to VIIA are mainly devoted to consolidating the knowledge acquired in IG-IIIG; and IVG to VIG address problems involving the use of indeterminate equations of degree higher than two.

Throughout the *Arithmetica*, Diophantus admits only positive rational solutions (that is, solutions of the form p/q where p and q are natural numbers). Although negative numbers are used in his work, he seems to make sense of them only with respect to some positive quantity and not as having a meaning on their own. For example, in VG.2 (where 2 refers to problem 2 of VG), the equation $4 = 4x + 20$ is considered absurd because the only solution is -4.

In IG are found many problems involving pure determinate equations, such as equations in which the unknown is present only in one power. The solution to IG.30, for example, requires solution of the equation $100 - x^2 = 96$, which gives $x = 2$. Note that Diophantus is not interested in the solution $x = -2$. Diophantus gives a general rule for solving pure equations:

> Next, if there results from a problem an equation in which certain terms are equal to terms of the same species, but with different coefficients, it will be necessary to subtract like from like on both sides until one term is found equal to one term. If perchance there be on either side or on both sides any negative terms, it will be necessary to add the negative terms on both sides, until the terms on both sides become positive, and again to subtract like from like until on each side only one term is left.

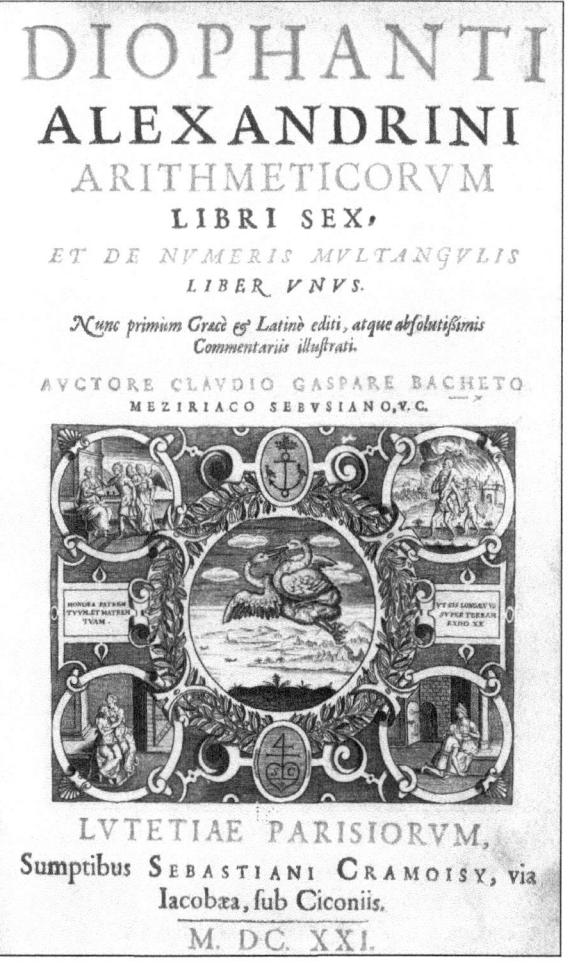

Title page from an edition of Diophantus's Arithmetica.
(Library of Congress)

In other words, Diophantus reduces the equation to the normal form $ax^m = c$. If the result were a mixed quadratic, however, such as $ax^2 + bx + c = 0$, Diophantus might have solved it by using a general method of solution similar to the one commonly learned in high school. As an example, problem VIG.9 can be reduced to finding the solution of $630x^2 - 73x = 6$, for which Diophantus merely states the solution to be $x = 6/35$. Although the possibility that Diophantus might have solved these problems by trial and error is open, internal evidence strongly suggests that he knew more than is relayed in the *Arithmetica*. In fact, the passage immediately following the above quote reads, "we will show you afterwards how, in the case also when two terms are left equal to a single term, such an equation can be solved." The promised solution may be in the lost three books.

Diophantus also solves problems involving equations (or systems of equations) of the form

(a) $a_n x^n + a_{n-1} x^{n-1} + \ldots + a_1 x - a_0 = y^2$
(where n is at most 6)
(b) $a_n x^n + a_{n-1} x^{n-1} + \ldots + a_1 x - a_0 = y^2$
(where n is at most 3)

The methods are seldom general, however, and rely on special cases of the above equations as found in VIG.19, where one finds the system given by the two equations $4x + 2 = y^3$ and $2x + 1 = z^2$. (The reader is reminded that Diophantus always works with numerical cases and so equations in abstract form are not to be found in his work.)

In many problems, Diophantus needs to find solutions that are subject to certain limits imposed by a condition of the problem at hand. He often uses some very interesting techniques to deal with such situations (so-called methods of limits and approximation to limits).

The tract on polygonal numbers has been transmitted in incomplete form. Whereas the *Arithmetica* used methods that could be called algebraic, the treatise on polygonal numbers follows the geometrical method, in which numbers are represented by geometrical objects.

Of the other two works, *Porismata* and *Moriastica*, virtually nothing is known. The *Moriastica* was mentioned by Iamblichus (fourth century C.E.) and seems to have been merely a compendium of rules for computing with fractions similar (or identical) to the one found in IG. The *Porismata* is referred to often by Diophantus himself. In the *Arithmetica*, he often appeals to some results of number theoretic nature and refers to the *Porismata* for their proofs. It is unclear, as in the case of the *Moriastica*, whether the *Porismata* was part of the *Arithmetica* or a different work. There are other number theoretic statements that are used by Diophantus in the *Arithmetica* and that might have been part of the *Porismata*. They concern the expressibility of numbers as sums of two, three, or four squares. For example, Diophantus certainly knew that numbers of the form $4n + 3$ cannot be odd and that numbers of the form $8n + 7$ cannot be written as sums of three squares. It was in commenting on these insights of Diophantus that the distinguished mathematician Pierre de Fermat (1601-1665) gave some of his most famous number theoretic statements.

Significance

Diophantus's *Arithmetica* represents the most extensive treatment of arithmetic problems involving determinate and indeterminate equations from Greek times. It is clear from the sources that Diophantus did not create the field anew but was heavily dependent on the older Greek tradition. Although it is difficult to assess how much he improved on his predecessors' results, his creativeness in solving so many problems by exploiting new stratagems to supplement the few general techniques at his disposal was impressive.

The *Arithmetica* was instrumental in the development of algebra in the medieval Islamic world and Renaissance Europe. The Arabic writers al-Khazin (fl. c. 940), Abul Wefa (940-998), and al-Karaji (fl. c. 1010), among others, were deeply influenced by Diophantus's work and incorporated many of his problems in their algebra textbooks. The Greek books have come to the West through Byzantium. The Byzantine monk Maximus Planudes (c. 1260-c. 1310) wrote a commentary on the first two Greek books and collected several extant manuscripts of Diophantus that were brought to Italy by Cardinal Bessarion. Apart from a few sporadic quotations, there was no extensive work on the *Arithmetica* until the Italian algebraist Rafael Bombelli ventured into a translation (with Antonio Maria Pazzi), which was never published, and used most of the problems found in IG-VIG in his *Algebra*, published in 1572. François Viète, the famous French algebraist, also made use of several problems from Diophantus in his *Zetetica* (1593). In 1575, the first Latin translation, by Wilhelm Holtzmann (who grecized his name as Xylander), appeared with a commentary. In 1621, the Greek text was published with a Latin translation by Claude-Gaspar Bachet. This volume became the standard edition until the end of the nineteenth century, when Paul Tannery's edition became available. A new French-Greek edition of the Greek books is planned since the Tannery edition is long outdated.

Wilbur R. Knorr and Paolo Mancuso

FURTHER READING

Bashmakova, Isabella G. *Diophantus and Diophantine Equations*. Translated by Abe Shenitzer. Washington, D.C.: Mathematical Association of America, 1997. A discussion of the methods of Diophantus, accessible to readers who have taken some university mathematics. It includes the elementary facts of algebraic geometry indispensable for its understanding. Examines the development of Diophantine methods during the Renaissance and in the work of Pierre de Fermat.

Heath, Thomas L. *Diophantos of Alexandria: A Study in the History of Greek Algebra*. Cambridge, England: Cambridge University Press, 1885. This volume is still the major reference work on Diophantus in English. It gives an extensive treatment of the sources, the works, and the influence of Diophantus. The appendix contains translations and a good sample of problems from IG-VIG of the *Arithmetica* and translations from the tract on polygonal numbers.

_____. *A History of Greek Mathematics*. 2 vols. 1921. Reprint. New York: Dover Press, 1981. The second volume of this classic study contains a thorough exposition of Diophantus's work with a rich analysis of types of problems from the *Arithmetica*.

Sesiano, Jacques. *Books IV to VII of Diophantus's Arithmetica: In the Arabic Translation Attributed to Qusta Ibn Luqa*. New York: Springer-Verlag, 1982. A detailed analysis of the Arabic books with a translation and a commentary on the text. The introduction presents a summary of the textual history of arithmetic theory in Greek and Arabic. The English translation and the commentary are followed by an edition of the Arabic text. Other features include an Arabic index, an appendix that gives a conspectus of the problems in the *Arithmetica*, and an extensive bibliography.

Thomas, Ivor, ed. *Greek Mathematical Works*. 2 vols. Cambridge, Mass.: Harvard University Press, 1980. Volume 2 of this work contains selections from the *Arithmetica* and the quotations from the *Greek Anthology*, Psellus, and Theon of Alexandria that are relevant for Diophantus's dates. Greek texts with English translation.

Vogel, Kurt. "Diophantus of Alexandria." In *Concise Dictionary of Scientific Biography*, vol. 4. New York: Scribner's, 2000. A survey of Diophantus's life and works, with an extensive selection of types of problems and solutions found in the *Arithmetica*.

EUCLID

Greek geometer

Euclid took the geometry known in his day and presented it in a logical system. His work on geometry became the standard textbook on the subject down to modern times.

Born: c. 330 B.C.E.; probably Greece
Died: c. 270 B.C.E.; Alexandria, Egypt

EARLY LIFE

Little is known about Euclid (YOO-klihd), and even the city of his birth is a mystery. Medieval authors often called him Euclid of Megara, but they were confusing him with an earlier philosopher, Eucleides of Megara, who was an associate of Socrates and Plato. It is virtually certain that Euclid came from Greece proper and probable that he received advanced education in the Academy, the school founded by Plato in Athens. By the time Euclid arrived there, Plato and the first generation of his students had already died, but the Academy was the outstanding mathematical school of the time. The followers of Aristotle in the Lyceum included no great mathematicians. The majority of the geometers who instructed Euclid were adherents of the Academy.

Euclid traveled to Alexandria and was appointed to the faculty of the Museum, the great research institution that was being organized under the patronage of Ptolemy Soter, who ruled Egypt from 323 to 283. Ptolemy, a boyhood friend of Euclid and then a lieutenant of Alexander the Great, had seized Egypt soon after the conqueror's death, become the successor of the pharaohs, and managed to make his capital, Alexandria, an intellectual center of the Hellenistic Age that outshone the waning light of Athens. Euclid presumably became the librarian, or head, of the Museum at some point in his life. He had many students, and although their names are not recorded, they carried on the tradition of his approach to mathematics. His influence can still be identified among those who followed in the closing years of the third century B.C.E. He was thus a member of the first generation of Alexandrian scholars, along with Demetrius of Phalerum and Strato of Lampsacus.

Two famous remarks are attributed to Euclid by ancient authors. On being asked by Ptolemy if there was any easier way to learn the subject than by struggling through the proofs in Euclid's work the *Stoicheia* (c. 300 B.C.E.; *The Elements of Geometrie of the Most Auncient Philosopher Euclide of Megara*, 1570, commonly known as the *Elements*), Euclid replied that there is no "royal road" to geometry. Then when a student asked him if geometry would help him get a job, he ordered his slave to give the student a coin, "since he has to make a profit from what he learns." In spite of this rejoinder, his usual temperament is described as gentle and benign, open, and attentive to his students.

LIFE'S WORK

Euclid's reputation rests on his greatest work, the *Elements*, consisting of thirteen books of his own and two spurious books added later by Hypsicles of Alexandria and others. This work is a systematic explication of geometry in which each brief and elegant demonstration rests on the axioms and postulates given previously. It embraces and systematizes the achievements of earlier mathematicians. Books 1 and 2 discuss the straight line, triangles, and parallelograms; books 3 and 4 examine the circle and the inscription and circumscription of triangles and regular polygons; and books 5 and 6 explain the theory of proportion and areas. Books 7, 8, and 9 introduce the reader to arithmetic and the theory of rational numbers, while book 10 treats the difficult subject of irrational numbers. The remaining three books investigate elementary solid geometry and conclude with the five regular solids (tetrahedron, cube, octahedron, dedecahedron, and icosahedron). It should be noted that the *Elements* discusses several problems that later came to belong to the field of algebra, but Euclid treated them in geometric terms.

The genius of the *Elements* lies in the beauty and compelling logic of its arrangement and presentation, not in its new discoveries. Still, Euclid showed originality in his development of a new proof for the Pythagorean theorem as well as his convincing demonstration of many principles that had been advanced less satisfactorily by others. The postulate that only one parallel to a

Euclid. (Library of Congress)

line can be drawn through any point external to the line is Euclid's invention. He found this assumption necessary in his system but was unable to develop a formal proof for it. Modern mathematicians have maintained that no such proof is possible, so Euclid may be excused for not providing one.

Other works by Euclid are extant in Greek. *Ta dedomena* (c. 300-270 B.C.E.; *Data in Euclid's Elements of Geometry*, 1661) is another work of elementary geometry and includes ninety-four propositions. The *Optika* (c. 300-270 B.C.E.; *The Optics of Euclid*, 1945), by treating rays of light as straight lines, makes its subject a branch of geometry. Spherical geometry is represented by the *Phainomena* (c. 300-270 B.C.E.; *Euclid's Phaenomena*, 1996), which is an astronomical text based in part on a work of Autolycus of Pitane, a slightly older contemporary. Euclid wrote on music, but the extant *Katatomē kanonos* (known by its Latin title, *Sectio canonis*) is at best a reworking by some later, inferior writer of a genuine text by Euclid, containing no more of his actual words than some excerpts. Discovered in Arabic translation was *Peri diaireseon biblion* (c. 300-270 B.C.E.; *On Divisions of Figures*, 1915), for which the proofs of only four of the propositions survive.

Also discovered have been the names of several lost books by Euclid on advanced geometry: The *Pseudaria* (fallacies) exposed fallacies in geometrical reasoning, and *Konika* (conics) laid some of the groundwork for the later book of the same title by Apollonius of Perga. There was a discussion of the relationships of points on surfaces titled *Topoi pros epiphaneia* (surface loci), and *Porismata* (porisms), a work of higher geometry, treated a kind of proposition intermediate between a theorem and a problem.

In addition to the last two books of the *Elements*, there are works bearing Euclid's name that are not genuinely his. These include the *Katoptrica* (catoptrica), a later work on optics, and *Eisagōgē armonikē* (*Introduction to Harmony*), which is actually by Cleonides, a student of Aristoxenus. None of Euclid's reputation, however, depends on these writings falsely attributed to him.

Significance

Euclid left as his legacy the standard textbook in geometry. There is no other ancient work of science that needs so little revision to make it current, although many modern mathematicians, beginning with Nikolay Lobachevski and Bernhard Riemann and including Albert Einstein, have developed non-Euclidean systems in reaction to the *Elements*, thus doing it a kind of honor. The influence of Euclid on later scientists such as Archimedes, Apollonius of Perga, Galileo Galilei, Sir Isaac Newton, and Christiaan Huygens was immense. Eratosthenes used his theorems to measure with surprising accuracy the size of the sphere of Earth, and Aristarchus attempted less successfully, but in fine Euclidean style, to establish the sizes and distances of the moon and the sun.

Other Hellenistic mathematicians, such as Hero of Alexandria, Pappus, Simplicius, and, most important, Proclus, produced commentaries on the *Elements*. Theon of Alexandria, father of the famous woman philosopher and mathematician Hypatia, introduced a new edition of the *Elements* in the fourth century C.E. The sixth century Italian Boethius is said to have translated the *Elements* into Latin, but that version is not extant. Many translations were made by early medieval Arabic scholars, beginning with one made for Harun al-Rashid near 800 C.E. by al-Hajjaj ibn Yusuf ibn Matar. Athelhard of Bath made the first surviving Latin translation from an Arabic text about 1120 C.E. The first printed version, a Latin translation by the thirteenth century scholar Johannes Campanus, appeared in 1482 in Venice. Bartolomeo Zamberti was the first to translate the *Elements* into Latin directly from the Greek, rather than Arabic, in 1505. The first English translation, printed in 1570, was done by Sir Henry Billingsley, later the lord mayor of London. The total number of editions of Euclid's *Elements* has been estimated to be more than a thousand, making it one of the most often translated and printed books in history and certainly the most successful textbook ever written.

J. Donald Hughes

Further Reading

Euclid. *The Thirteen Books of Euclid's "Elements."* Translated by Thomas Little Heath. 3 vols. 1925. Reprint. New York: Dover, 1956. This English translation contains extensive commentary on Euclid's *Elements*. This admirable work supersedes all previous translations. It contains a full introduction, 151 pages in length, touching on all the major problems.

Fraser, P. M. *Ptolemaic Alexandria*. 3 vols. Oxford, England: Clarendon Press, 1972. Has a useful section on the intellectual background and influences of Euclid but is primarily valuable in providing a study of the cultural setting of Alexandria in Euclid's day.

Heath, Thomas Little. *From Thales to Euclid*. Vol. 1 in *A History of Greek Mathematics*. New York: Dover, 1981. Places Euclid in the context of the development of ancient mathematics. A thoroughly dependable treatment.

Mlodinow, Leonard. *Euclid's Window: The Story of Geometry from Parallel Lines to Hyperspace*. New York: Free Press, 2001. In this history of geometry, reason, and abstraction, Euclid is represented as a major figure.

Mueller, Ian. *Philosophy of Mathematics and Deductive Structure in Euclid's "Elements."* Cambridge, Mass.: MIT Press, 1981. A study of the Greek concepts of mathematics found in the *Elements*, emphasizing philosophical, foundational, and logical rather than historical questions, although the latter are not totally neglected. Attention is directed to Euclid's work, not that of his predecessors. This monograph requires mathematical literacy, and the general reader may find it overly technical.

Reid, Constance. *A Long Way from Euclid*. New York: Thomas Y. Crowell, 1963. An explanation of how modern mathematical thought has progressed beyond Euclid, written for those whose introduction to mathematics consisted mainly of studying the *Elements*. Accessible to the general reader, this study takes Euclid as its starting point and shows that he did not provide the reader with all the answers, or even all the questions, with which mathematicians concern themselves.

Szabo, Arpad. *The Beginnings of Greek Mathematics*. Translated by A. M. Ungar. Boston: D. Reidel, 1978. Places Euclid within the context of the development of the Greek mathematical tradition.

Eudoxus of Cnidus
Greek geometer

Eudoxus and his disciples resolved classical difficulties in the fields of geometry and geometric astronomy. Their approach became definitive for later research in these fields.

Born: c. 390 B.C.E.; Cnidus, Asia Minor (now in Turkey)
Died: c. 337 B.C.E.; Cnidus, Asia Minor

Early Life

As for so many ancient figures, little is known about the life of Eudoxus (yew-DAHK-suhs) of Cnidus. If one follows the account of the ancient biographer Diogenes Laertius (c. 250 C.E.), Eudoxus first visited Athens at age twenty-three to study medicine and philosophy. He soon returned home to Cnidus, however, and from there, joining the company of the Cnidian physician Chrysippus, he moved on to Egypt, where for more than a year he studied among the priests and engaged in astronomical investigations. Later, as he traveled and lectured in the wider Aegean area (specifically, Cyzicus and the Propontis), he built up a following and thus returned to Athens a man of considerable distinction. His main subsequent activity seems to have centered on Cnidus, where he was honored as a lawgiver. His renown extended to many areas, including astronomy, geometry, medicine, geography, and philosophy.

There is disagreement over his dates. The ancient chronicler Apollodorus sets Eudoxus's prime activity in 368-365 B.C.E. In general, the prime means age forty; if that holds here, Eudoxus's birth would be set c. 408. There is reason for doubt, however, because this early a date conflicts with other biographical data. G. L. Huxley favors c. 400; G. de Santillana and others argue for c. 390. Eudoxus is reported to have died at the age of fifty-three; the corresponding date would be 355, 347, or 337.

Life's Work

None of Eudoxus's writings survives, but fragments cited by ancient authors offer a reasonable impression of their diversity and significance. His principal efforts were in the areas of mathematics and astronomy, the former best represented in portions of Euclid's *Stoicheia* (c. 300 B.C.E.; *The Elements of Geometrie of the Most Auncient Philosopher Euclide of Megara*, 1570, commonly known as the *Elements*), the latter in astronomical discussions of the fourth-century cosmology of Aristotle and the ancient commentaries on it.

According to Archimedes (287-212 B.C.E.), Eudoxus was the first to set out a rigorous proof of the theorems that any pyramid equals one-third of the associated prism (that is, having the same height and base as the pyramid), and any cone equals one-third of the associated cylinder. Eudoxus also appears to have proved two other theorems: that circles are as the squares of their diameters and that spheres are as the cubes of their diameters. The proofs of these four theorems constitute the main part of book 17 of *Elements*, and the technique used there is likely to derive from Eudoxus.

To take the circle theorem as an example, one could imagine a regular polygon having so many sides that it seems practically indistinguishable from a circle. As two such polygons (with equally numerous sides) are proportional to the squares of their diameters, the same could be supposed for the corresponding circles. Presumably, an argument of this sort was assumed by geometers who used the circle theorem in the time before Eudoxus. In the strict sense, however, the reasoning would be invalid, for only by an infinite process can rectilinear figures eventuate in the circle.

In the Eudoxean scheme, one assumes the stated proportion to be false: If two circles are not in the ratio of the squares of their diameters, then one can construct two similar regular polygons, one inscribed in each circle, and one can make the difference between the polygon and its circle so small that the polygon is found to be simultaneously greater and less than a specified amount. Because that is clearly impossible, the stated theorem must be true. (This indirect manner for proving theorems on curved figures is often called, if somewhat misleadingly, the "method of exhaustion.")

50

A key move in this proof is making the polygon sufficiently close to the circle. To this end, one observes that as the number of sides is doubled, the difference between the polygon and the circle is reduced by more than half. The procedure of successively bisecting a given quantity will eventually make it less than any preassigned amount, however, as Euclid proves in *Elements*. According to Archimedes, however, it seems that Eudoxus took this bisection principle as an axiom.

The notion of proportion itself runs into a similar difficulty with the infinite. As long as quantities are related to one another in terms of whole or fractional numbers (for example, if one area is 7/5 of another area), their ratios can be specified from these same numbers (that is, the ratio of the one area to the other will be 7 to 5). Yet what if no such numbers exist? For example, it was found, a century or so before Eudoxus, that the diagonal and side of a square cannot equal a ratio of whole numbers. (In modern terms, one calls the associated number $\sqrt{2}$ "irrational"; its decimal equivalent 1.414 . . . will be nonterminating and nonrepeating.) Only by means of some form of infinite sequence can "commensurable" quantities (those whose ratio is expressible by two integers) equal the ratio of incommensurable ones. Geometers in the generation before Eudoxus had pursued the study of incommensurables with considerable interest, but Eudoxus was the first to see how the theorems on ratios could be rigorously proved when their terms were incommensurable.

It is usually supposed that Eudoxus's approach was identical to that given by Euclid in book 5 of *Elements*. Other writers, in particular Archimedes, however, knew of a different technique of proportions that seems more like what Eudoxus would have proposed. By this technique, one first establishes the stated theorem for the case of commensurable quantities. For the incommensurable case, one uses an indirect argument: If the proportionality does not hold, one can find commensurable terms whose ratio differs by less than a specified amount from the ratio of the given incommensurable terms—this is done by successively bisecting one of the givens until it is less than the difference between two others; when the commensurable case of the theorem (already proved) is applied, a certain term will be found to be simultaneously greater and less than another. Because that is impossible, the theorem must be true.

The defining notions of the proportion theory in Euclid's book 5 can be derived as a simple modification of this technique, for the role that the intermediate commensurable terms play in it is assumed by the Euclidean definition of proportion: that A:B = C:D means that for all positive integers m, n, if $mA > nB$, then also $mC > nD$; the same holds true if = or < is substituted for >. Proofs in this Euclidean manner do not require a division into commensurable and incommensurable cases, nor do they make use of the bisection principle; in general, they are rather easier to set up than in the alternative technique. It is thus possible to see Euclid's approach as an intended refinement of the Eudoxean.

In either the Eudoxean or Euclidean form, this manner of proportion theory can be made to correspond to the modern definition of real number, as formulated by the German mathematician Julius Wilhelm Richard Dedekind. In each example, the real term (possibly irrational) is considered to separate all the rationals into those greater and those less than it.

It seems likely that Eudoxus also contributed to the study of incommensurable lines. His predecessor Theaetetus (d. c. 369) had shown that if the squares of two lines are commensurable with each other but do not have the ratio of square integers, then the lines themselves will be incommensurable with each other; further, if two such lines A, B are taken, the lines A ± B will be incommensurable with them, not only as lines but also in square (lines of this latter type were called *alogoi*, literally, "without ratio"). The further study of the *alogoi* lines, as collected in book 10 of Euclid's *Elements*, divided into twelve classes all the *alogoi* formed as the square roots of R(A ± B), where R is a unit line, and A and B are commensurable with each other in square only. Presumably, Eudoxus and his followers played a part in this investigation.

Eudoxus's efforts are rooted in a concern for logical precision in geometry, and this interest may reflect his close association with the Platonic Academy at Athens. Two anecdotes (of questionable historicity) celebrate this connection. The first explains how Eudoxus came to be involved in seeking the cube duplication, the so-called Delian problem. To allay a plague, the citizens of Delos were commanded by the oracle to double a cube-shaped altar. When their attempts failed, they sent to Plato, who directed his mathematical associates Archytas, Menaechmus, and Eudoxus to solve it. When they did so, however, Plato criticized their efforts for being too mechanical. The solutions of a dozen different ancient geometers are known, but that of Eudoxus has not been preserved. It supposedly employed "curved lines" of some sort, and reconstructions have been proposed.

In a second story, Plato is said to have posed to Eudoxus the problem of "saving the phenomena" of planetary motion on the restriction to uniform circular motion. An account of Eudoxus's scheme is transmitted by Simplicius of Cilicia (sixth century C.E.) in his commentary on Aristotle's *De caelo* (before 335 B.C.E.; *On the Heavens*, 1777). From this account a reconstruction was worked out by the Italian historian of astronomy Giovanni Virginio Schiaparelli in 1875. The Eudoxean system reproduces the apparent motion of a planet by combining the rotations of a set of homocentric spheres. The planet is set on the equator of a uniformly rotating sphere. If a second sphere is set about the first, rotating with equal speed to the first but in the opposite direction and having its axis inclined, then the planet will trace out an eight-shaped curve (which the ancients called the *hippopede*, or horse fetter), so as to complete the full double loop once for every full revolution of the spheres. One superimposes over this a third spherical rotation, corresponding to the general progress of the planet in the ecliptic, and finally over this a fourth rotation, corresponding to the daily rotation of the whole heaven. In this way, each of the five planets requires four spheres, while the sun and the moon each take three.

Schiaparelli's exposition thus revealed the ingenuity of Eudoxus's scheme for reproducing geometrically the seemingly erratic forward and backward (retrograde) motion of the planets. Nevertheless, the model proves unsuccessful in some respects: Because the planets do not vary in distance from the earth (the center of their spheres), Eudoxus cannot account for their variable brightness or for asymmetries in the shape of their retrograde paths. Even worse, the values that Eudoxus had to assign for the rotations of the spheres do not produce retrogrades for Mars or Venus, and the sun and the moon are given uniform motions, contrary to observation. Apparently, the latter two defects were recognized, for Eudoxus's follower Callippus introduced seven additional spheres (two each for sun and moon, one each for Mercury, Venus, and Mars) to make the needed corrections.

The Eudoxean-Callippean scheme is enshrined in Aristotle's *Metaphysica* (c. 335-323 B.C.E.; *Metaphysics*, 1801), in which it serves as the mathematical basis of a comprehensive picture of the entire physical cosmos. Doubtless, Eudoxus proposed his geometric model without specific commitments on physical and cosmological issues. Nevertheless, it suited well the basic Aristotelian principles—for example, that the cosmos separates into two spherical realms, the celestial and, at its center, the terrestrial, and that the natural motions of matter in the central realm (for example, earthy substances moving in straight lines toward the center of the cosmos) differ from those in the outer (where motion is circular, uniform, and eternal). Ironically, these Aristotelian principles persisted in later cosmology, even after astronomers had switched from the homocentric spheres to eccentrics, epicycles, and other geometric devices.

Eudoxus also produced works of a descriptive and empirical sort in astronomy and geography. His *Phaenomena* (fourth century B.C.E.; phenomena) and *Enoptron* (fourth century B.C.E.; mirror) recorded observations of the stars—the basis, one would suppose, of a systematic almanac of celestial events (for example, solstices and equinoxes, lunar phases, heliacal risings of stars). He adopted, as Diogenes and others report, an *oktaeteris*, or eight-year calendar cycle. As known to later authors, an *oktaeteris* is one of the cycles found to reconcile the solar year of 365.25 days with the period of the moon's phases (somewhat over 29.5), by parsing out the 2,922 days in eight years into ninety-nine lunar periods (fifty-one of thirty days and forty-eight of twenty-nine). However, it is unclear whether this was the arrangement used by Eudoxus. His geographical treatise, the *Gēs periodos* (fourth century B.C.E.; circuit of the Earth), systematically described the lands and peoples of the known world, from Asia in the east to the western Mediterranean region. A connection with his astronomical studies can be seen in the use of the ratio of longest to shortest periods of daylight for designating the latitudes of places.

Significance

However interesting Eudoxus's contributions to calendarics, geography, and philosophy may be, they are secondary to his achievement in mathematics, for he may justly be viewed as the most significant geometer in the pre-Euclidean period. He advanced the study of incommensurables, introduced a new technique for generalizing the theory of proportion, and made exact the theory of limits with his new method of "exhaustion." Remarkable for the logical precision of his proofs, Eudoxus here set the standard against which even the foremost of the later geometers, such as Euclid and Archimedes, measured their own efforts.

Eudoxus's influence on geometric astronomy is more subtle. Already, early in the third century B.C.E., astronomers had discarded his system of homocentric spheres in their pursuit of viable geometric models for the planetary motions. If the shortcomings of Eudoxus's model were evident, however, it nevertheless defined for later astronomers the essence of their task: to represent the planetary phenomena by means of uniform circular motion. Eudoxus's success thus remains implicit in the later development of astronomy, from Apollonius and Hipparchus to Ptolemy.

Wilbur R. Knorr

FURTHER READING

Charles, David. *Aristotle on Meaning and Essence*. Oxford, England: Clarendon Press, 2000. In this work on Aristotle, Charles is critical of his subject's analysis of Eudoxus and Callippus's astronomical conclusions.

De Santillana, G. "Eudoxus and Plato: A Study in Chronology." In *Reflections on Men and Ideas*. Cambridge, Mass.: MIT Press, 1968. A revised chronology of Eudoxus's life is argued on the basis of a detailed examination of the ancient biographical data and collateral historical evidence.

Knorr, W. R. *The Ancient Tradition of Geometric Problems*. Boston: Birkhäuser, 1986. Chapter 3 considers Eudoxus's studies of exhaustion and cube duplication, discussed in the wider context of pre-Euclidean geometry.

Neugebauer, O. *A History of Ancient Mathematical Astronomy*. New York: Springer-Verlag, 1975. All facets of Eudoxus's contributions to astronomy are covered; particularly detailed is the discussion of his planetary models. Includes an index.

Waerden, B. L. van der. *Science Awakening*. 4th ed. Princeton Junction, N.J.: Scholar's Bookshelf, 1988. The author provides a brief, insightful review of Eudoxus's mathematical work.

LEONHARD EULER

Swiss mathematician and physicist

Euler had a tremendous impact on almost all fields of mathematics, opening new and more fruitful courses of inquiry. One of the most prolific mathematical writers ever, his founding of the field of analysis was particularly important, and his notations remain in common use in mathematics.

Born: April 15, 1707; Basel, Switzerland
Died: September 18, 1783; St. Petersburg, Russia

EARLY LIFE

Leonhard Euler (LAY-awn-hahrt OY-luhr) was born to a Calvinist minister, Paul Euler, and his wife, Marguerite Brucker, the daughter of a minister, in Basel, Switzerland, on April 15, 1707. The family soon moved to the suburb of Riehen. Little is known of Euler's childhood, but the information that is available indicates that his interest in mathematics was quite logical, because his father was an excellent mathematician in his own right.

Paul Euler had studied under Jakob I Bernoulli, a member of the famous Bernoulli family of mathematicians, while he was studying for his degree in theology. The elder Euler gave Leonhard his first instruction at home. During this period, the younger Euler studied some of the most difficult texts in mathematics available at the time. He later went to live with his grandmother in Basel, where he went to the local school (*Gymnasium*). Euler was not satisfied with the mathematics instruction offered there and received private tutoring from Johann Burckhardt.

When Euler was almost fourteen, at his father's wish he entered the University of Basel to study theology. Although Leonhard was quite devout and worked dutifully, he had no desire to become a minister, and he filled his free time with mathematics. In fact, in time he received limited tutoring from Johann I Bernoulli. Bernoulli suggested texts for Euler and agreed to explain any difficulties during his free time on Saturdays. Since Euler did not want to disappoint Bernoulli, he worked very hard to ensure that he did not waste the professor's time.

LIFE'S WORK

In 1724, at age seventeen, Euler received a master's degree. His father was concerned about the progress Leonhard was making in theology, but the Bernoullis convinced Paul Euler that the young man was extremely gifted in mathematics and that the gift should not be wasted. Thus, Euler was free at a very young age to pursue a career in mathematics. Euler began working independently and submitted a solution to a problem in navigation proposed by the Academy of Sciences in Paris in 1727. Although he received only an honorable mention from the academy, his name was placed before many of the people who could influence his career.

Unfortunately, there were many mathematicians who were not ready to accept one so young. When Euler applied for a post as professor of mathematics at the University of Basel, his name was not forwarded, probably because of his age. As such positions were rare in his home country, Euler was very discouraged, but the Bernoullis encouraged him with news of the newly formed Academy at St. Petersburg in Russia. This institution had a twofold purpose: Its members received a stipend to continue their own work, and, from time to time, the czar might pose practical problems to be solved by the members. Both Nikolaus and Daniel Bernoulli held positions there, and they wrote to Euler that there would soon be an opening in medicine. Therefore, Euler began to study anatomy so that he would be qualified for the position when it became available. He received the appointment in 1727, and he traveled to Russia, intending to accept this medical position. The reigning monarch, Catherine I, died before he could take up his appointment, however, and he instead joined the mathematical group, unnoticed in the change of regimes.

Although the political situation in Russia was not entirely satisfactory, the Russian academy offered Euler security and a comfortable lifestyle, and he was able to marry Catharina Gsell and begin a family. Except for a two-decade stint in Berlin, he made St. Petersburg his permanent home. His work for the first six years was fairly routine as a member of the physics staff, but in 1733 he became the leading member in mathematics when Daniel Bernoulli left to return home.

Euler threw himself into his work with fervor. (Indeed, when the Swiss prepared to publish Euler's writings in the twentieth century, the project's editors were stunned by the amount of material found in St. Petersburg.) His work at the Russian academy was diverse, spanning navigation, cartography, ballistics, mechanics, measurement, and especially mathematics. During this first, fourteen-year period in Russia, Euler wrote nearly one hundred articles and memoirs for publication. He also maintained correspondence with the most widely known European mathematicians, both for himself and in the name of the Academy of Sciences. Indeed, as the result of his strenuous pursuit of a Parisian prize in the field of astronomy, Euler developed an illness that resulted in the loss of sight in one eye.

In 1741, Euler was invited to Berlin by Frederick the Great of Prussia as part of the reorganization and refurbishing of the Berlin Academy. Euler accepted this position, which he filled from 1741 until 1766. He also maintained his membership in the St. Petersburg Academy, as well as in the Royal Society of London, to which he was elected in 1749. He continued to write for the Russian academy during this time, as he was still in their employ. While living abroad, Euler received a stipend from Russia in addition to his recompense for his post in Berlin. While in Russia, he was supported well enough to have several servants. Euler and Frederick got along so poorly that at least once Frederick tried to remove him as the director of the academy but was convinced by others that this would be a mistake. Nevertheless, Euler more than earned his pay. He worked on many applications for Frederick, including coinage, insurance, and pensions, and he held several administrative posts. In addition, he produced almost three hundred mathematical papers and tutored some of Frederick's relatives.

By 1766, however, the situation with Frederick had become so bad that Euler, then fifty-nine, decided to accept the invitation of Catherine the Great to return to Russia. Because he had regularly sent memoirs back to that country and had enjoyed its financial support while in Berlin, the move seemed logical. He was to live there for the remainder of his life.

Soon after his return, Euler began to develop a cataract in his remaining eye. For a man of lesser gifts,

Leonhard Euler. (Library of Congress)

blindness would have been a career-ending disability. Euler, however, began to train himself to solve problems mentally and dictate the results to others, principally his sons Johann Albrecht and Christoph, who were also mathematicians. He succeeded so completely that he was able to work in this fashion for another fifteen years, holding his post at the academy and actually producing more papers than ever before. During this time, Euler produced some of his best work, including his analysis of the effects of the gravitational pull of the Earth and the Sun upon the motion of the Moon. Although he did benefit from discussions with his peers, all the work had to be done without the aid of writing partial results or ideas. He also produced a monograph on integral calculus, work on fluid mechanics, and won a prize for work in astronomy.

Despite a lifetime of work, Euler was most comfortable with his family. A devoted family man, he even held his children while working on mathematics when they were still small. Working with his sons in his later life was also quite fulfilling for him. Although he had chosen not to pursue theology as a youth, Euler never left his church, and he held daily services with his family. Euler's wife died in 1776, and he soon was married to his first wife's sister, with whom he lived until his death in St. Petersburg on September 18, 1783.

Significance

The extent of Leonhard Euler's work was vast. In mathematics, he developed much of the notation in current use, in addition to a considerable amount of theory. Euler was the first to treat trigonometry as a field in itself rather than a branch of geometry, and he developed spherical trigonometry. Thus, he led the way in its development as a discipline. He made great progress in calculus, writing two texts, *Institutiones calculi differentialis* (1755) and *Institutiones calculi integralis* (1768-1770), that are still used by mathematicians as reference works. Included in these books are several discoveries Euler made concerning differential equations and partial differential equations. Euler made significant refinements to the fundamental theorem of algebra. He was also extremely interested in summation of infinite series and developed much of the basis upon which convergence theories would later be founded.

Although he produced a great quantity of work in physics, in part in response to requests by monarchs, Euler's major contribution in this field was his imposing analysis of mechanics. He was far more interested in the mathematical aspects of physical problems and thus was able to systematize his study. Euler published his results in *Mechanica sive motus scientia analytice exposita* (1736) and *Theoria motus corporum solidorum seu rigidorum* (1765). The former work was the first attempt to establish clear solutions to mechanical problems. Other sciences in which Euler worked include astronomy, navigation, and optics, yet Euler's foremost field was mathematical analysis, a field that owes its foundation to Euler's book *Introductio in analysin infinitorum* (1748; *Introduction to Analysis of the Infinite*, 1988-1990). Of particular interest to mathematicians is his development of function theory and notation.

The republication of Euler's work began in 1911 in Leipzig, Germany, and moved to Lausanne, Switzerland, in 1942. Three series have been produced, *Opera mathematica, Opera mechanica et astronomica*, and *Opera physica*, in which each work is reproduced in the original language of publication. Although only those papers that Euler personally prepared for publication are included, it is estimated that to include them all would take more than fifty volumes. Euler ranks as one of the most prolific mathematicians in history.

Celeste Williams Brockington

Further Reading

Bell, Eric T. "Analysis Incarnate." In *Men of Mathematics*. New York: Simon & Schuster, 1937. Each chapter in this book deals with a major mathematician, dating from ancient Greece to the early twentieth century. This chapter on Euler includes biographical information and a limited discussion of his work.

Boyer, Carl B. *A History of Mathematics*. 2d ed., rev. New York: Wiley, 1989. Boyer's book is a standard though extensive history, and his discussion of Euler and his work is both interesting and clear.

Dunham, William. *Euler: The Master of Us All*. Washington, D.C.: Mathematical Association of America, 1999. In this book aimed at readers with a knowledge of mathematics, Dunham describes Euler's many contributions to the field. Topics include number theory, logarithms, infinite series, complex variables, algebra, and geometry.

Eves, Howard. *An Introduction to the History of Mathematics*. 5th ed. Philadelphia: Saunders College, 1983. Although the treatment of Euler is extremely brief, Eves is excellent in placing Euler within the evolution of mathematics.

Havil, Julian. *Gamma: Exploring Euler's Constant.* Princeton, N.J.: Princeton University Press, 2003. Euler first described how gamma was a constant in many areas of mathematics. Almost three hundred years later, however, the nature of this constant remains a mystery. Havil examines Euler's discovery and subsequent developments in the understanding of gamma.

Struik, Dirk J. *A Concise History of Mathematics.* 4th rev. ed. New York: Dover, 1987. A standard history of mathematics; the treatment of Euler and his work is concise yet informative.

Youschkevitch, A. P. "Leonhard Euler." In *Dictionary of Scientific Biography.* Vol. 4. New York: Charles Scribner's Sons, 1971. This article is of particular note for at least two reasons. First, Youschkevitch was a fellow of the Soviet Academy of Sciences, an outgrowth of the St. Petersburg Academy. As such, he had easy access to Euler's work. Second, the article contains an extensive bibliography (seventy entries).

Pierre de Fermat

French mathematician

Fermat made several pivotal discoveries in the foundations of analytical geometry, differential calculus, and probability theory, serving as an intellectual catalyst for René Descartes, Gottfried Wilhelm Leibniz, and Sir Isaac Newton. His main achievements, however, were in number theory, in which he established the basis of the modern theory and formulated two fundamental theorems that still bear his name.

Born: August 17, 1601; Beaumont-de-Lomagne, France
Died: January 12, 1665; Castres, France

Early Life

Pierre de Fermat (pyehr deh fehr-mah) was born in a provincial village northwest of Toulouse in the Gascony region of southern France. His father, Dominique, was a well-to-do leather merchant and petty official; his mother, Claire, née de Long, belonged to a prominent family of jurists. Pierre, his brother Clement, and his two sisters acquired primary and secondary education at the local monastery of Grandselve. Pierre then attended the University of Toulouse, from which, having decided on a legal career, he entered the University of Law at Orléans, where he earned a bachelor of civil laws degree in 1631. Shortly before graduation, he purchased an office in the *parlement* of Toulouse; shortly after, he married a distant cousin, Louise de Long, and settled down to a long and apparently uneventful career as a civil official and legislator. For the next thirty-four years, he fathered five children, served capably in office, and overtly did little to distinguish himself. Few records remain of his life, beyond the normal transactions of the bourgeois.

The single remaining portrait of Fermat, apparently done when he was around forty-five, shows a round, somewhat fleshy face, with arched brows, a large straight nose, and a small, rather delicate mouth. The large eyes, the most prominent feature of his face, seem unfocused, as though staring at something deep within. On the whole, he looks remote, withdrawn, aloof, and a bit patrician—a proper image for a provincial jurist.

He looked undistinguished largely because he wanted to. His life spanned a turbulent period in French history, when distinction often led to disgrace or at least to difficulty. Fermat avoided this adroitly, and he therefore gained stability and a measure of leisure, allowing him to pursue his real interest: mathematics. Mathematics had not yet become a profession, hence, it could be pursued as a hobby. Fermat became one of the greatest mathematical hobbyists of his or of any other time. His correspondence is filled with the most daring mathematical speculations ever recorded, all the more striking because he strenuously resisted publication or any kind of public recognition. Publication might have jeopardized his stability, his security, and his serenity.

Life's Work

Fermat's major achievements lie in the field of his great love, number theory; but he anticipated these with striking discoveries in other analytical areas, which, characteristically, he neglected to publish, thereby allowing others—notably René Descartes and Blaise Pascal—to gain credit that was properly his. Thus, for example, he anticipated the fundamental discovery underlying the differential calculus thirteen years before the birth of Sir Isaac Newton and seventeen years before that of Gottfried Wilhelm Leibniz; yet they are commonly given independent credit for that finding. He did this in a characteristic way.

The basic problem of differentiation is to determine the rate of change of a system at a particular instant in time. This is commonly represented by the attempt to draw the straight-line tangent to the graph of a continuous function—that is, to discover how to construct a line tangent to any point of a given curve. After Descartes had invented a coordinate system, constructing the graph itself was relatively easy. The difficulty lay in determining the tangent, for it changed at every point in the curve. The inventors of the calculus solved this simply by visualizing what would happen to a given tangent as it approached a given point—that is, by seeing how the tangent changed as the distance between it and the point dwindled to nothing. Once they had seen this in their imagination, they could proceed to create algebraic or graphical means of specifying it; this was done by determining the limiting values of the y-component divided by the x-component as both approached zero simultaneously. This sounds complicated and is extremely difficult to visualize without graphic demonstration and some knowledge of trigonometry, but these problems had real physical applications that had to be determined before modern physics and the technology based on it could be carried out.

Fermat made a second discovery in this new differential calculus closely related to the first. Basic algebra presents equations in which one quantity is expressed in terms of another: $y = 4t$ for example. This means that the value of y can be determined by calculating the value of $4t$. From another point of view, the value of y depends on t, or y depends on t, or y is a function of t: $y = f(t)$, in algebraic notation. In this particular instance, the function can be graphed as a straight line, and the calculations for applications are quite simple. When the graph is a complex curve, however, in many practical applications it is necessary to find the maximum and minimum values of the function. Fermat derived a way to do this simply, both graphically and algebraically. Beginning with his earlier observations about tangents, Fermat reasoned that the highest and lowest values of any function would be found at the highest and lowest points of the curve. Observation would show this easily on a graph. Furthermore, these tops and bottoms would occur only where the tangents became parallel to the horizontal axis—that is, where the equation of the tangent became zero. To find them, he had only to set the tangent equation equal to zero and calculate the point. This discovery, relatively simple, had vital and far-reaching effects.

Pierre de Fermat. (Library of Congress)

Fermat himself occasionally dabbled in particular applications of his general theorems. In one case, he turned his attention to optics and the problems of determining how a ray of light will behave when it reflects from or is refracted through a surface. In the process of studying this, he discovered what has come to be called the principle of least time, which is the fundamental principle of quantum theory, particularly in its mathematical aspect of wave mechanics. A ray of light passes from point A to point B, undergoing several reflections and passing through several surfaces. Fermat proved that regardless of deviations, the path the ray must take can be calculated by a single factor: The time spent in passage must be a minimum. On the strength of this theorem, Fermat deduced that, in reflection, the angle of incidence equals the angle of reflection and, in refraction, the sine of the angle of incidence equals a constant multiple of the angle of refraction in moving from one medium to another.

Fermat was also an innovator in analytic geometry, anticipating Descartes in the process but characteristically refusing to declare his precedence by publishing his findings. In fact, he went beyond Descartes and made the crucial applications on which all further progress in the discipline depended. Fermat was the first to postulate a space of three dimensions, thereby laying the basis for modern multidimensional analytic geometry. Like most of his other discoveries, this marked a true turning point, for the great difficulty in this method of analysis is going from two to three dimensions. Moreover, in making this transition, Fermat also corrected Descartes in the classification of curves by degrees of equation. Descartes, assuming proprietary rights as the assumed inventor of the system, at first balked at accepting the corrections of an amateur but had to concede in the end. Yet true credit for this invention was denied Fermat for centuries. Similarly, not until 1934 did anyone discover that Newton had borrowed the fundamental theorem of the differential calculus from Fermat.

Any of these discoveries alone would have sufficed for the life's work of any mathematical genius. For Fermat, however, these were mere incidents; his principal mathematical occupation was the theory of numbers, a field in which he made his major achievements. This field concerns itself with the most basic of all topics in mathematics: the simple whole numbers and their common relationships and properties. Although basic to mathematics and simple in the beginning, the problems presented here have led to the most abstract theories.

Fermat began by concentrating on prime numbers—those numbers greater than one that have no divisors other than 1 and the number itself: 2, 3, 5, 7, 11, and the like. In working with these numbers, Fermat routinely presented his theorems without proofs, or without proving them completely, or simply with hints about the methods he used to discover them. Furthermore, he sometimes happened to be completely—or partly—wrong. Something like that took place in his formulation of what came to be known as Fermat's numbers: the series 3, 5, 17, 257, 65537. All of these numbers are found by the same process of raising 2 to a further power of 2 itself raised to a sequential power of 2. Fermat asserted that every number so found is a prime. He was right for the first five numbers, but the next two that follow are not primes. Thereafter in the sequence, there seems to be no general rule, though that could not be determined until the development of modern computers. The amazing point is this: Fermat was wrong, but these numbers still turned out to have significant applications in physics.

Fermat's greatest accomplishments in number theory are found in two theorems that still bear his name: Fermat's theorem and his last theorem. The first can be stated simply: If n is any whole number and p is any prime, then $n^p - n$ is divisible by p. Typically, he gave this without proof, and one was not presented until fifty years after his death. Yet the proof depends on only two facts: that a given whole number can be made only by multiplying primes, and that if a prime divides a product of two numbers then it divides at least one of them. Yet it also depends on the use of the principle of mathematical induction, which was first formulated during Fermat's lifetime. That Fermat had formulated this principle independently is clear from this theorem and from his description of a method he called "infinite descent," already suggested in the account given of his method of tangents.

Fermat's so-called last theorem, which states that there are no solutions to equations of the type $x^n + y^n = z^n$ when n is greater than 2, grew out of his fascination with the kind of equations called Diophantine—equations with two or more unknowns requiring whole number solutions. Fermat accomplished much with these equations. For example, he asserted that the equation $y^3 = x^2 + 2$ has only one solution: $y = 3$, $x = 5$. As usual, he gave no proof; yet he must have had one, since one eventually emerged.

Significance

The significance of someone such as Fermat is difficult to state directly or simply, since he worked solely in the area of abstract mathematics and produced few tangible or readily measurable results. It is even more difficult with him than with other mathematicians because he refused publicity; at his death, few could have been aware that one of the world's truly seminal minds was passing away. Furthermore, many of his discoveries were paralleled by other workers; it would be easy to dismiss him as interesting but not particularly significant, but such a judgment would not do him justice.

So many of his discoveries were pivotal, providing a necessary impetus to the opening of several new and rich fields of inquiry. While it is true that others paralleled some of his work, in most cases he provided the catalyst. In analytic geometry, for example, Descartes preceded him in print, but Fermat made the absolutely necessary transition to the third dimension; he also corrected Descartes's formulation of the degrees of the equations. Without these contributions, Cartesian analysis could not have become a formidable instrument in the development of mechanics and the incipient engineering of the Industrial Revolution. Similarly, Newton and Leibniz could not have begun differential calculus without Fermat's work on tangents and slopes. Finally, whole areas of analysis in physics remain indebted to Fermat.

James Livingston

FURTHER READING

Beiler, Albert H. *Recreations in the Theory of Numbers: The Queen of Mathematics Entertains.* 2d ed. New York: Dover, 1966. Beiler provides a solid introduction to the problems that Fermat attacked and his methods of solution. This work is less technical in approach than many books in number theory, but it requires some knowledge of advanced mathematics.

Bell, Eric T. *Men of Mathematics.* New York: Simon & Schuster, 1986. An excellent general introduction to the major figures in classical mathematics. Bell's discussion of Fermat covers all major topics with style and wit.

Burton, David M. *The History of Mathematics: An Introduction.* Boston: Allyn & Bacon, 1984. A standard text, written for readers with some knowledge of advanced mathematics. Burton re-creates the process of problem solving, which is particularly good for understanding Fermat.

Kline, Morris. *Mathematical Thought from Ancient to Modern Times.* New York: Oxford University Press, 1972. An excellent general history of mathematics, with a clear, succinct account of Fermat's contributions.

Krizek, Michal, Florian Luca, and Lawrence Somer. *Seventeen Lectures on Fermat Numbers.* New York: Springer, 2001. These lectures provide an overview of the properties of Fermat numbers and their various mathematical applications.

Mahoney, Michael Sean. *The Mathematical Career of Pierre de Fermat (1601-1665).* Princeton, N.J.: Princeton University Press, 1973. The quality of writing and clarity of exposition makes this book a good work for general readers as well as historians of mathematics or science. The mathematical explanations are fully detailed and not overly technical.

Ribenboim, Paulo. *Fermat's Last Theorem for Amateurs.* New York: Springer, 1999. Intended for students, teachers, and amateur mathematicians, the book explains the proofs related to Fermat's last theorem.

Simmons, George Finlay. *Calculus Gems: Brief Lives and Memorable Mathematics.* New York: McGraw-Hill, 1992. This collection of biographies includes a chapter about Fermat's discovery of analytic geometry and his founding of modern numbers theory.

Singh, Simon. *Fermat's Enigma: The Epic Quest to Solve the World's Greatest Mathematical Problem.* New York: Walker, 1997. Recounts how scientists and mathematicians during a 350-year period attempted to solve Fermat's last theorem, concluding in Andrew Wiles's eventual solution.

JOSEPH FOURIER

French mathematician and physicist

In deriving and solving equations representing the flow of heat in bodies, Fourier developed analytical methods that proved to be useful in the fields of pure mathematics, applied mathematics, and theoretical physics.

Born: March 21, 1768; Auxerre, France
Died: May 16, 1830; Paris, France
Also known as: Baron Fourier; Jean-Baptiste-Joseph Fourier (full name)

EARLY LIFE

The twelfth child of master tailor Joseph Fourier and the ninth child of Édmie Fourier, Jean-Baptiste-Joseph Fourier (few-ree-ay) became an orphan at the age of nine. He was placed in the local Royal Military School run by the Benedictine Order and soon demonstrated his passion for mathematics. Fourier and many biographers after him attribute the onset of his lifelong poor health to his habit of staying up late, reading mathematical texts in the empty classrooms of the school. He completed his studies in Paris. He was denied entry into the military and decided to enter the Church and teach mathematics.

Fourier remained at the Benedictine Abbey of St. Benoit-sur-Loire from 1787 to 1789, occupied with teaching and frustrated that he had little time for mathematical research. Whether he left Paris because of the impending revolution or because he did not want to take his vows is uncertain. He returned to Auxerre and from 1789 to 1794 served as professor and taught a variety of subjects at the Royal Military School. The school was run by the Congregation of St. Maur, the only religious order excluded from the postrevolutionary decree confiscating the property of religious orders.

Fourier became involved in local politics in 1793 and was drawn deep into the whirlpool as internal unrest and external military threats turned the committees on which he served into agents of the Terror. Fourier made the mistake of defending a group of men who turned out to be enemies of Robespierre. He was arrested and nearly guillotined, spared only by the death of Robespierre. He became a student at the short-lived École Normale, mainly to have the opportunity to go to Paris and meet Pierre-Simon Laplace, Joseph-Louis Lagrange, and Gaspard Monge, the foremost mathematicians in France. In 1795, the École Polytechnique was opened, and Fourier was invited to join the faculty, but he was arrested once again, this time by the extreme reactionaries who hated him for his role in the Terror, even though he did much to moderate the excesses of the Terror in Auxerre. As with many other aspects of Fourier's life during the Revolution, the exact reason for his release is unknown. In any case, he was released and occupied himself with teaching and administrative duties at the École Polytechnique.

In 1798, Fourier was chosen to be part of Napoleon I's expedition to Egypt. Fourier was elected permanent secretary of the newly formed Institute of Egypt, held a succession of administrative and diplomatic posts in the French expedition, and conducted some mathematical research. Upon his return to France in 1801, Fourier was named by Napoleon to be the prefect of Isère, one of the eighty-four newly formed divisions of France. It is during his prefecture that Fourier began his life's work.

Life's Work

Fourier's work in the development of an analytical theory of heat diffusion dates from the early nineteenth century, when he was in his early thirties, and after he had distinguished himself in administration of scientific and political institutions in Egypt. He had demonstrated a talent and passion for mathematics early, but he had not yet made significant contributions to the field. It was during whatever time he could spare from his administrative duties as prefect that he made his lasting contribution to physics and mathematics.

Fourier remained at Grenoble until Napoleon's downfall in 1814. He turned a poorly managed department into a well-managed one in a short time. It is not clear why Fourier began to study the diffusion of heat, but in 1804 he began with a rather mathematically abstract derivation of heat flow in a metal plate. He conducted numerous experiments in an attempt to establish the laws regulating the flow of heat. He expanded the scope of problems addressed, polished the mathematical formalism, and infused physical concepts into the derivation of the equations that expressed heat flow. In 1807, he presented a long paper to the French Academy of Sciences, but opposition from Laplace and others prevented its publication. At issue was a fundamental disagreement over mathematical rigor and the underlying physical concepts.

Laplace's first objection was that Fourier's methods were not mathematically rigorous. Fourier claimed that any function could be represented by an infinite trigonometric series—a sum of an infinite number of sine and cosine functions each with a determinable coefficient. Such series were instrumental in Fourier's formulation and solution of the problems of the diffusion of heat. Only later were Fourier's methods shown to be strictly rigorous mathematically. The second objection concerned the method of derivation. Laplace preferred to explain phenomena by the action of central forces acting between particles of matter. Fourier, while not denying the correctness or the usefulness of that approach, took a different approach. Heat, for Fourier, was the flow of a substance and not some relation between atoms and their motions. He attempted, successfully, to account for the phenomenon of heat diffusion through mathematical analysis. His paper of 1807 languished in the archives of the Academy of Sciences, unpublished.

As a result of his work in Egypt and his position as permanent secretary of the Institute of Egypt, Fourier edited and wrote the historical introduction to the *Description de l'Égypte* (1809-1828; description of Egypt). He worked on this project from around 1802 until 1810.

A prize was offered in 1810 by the Academy of Sciences on the subject of heat diffusion, and Fourier slightly revised and expanded his 1807 paper to include discussion of diffusion in infinite bodies and terrestrial and radiant heat. Fourier won the prize, but he had faced no serious competition. The jury criticized the paper in much the same way as the 1807 paper had been criticized, and again Fourier's work was not published. Eventually, after years of prodding, the work was published in 1815.

Fourier was probably not happy being virtually exiled from Paris, the scientific capital of France. He seemed destined to live out his days in Grenoble. With Napoleon's abdication in April, 1814, Fourier provisionally retained his job as prefect during the transfer of power to Louis XVIII. He also managed to alter the route Napoleon took from Paris to exile in Elba, bypassing Grenoble, in order to avoid a confrontation between himself and Napoleon. Upon Napoleon's return in March, 1815, Fourier prepared the defenses of the town and made a diplomatic retreat to Lyons. Fourier returned before completing the journey upon learning that Napoleon had made him prefect of the Rhône department. He was dismissed before Napoleon fell once again.

Fourier's scientific work began again after 1815. One of his former pupils was now a prefect and appointed Fourier director of the Bureau of Statistics for the Seine department, which included Paris. He now had a modest income and few demands on his time. Fourier was named to the Academy of Sciences in 1817. During the next five years, he actively participated in the affairs of the Academy, sitting on commissions, writing reports, and conducting his own research. His administrative duties increased in 1822, when he was elected to the powerful position of permanent secretary of the mathematical section of the Academy. His *Théorie analytique de la chaleur* (1822; *The Analytical Theory of Heat*, 1878) differs only slightly from his 1810 essay. The papers he wrote in his later years contained little that was new. He led a satisfying academic life in his last years, but his health began to deteriorate. His rheumatism had returned, he had trouble breathing, and he was sensitive to cold. Fourier died from a heart attack in May, 1830.

Significance

The core of Joseph Fourier's scientific work is *The Analytical Theory of Heat*. This work is basically a textbook describing the application of theorems from pure mathematics applied to the problem of the diffusion of heat in bodies. Fourier was able to express the distribution of heat inside and on the surface of a variety of bodies, both at equilibrium and when the distribution was changing because of heat loss or gain.

Fourier significantly influenced three different fields: pure mathematics, applied mathematics, and theoretical physics. In pure mathematics, Fourier's most lasting influence has been the definition of a mathematical function. He realized that any mathematical function can be represented by a trigonometric series, no matter how difficult to manipulate the function may appear. Some scholars single out this concept as the stepping-stone to the work of pure mathematicians later in the century, which resulted in the modern definition of a function. Additional influences are that of a clarification of a notational issue involving integral calculus and properties of infinite trigonometric series.

Applied mathematics has been influenced to a great extent by Fourier's use of trigonometric series and techniques of integration. The class of problems that Fourier series and Fourier integrals can solve extends far beyond diffusion of heat. His methods form the foundation of applied mathematics techniques taught to undergraduates. Some mathematicians before him had used trigonometric series in the solutions of problems, but the clarity, scope, and rigor that he brought to the field were significant.

Fourier's influence in theoretical physics is more subdued, perhaps because of the completeness of his results. There was little room for others to extend the physical aspects of Fourier's work—his results did not need extending. Other branches of physics appear to have been influenced by his approach, and a direct influence on the issue of determining the age of the earth by calculating its heat loss has been documented.

Roger Sensenbaugh

Further Reading

Bell, Eric T. *The Development of Mathematics*. 2d ed. New York: McGraw-Hill, 1945. Presents a narrative history of the decisive epochs in the development of mathematics without becoming overly technical. The majority of references to Fourier appear in chapter 13.

_____. *Men of Mathematics*. New York: Simon & Schuster, 1986. First published in 1937, this book remains one of the best accounts of the history of mathematics. The contributions of Fourier are discussed in chapter 12, "Friends of an Emperor."

Fourier, Joseph. *The Analytical Theory of Heat*. Translated by Alexander Freeman. 1878. Reprint. Mineola, N.Y.: Dover, 2003. Fourier's preliminary discourse to his most famous work explains in clear terms what he is attempting in the work. Devoid of technical matters, this book offers the reader a glimpse of why Fourier has achieved the status he has.

Fox, Robert. "The Rise and Fall of Laplacian Physics." *Historical Studies in the Physical Sciences* 4 (1974): 89-136. This paper presents a description of the research program of Laplace, which dominated French science at one of its most successful periods, from 1805 to 1815. Fourier led the revolt against this program.

Friedman, Robert Marc. "The Creation of a New Science: Joseph Fourier's Analytical Theory of Heat." *Historical Studies in the Physical Sciences* 8 (1977): 73-100. This paper concentrates on conceptual and physical issues rather than the mathematical aspects stressed in most older works. Also discusses how Fourier's philosophy of science compared to that of his contemporaries.

Grattan-Guinness, Ivor, with J. R. Ravetz. *Joseph Fourier, 1768-1830: A Survey of His Life and Work, Based on a Critical Edition of His Monograph on the Propagation of Heat, Presented to the Institute de France in 1807.* Cambridge, Mass.: MIT Press, 1972. Intertwines a close study of Fourier's life and work with a critical edition of his 1807 monograph. The 1807 monograph is in French, but everything else is in English. Contains a bibliography of Fourier's writings, a list of translations of his works, and a secondary bibliography.

Herivel, John. *Joseph Fourier: The Man and the Physicist.* Oxford, England: Clarendon Press, 1975. Although this work does not claim to be the definitive biography of Fourier, it goes much further than any other work written in English. Although the book is almost devoid of technical detail, the prospective reader would benefit from a knowledge of the history of France from 1789 to 1830.

James, Ioan. *Remarkable Physicists: From Galileo to Yukawa.* New York: Cambridge University Press, 2004. This collection of brief biographies of famous physicists contains a five-page biography of Fourier. Written in a nontechnical style for readers with limited knowledge of science.

Purrington, Robert D. *Physics in the Nineteenth Century.* New Brunswick, N.J.: Rutgers University Press, 1997. Fourier is mentioned in several places, and his work is placed in a historical context, in this survey of nineteenth century physics; references to Fourier are listed in the index. Chapter 4, "Heat and Thermodynamics," contains information about the Fourier series.

GOTTLOB FREGE
German mathematician

Frege's writings were never widely read or appreciated during his own time, and his complex system of symbols and functions was forbidding even to the best minds in mathematics. Nevertheless, he is recognized as the founder of modern symbolic logic and the creator of the first system of notations and quantifiers of modern logic.

Born: November 8, 1848; Wismar, Mecklenburg-Schwerin (now in Germany)
Died: July 26, 1925; Bad Kleinen, Germany
Also known as: Friedrich Ludwig Gottlob Frege (full name)

EARLY LIFE
Friedrich Ludwig Gottlob Frege (FREH-gah) was the son of the principal of a private girls' high school. While Frege was in high school in Wismar, his father died. Frege was devoted to his mother, who was a teacher and later principal of the girls' school. He may have had a brother, Arnold Frege, who was born in Wismar in 1852. Nothing further is known about Frege until he entered the university at the age of twenty-one. From 1869 to 1871, he attended the University of Jena and proceeded to the University of Göttingen, where he took courses in mathematics, physics, chemistry, and philosophy. By 1873, Frege had completed his thesis and had received his doctorate from the university. Frege returned to the University of Jena and applied for an unsalaried position. His mother wrote to the university that she would support him until he acquired regular employment. In 1874, as a result of publication of his dissertation on mathematical functions, he was placed on the staff of the university. He spent the rest of his life at Jena, where he investigated the foundations of mathematics and produced seminal works in logic.

Frege's early years at Jena were probably the happiest period in his life. He was highly regarded by the faculty and attracted some of the best students in mathematics. During these years, he taught an extra load as he assumed the courses of a professor who had become ill. He also worked on a volume on logic and mathematics. Frege's lectures were thoughtful and clearly organized, and were greatly appreciated by his students. Much of Frege's personal life, however, was beset by tragedies. Not only did his father die while he was a young man but his children also died young, as did his wife. He dedicated twenty-five years to developing a formal system, in which all of mathematics could be derived from logic, only to learn that a fatal paradox destroyed the system. During his life, he received little formal recognition of his monumental work and, with his death in 1925, passed virtually unnoticed by the academic world.

LIFE'S WORK

Frege's first major work in logic was published in 1879. Although this was a short book of only eighty-eight pages, it has remained one of the most important single works ever written in the field. *Begriffsschrift: Eine der Arithmetischen Nachgebildete Formelsprache des reinen Denkens* (conceptual treatise: a formal language, modeled upon that of arithmetic, for pure thought) presented for the first time a formal system of modern logic. He created a system of formal symbols that could be used more regularly than ordinary language for the purposes of deductive logic. Frege was by no means the first person to use symbols as representations of words, because Aristotle had used this device and was followed by others throughout the history of deductive logic.

Earlier logicians, however, had thought that in order to make a judgment on the validity of sentences, a distinction was necessary between subject and predicate. For the purposes of rhetoric, there is a difference between the statements "The North defeated the South in the Civil War" and "The South was defeated by the North in the Civil War." For Frege, however, the content of both sentences conveyed the same concept and hence must be given the same judgment. In this work, Frege achieved the ideal of nineteenth century mathematics: that if proofs were completely formal and no intuition was required to judge the correctness of the proofs, then there could be complete certainty that these proofs were the result of explicitly stated assumptions. During this period, Frege began to use universal quantifiers in his logic, which cover statements that contain "some" or "every." Consequently, it was now possible to cover a range of objects rather than a single object in a statement.

In 1884, Frege published *Die Grundlagen der Arithmetik* (*The Foundations of Arithmetic*, 1950), which followed his attempt to apply similar principles to arithmetic as his earlier application to logic. In this work, he first reviewed the works of his predecessors and then raised a number of fundamental questions on the nature of numbers and arithmetic truth. This work was more philosophical than mathematical.

Throughout the work, Frege enunciated three basic positions concerning the world of philosophical logic. Mental images of a word as perceived by the speaker are irrelevant to the meaning of a word in a sentence in terms of its truth or falsity. The word "grass" in the sentence "the grass is green" does not depend on the mental image of "grass" but on the way in which the word is used in the sentence. Thus the meaning of a word was found in its usage. A second idea was that words have meaning only in the context of a sentence. Rather than depending on the precise definition of a word, the sentence determined the truth-value of the word. If "all grubs are green," then it is possible to understand this sentence without necessarily knowing anything about "grubs." Also, it is possible to make a judgment about a sentence that contains "blue grubs" as false, because "all grubs are green."

Frege's third idea deals with the distinction between concepts and objects. This distinction raises serious questions concerning the nature of proper names, identity, universals, and predicates, all of which were historically troublesome philosophical and linguistic problems.

After the publication of *The Foundations of Arithmetic*, Frege became known not only as a logician and a mathematician but also as a linguistic philosopher. Although the notion of proper name is important for his system of logic, it also extends far beyond those concerns. There had existed an extended debate as to whether numbers such as "1,2,3, . . ." or directions such as "north" were proper names. Frege argued that it was not appropriate to determine what can be known about these words and then see if they can be classified as objects. Rather, like his theory of meaning, in which the meaning of a word is determined by its use in a sentence, if numbers are used as objects they are proper names.

Frege's insistence on the usage of words extended to the problem of universals. According to tradition, something that can be named is a particular, while a universal is predicated on a particular. For example, "red rose" comprises a universal "red" and a particular "rose." Question arose as to whether universals existed in the sense that the "red" of the "red rose" existed independently of the "rose." Frege had suggested that universals are used as proper names in such sentences as "The rose is red."

Between 1893 and 1903, Frege published two volumes of his unfinished work *Grundgesetze der Arithmetik* (the basic laws of arithmetic). These volumes contained both his greatest contribution to philosophy and logic and the greatest weaknesses of his logical system.

Frege made a distinction between sense and reference, in that words frequently had the same reference, but may imply a different sense. Words such as "lad," "boy," and "youth" all have the same reference or meaning, but not in the same sense. As a result, two statements may be logically identical, yet have a different sense. Hence, $2 + 2 = 4$ involves two proper names of a number, namely "$2 + 2$" and "4," but are used in different senses. Extending this idea to a logical system, the meaning or reference of the proper names and the truth-value of the sentence depend only on the reference of the object and not its sense. Thus, a sentence such as "The boy wore a hat" is identical to the sentence "The lad wore a hat." Because the logical truth-value of a sentence depends on the meaning of the sentence, the inclusion of a sentence without any meaning within a complex statement means that the entire statement lacks any truth-value. This proved to be a problem that Frege could not resolve and became a roadblock to his later work.

A further problem that existed in *Grundgesetze der Arithmetik*, which was written as a formal system of logic including the use of terms, symbols, and derived proofs, was the theory of classes. Frege wanted to use logic to derive the entire structure of mathematics to include all real numbers. To achieve this, Frege included, as part of his axioms, a primitive theory of sets or classes.

While the second volume of *Grundgesetze der Arithmetik* was being prepared for publication, Frege received a letter from Bertrand Russell describing a contradiction that became known as the Russell Paradox. This paradox, sometimes known as the Stranger Loop, asks, is "the class of all classes that are not members of itself" a member of itself or not? For example, the "class of all dogs" is not a dog; the "class of all animals" is not an animal. If the class of all classes is a member of itself, then it is one of those classes that are not members of themselves. However, if it is not a member of itself, then it must be a member of all classes that are members of themselves, and the loop goes on forever. Frege replaced the class axiom with a modified and weaker axiom, but his formal system was weakened, and he never completed the third volume of the work.

Between 1904 and 1917, Frege added few contributions to his earlier works. During these years, he attempted to work through those contradictions that arose in his attempt to derive all of mathematics from logic. By 1918, he had begun to write a new book on logic, but he completed only three chapters. In 1923, he seemed to have broken through his intellectual dilemma and no longer believed that it was possible to create a foundation of mathematics based on logic. He began work in a new direction, beginning with geometry, but completed little of this work before his death.

Significance

In *Begriffsschrift*, Gottlob Frege created the first comprehensive system of formal logic since the ancient Greeks. He provided some of the foundations of modern logic with the formulation of the principles of noncontradiction and excluded middle. Equally important, Frege introduced the use of quantifiers to bind variables, which distinguished modern symbolic logic from earlier systems.

Frege's works were never widely read or appreciated. His system of symbols and functions was forbidding even to the best minds in mathematics. Russell, however, made a careful study of Frege and was clearly influenced by his system of logic. Also, Ludwig Wittgenstein incorporated a number of Frege's linguistic ideas, such as the use of ordinary language, into his works. Frege's distinction between sense and reference later generated a renewed interest in his work, and a number of important philosophical and linguistic studies are based on his original research.

Victor W. Chen

Further Reading

Bynum, Terrell W. Introduction to *Conceptual Notations*, by Gottlob Frege. Oxford, England: Clarendon Press, 1972. An eighty-page introduction to the logic of Frege. Although sections of the text on the logic are not suited for the general reader, the introductory text is clear, concise, and highly accessible. The significant works by Frege are outlined in simple terms, and the commentary is useful.

Currie, Gregory. *Frege: An Introduction to His Philosophy*. Brighton, England: Harvester Press, 1982. Discusses all the major developments in Frege's thought from a background chapter to the *Begriffsschrift*, theory of numbers, philosophical logic and methods, basic law, and the fatal paradox. Some parts of this text are accessible to the general reader; other parts require a deeper understanding of philosophical issues.

Dummett, Michael A. E. *Frege: Philosophy of Language*. London: Duckworth, 1973. One of the leading authorities on the philosophy of Frege. The advantage of this text over others is that Dummett is in part responsible for the idea that Frege is a linguistic philosopher. Somewhat difficult but good introduction to Frege.

Grossmann, Reinhardt. *Reflections on Frege's Philosophy*. Evanston, Ill.: Northwestern University Press, 1969. Delineates three major areas of Frege's thoughts as found in *Begriffsschrift* and *The Foundations of Arithmetic*, and describes the distinction between meaning, sense, and reference. Within these areas the author writes an exposition on a few selected problems that are of current interest.

Hill, Claire O. *Rethinking Identity and Metaphysics: The Foundations of Analytic Philosophy*. New Haven, Conn.: Yale University Press, 1997. The author provides a reassessment of twentieth century analytic philosophy by examining the writings of Frege, Bertrand Russell, and Willard Quine. Hill concludes that the lack of clarity inherent in the abstract issue of identity has implications for solid subjects, such as medical ethics.

Kneale, William, and Martha Kneale. *The Development of Logic*. Oxford, England: Clarendon Press, 1962. Three chapters in this work are useful. Chapter 7 covers Frege and his contemporaries, Frege's criticism of his predecessors, and Frege's definition of natural numbers. Chapter 8 covers Frege's three major works and outlines his contributions to the world of logic. Chapter 9 covers formal developments in logic after Frege and reveals his pivotal position in the development of modern symbolic logic.

Noonan, Harold W. *Frege: A Critical Introduction*. Malden, Mass.: Blackwell, 2001. An overview of Frege's ideas, emphasizing his logic, theory of meaning, distinctions between sense and reference, and ideas about object, concept, and function.

Reck, Erich H., ed. *From Frege to Wittgenstein: Perspectives on Early Analytic Philosophy*. New York: Oxford University Press, 2002. Considers how the two men developed an analytic philosophy, including a discussion of Wittgenstein's debt to Frege, an explanation of the roots of analytic philosophy, and interpretations of some of Frege's writings.

Salerno, Joseph. *On Frege*. Belmont, Calif.: Wadsworth/Thomson Learning, 2001. Brief, nontechnical overview of Frege's ideas aimed at students seeking a better understanding of his philosophy.

Weiner, Joan. *Frege Explained: From Arithmetic to Analytic Philosophy*. Chicago: Open Court, 2004. A summary of Frege's philosophy, tracing the development of his ideas about logic, sense, meaning, and other subjects. Includes a chapter about his life and character.

ÉVARISTE GALOIS
French mathematician

With the aid of group theory, Galois produced a definitive answer to the problem of the solvability of algebraic equations, a problem that had preoccupied mathematicians since the eighteenth century. Consequently, he laid one of the foundations of modern algebra.

Born: October 25, 1811; Bourg-la-Reine, near Paris, France
Died: May 31, 1832; Paris, France

EARLY LIFE
Évariste Galois (gahl-wah) was the son of Nicolas-Gabriel Galois, a friendly and witty liberal thinker who headed a school that accommodated about sixty boarders. Elected mayor of Bourg-la-Reine during the Hundred Days after Napoleon's escape from Elba, the elder Galois retained office under the second Restoration. Galois's mother, Adélaïde-Marie Demante, was from a long line of jurists and had received a more traditional education. She had a headstrong and eccentric personality. Having taken control of her son's early education, she attempted to implant in him, along with the elements of classical culture, strict religious principles as well as respect for a stoic morality. Influenced by his father's imagination and liberalism, the eccentricity of his mother, and the affection of his elder sister Nathalie-Théodore, Galois seems to have had a childhood that was both happy and studious.

Galois continued his studies at the Collège Louis-le-Grand in Paris, entering in October, 1823. He found it difficult to adjust to the harsh discipline imposed by the school during the Restoration at the orders of the political authorities and the Church, and, although a brilliant student, he was rebellious. During the early months of 1827, he attended the first-year preparatory mathematics courses taught by H. J. Vernier; this first exposure to mathematics was a revelation for him. He rapidly became bored with the elementary nature of this instruction and with the inadequacies of some of his textbooks and began reading the original works themselves.

After appreciating the difficulty of Adrien-Marie Legendre's geometry, Galois acquired a solid background from the major works of Joseph-Louis Lagrange. During the next two years, he attended Vernier's second-year preparatory mathematics courses, then the more advanced ones of L. P. E. Richard, who was the first to recognize Galois's superiority in mathematics. With this perceptive teacher, Galois excelled in his studies, even though he was already devoting much more of his time to his personal work than to his classwork. In 1828, he began to study some then-recent works on the theory of equations, on number theory, and on the theory of elliptic functions.

This was the time period in which Galois's first memoir appeared. Published in March, 1829, in the *Annales de mathématiques pures et appliquées* (annals of pure and applied mathematics), it demonstrated and clarified a result of Lagrange concerning continuous fractions. Although it revealed a certain astuteness, it did not demonstrate exceptional talent.

Life's Work

In 1828, by his own admission Galois falsely believed—as Niels Henrik Abel had eight years earlier—that he had solved the general fifth-degree equation. Quickly enlightened, he resumed with a new approach the study of the theory of equations, a subject that he pursued until he elucidated the general problem with the aid of group theory. The results he obtained in May, 1829, were sent to the Academy of Sciences by a particularly competent judge, Augustin-Louis Cauchy. Fate was to frustrate these brilliant beginnings, however, and to leave a lasting impression on the personality of the young mathematician.

First, at the beginning of July, his father, a man who had been persecuted for his liberal beliefs, committed suicide. A month later, Galois failed the entrance examination for the École Polytechnique, because he refused to use the expository method suggested by the examiner. Barred from entering the school that attracted him because of its scientific prestige and liberal tradition, he took the entrance examination for the École Normale Supérieure (then called the École Préparatoire), which trained future secondary school teachers. He entered the institution in November, 1829.

At this time he learned of Abel's death and, at the same time, that Abel's last published memoir contained several original results that Galois himself had presented as original in his memoir to the Academy. Cauchy, assigned to supervise Galois's work, advised his student to revise his memoir, taking into account Abel's research and new results. Galois wrote a new text that he submitted to the Academy in February, 1830, that he hoped would win for him the grand prix in mathematics. However, this memoir was lost upon the death of Joseph Fourier, who had been appointed to study it. Eliminated from the competition, Galois believed himself to be the object of a new persecution by both the representatives of institutional science and society in general. His manuscripts preserve a partial record of the revision of this memoir of February, 1830.

In June, 1830, Galois published in *Bulletin des sciences mathématiques* (bulletin of mathematical sciences) a short note on the resolution of numerical equations, as well as a much more significant article, "Sur la théorie des nombres" (on number theory). The fact that this same issue contained original works by Cauchy and Siméon-Denis Poisson sufficiently confirms the reputation that Galois had already acquired. The July Revolution of 1830, however, was to initiate a drastic change in his career.

Galois became politicized. Before returning for a second year to the École Normale Supérieure in November, 1830, he had already developed friendships with several republican leaders. Even less able to tolerate his school's strict discipline than before, he published a violent article against its director in an opposition journal. For this action he was expelled on December 8, 1830.

Left alone, Galois devoted most of his time to political propaganda. He participated in the riots and demonstrations then agitating Paris and was even arrested (but was eventually acquitted). Meanwhile, to a limited degree, he continued his mathematical research. His last two publications were a short note on analysis in the *Bulletin des sciences mathématiques* of December, 1830, and "Lettre sur l'enseignement des sciences" (letter on the teaching of the sciences), which appeared on January 2, 1831, in the *Gazette des écoles*. On January 13, he began to teach a public course on advanced algebra in which he planned to present his own discoveries; this project appears not to have been successful.

On January 17, 1831, Galois presented the Academy a new version of his memoir, hastily written at Poisson's request. However, in Poisson's report of July 4, 1831, on this, Galois's most important piece of work, Poisson suggested that a portion of the results could be found in several posthumous writings of Abel and that the rest was incomprehensible. Such a judgment, the profound injustice of which would become apparent in the future, only encouraged Galois's rebellion.

Arrested again during a republican demonstration on July 14, 1831, and imprisoned, Galois nevertheless continued his mathematical research, revised his memoir on equations, and worked on the applications of his theory and on elliptic functions. After the announcement of a cholera epidemic on March 16, 1832, he was transferred to a nursing home, where he resumed his investigations, wrote several essays on the philosophy of science, and became immersed in a love affair that ended unhappily. Galois sank into a deep depression.

Provoked into a duel under unclear circumstances following this breakup, Galois sensed that he was near death. On May 29, he wrote desperate letters to his republican friends, hastily sorted his papers, and addressed to his friend Auguste Chevalier—but intended for Carl Friedrich Gauss and Carl Gustav Jacob Jacobi—a testamentary letter, a tragic document in which he attempted to outline the principal results that he had attained. On May 30, fatally wounded by an unknown opponent, he was hospitalized; he died the following day, not even twenty-one years of age.

SIGNIFICANCE

Èvariste Galois's work seems not to have been fully appreciated by anyone during his lifetime. Cauchy, who would have been able to understand its significance, left France in September, 1830, having seen only its initial outlines. In addition, the few fragments published during his lifetime did not give an overall view of his achievement and, in particular, did not provide a means of judging the exceptional interest of the results regarding the theory of equations rejected by Poisson. Also, the publication of the famous testamentary letter does not appear to have attracted the attention it deserved.

It was not until September, 1843, that Joseph Liouville, who prepared Galois's manuscripts for publication, announced officially that the young mathematician had effectively solved the problem, already investigated by Abel, of deciding whether an irreducible first-degree equation is solvable with the use of radicals. Although announced and prepared for the end of 1843, the memoir of 1831 did not appear until the October/November, 1846, issue of the *Journal de mathématiques pures et appliquées*, when it was published with a fragment on the primitive equations solvable by radicals.

Beginning with Liouville's edition, which appeared in book form in 1897, Galois's work became progressively known to mathematicians and subsequently exerted a profound influence on the development of modern mathematics. Also important, although they came too late to contribute to the advancement of mathematics, are the previously unpublished texts that appeared later.

Although he formulated more precisely essential ideas that were already being investigated, Galois also introduced others that, once stated, played an important role in the genesis of modern algebra. Furthermore, he boldly generalized certain classic methods in other fields and succeeded in providing a complete solution and a generalization of problems by systematically drawing upon group theory—one of the most important structural concepts that unified the multiplicity of algebras in the nineteenth century.

Genevieve Slomski

FURTHER READING

Bell, Eric T. *Men of Mathematics*. New York: Simon & Schuster, 1937. Historical account of the major figures in mathematics from the Greeks to Georg Cantor, written in an interesting if at times exaggerated style. In a relatively brief chapter, "Genius and Stupidity," Bell describes the life and work of Galois in a tone that both worships and scorns the young mathematician and mixes fact with legend in his discussion.

Boyer, Carl B. *A History of Mathematics*. New York: John Wiley & Sons, 1968. In this standard and reputable history of mathematics, Boyer devotes a brief section to Galois. Galois is described as the individual who most contributed to the vital discovery of the group concept. The author also assesses Galois's impact on future generations of mathematicians.

Infeld, Leopold. *Whom the Gods Love: The Story of Èvariste Galois*. New York: Whittlesey House, 1948. This biography takes great license with the facts (many of which are unknown) of Galois's life and creates an interesting, if fictional, account. The author, maintaining that biography always mixes truth and fiction, puts Galois's life in the historical context of nineteenth century France by creating scenes and dialogues that might have occurred.

Kline, Morris. *Mathematical Thought from Ancient Times to Modern Times*. New York: Oxford University Press, 1972. In this voluminous work, the author surveys the major mathematical creators and developments through the first few decades of the twentieth century. The emphasis is on the leading mathematical themes rather than on the men. The brief section on Galois gives some biographical information and discusses the mathematician's work in finite fields, group theory, and the theory of equations.

Struik, Dirk J. *A Concise History of Mathematics*. Vol. 2 in *The Seventeenth Century-Nineteenth Century*. New York: Dover, 1948. In this book devoted to a concise overview of the major figures and trends in mathematics during the time period covered, a brief section is devoted to Galois. The author spends approximately equal time discussing Galois's life and major achievements, and views the mathematician both as a product of his times and as a unique genius.

Tota Rigatelli, Laura. *Evariste Galois, 1811-1832*. Translated from the Italian by John Denton. Boston: Birkhäuser Verlag, 1996. Brief biography based on new research offering a more accurate account of Galois's life than previous biographies. Includes a chapter describing Galois's mathematical work and a comprehensive bibliography.

CARL FRIEDRICH GAUSS
German scientist

A great scientific thinker who is often ranked with Archimedes and Isaac Newton, Gauss made significant contributions in many branches of science. His most notable achievement was the articulation of the two most revolutionary mathematical ideas of the nineteenth century: non-Euclidean geometry and non-commutative algebra.

Born: April 30, 1777; Brunswick, Duchy of Brunswick (now in Germany)
Died: February 23, 1855; Göttingen, Hanover (now in Germany)
Also known as: Johann Friedrich Carl Gauss (full name)

EARLY LIFE
Carl Friedrich Gauss (gowz) was born into a family of town workers who were struggling to achieve lower-middle-class status. Without assistance, Gauss learned to calculate before he could talk; he also taught himself to read. At the age of three, he corrected an error in his father's wage calculations. In his first arithmetic class, at the age of eight, he astonished his teacher by instantly solving a word problem that involved finding the sum of the first hundred integers. However, his teacher had the insight to furnish the child with books and encourage his intellectual development.

When he was eleven, Gauss studied with Martin Bartels, then an assistant in the school and later a teacher of Nikolay Ivanovich Lobachevsky at Kazan. Gauss's father was persuaded to allow his son to enter the gymnasium in 1788. At the gymnasium, Gauss made rapid progress in all subjects, especially in classics and mathematics, largely on his own. E. A. W. Zimmermann, then professor at the local Collegium Carolinum and later privy councillor to the duke of Brunswick, encouraged Gauss; in 1792, Duke Carl Wilhelm Ferdinand began the stipend that would assure Gauss's independence.

When Gauss entered the Brunswick Collegium Carolinum in 1792, he possessed a scientific and classical education far beyond his years. He was acquainted with elementary geometry, algebra, and analysis (often having discovered important theorems before reaching them in his books), but he also possessed much arithmetical information and number-theoretic insights. His lifelong pattern of research had become established: Extensive empirical investigation led to conjectures, and new insights guided further experiment and observation. By such methods, he had already discovered Johann Elert Bode's law of planetary distances, the binomial theorem for rational exponents, and the arithmetic-geometric mean.

During his three years at the Collegium, among other things, Gauss formulated the principle of least squares. Before entering the University of Göttingen in 1795, he had rediscovered the law of quadratic reciprocity, related the arithmetic-geometric mean to infinite series expansions, and conjectured the prime number theorem (first proved by Jacques-Salomon Hadamard in 1896).

While Gauss was in Brunswick, most mathematical classics had been unavailable to him. At Göttingen, however, he devoured masterworks and back issues of journals and often found that his discoveries were not new. Attracted more by the brilliant classicist Christian

Gottlob Heyne than by the mediocre mathematician A. G. Kästner, Gauss planned to be a philologist, but in 1796 he made a dramatic discovery that marked him as a mathematician. As a result of a systematic investigation of the cyclotomic equation (whose solution has the geometric counterpart of dividing a circle into equal arcs), Gauss declared that the regular seventeen-sided polygon was constructible by ruler and compasses, the first advance on this subject in two thousand years.

The logical aspect of Gauss's method matured at Göttingen. Although he adopted the spirit of Greek rigor, it was without the classical geometric form; Gauss, rather, thought numerically and algebraically, in the manner of Leonhard Euler. By the age of twenty, Gauss was conducting large-scale empirical investigations and rigorous theoretical constructions, and during the years from 1796 to 1800 mathematical ideas came so quickly that Gauss could hardly write them down.

LIFE'S WORK

In 1798, Gauss returned to Brunswick, and the next year, with the first of his four proofs of the fundamental theorem of algebra, earned a doctorate from the University of Helmstedt. In 1801, the creativity of the previous years was reflected in two extraordinary achievements, the *Disquisitiones arithmeticae* (1801; *Arithmetical Inquisitions*, 1966) and the calculation of the orbit of the newly discovered planet Ceres.

Although number theory was developed from the earliest times, during the late eighteenth century it consisted of a large collection of isolated results. In *Arithmetical Inquisitions*, Gauss systematically summarized previous work, solved some of the most difficult outstanding questions, and formulated concepts and questions that established the pattern of research for a century. The work almost instantly won for Gauss recognition by mathematicians, although readership was small.

In January, 1801, Giuseppi Piazzi had briefly discovered but lost track of a new planet he had observed, and during the rest of that year astronomers unsuccessfully attempted to relocate it. Gauss decided to pursue the matter. Applying both a more accurate orbit theory and improved numerical methods, he accomplished the task by December. Ceres was soon found in the predicted position. This feat of locating a distant, tiny planet from apparently insufficient information was astonishing, especially because Gauss did not reveal his methods. Along with *Arithmetical Inquisitions*, it established his reputation as a first-rate mathematical and scientific genius.

The decade of these achievements (1801-1810) was decisive for Gauss. Scientifically it was a period of exploiting ideas accumulated from the previous decade, and it ended with a work in which Gauss systematically developed his methods of orbit calculation, including a theory of and use of least squares. Professionally this decade was one of transition from mathematician to astronomer and physical scientist. Gauss accepted the post of director of the Göttingen Observatory in 1807.

This decade also provided Gauss with his one period of personal happiness. In 1805, he married Johanna Osthoff, with whom he had a son and a daughter. She created a happy family life around him. When she died in 1809, Gauss was plunged into a loneliness from which he never fully recovered. Less than a year later, he married Minna Waldeck, his deceased wife's best friend. Although she bore him two sons and a daughter, she was unhealthy and often unhappy. Gauss did not achieve a peaceful home life until his youngest daughter, Therese, assumed management of the household after her mother's death in 1831 and became his companion for the last twenty-four years of his life.

In his first years as director of the Göttingen Observatory, Gauss experienced a second burst of ideas and publications in various fields of mathematics and

Carl Friedrich Gauss.

matured his conception of non-Euclidean geometry. However, astronomical tasks soon dominated Gauss's life.

By 1817, Gauss moved toward geodesy, which was to be his preoccupation for the next eight years. The invention of the heliotrope, an instrument for reflecting the sun's rays in a measured direction, was an early by-product of fieldwork. The invention was motivated by dissatisfaction with the existing methods of observing distant points by using lamps or powder flares at night. In spite of failures and dissatisfactions, the period of geodesic investigation was one of the most scientifically creative of Gauss's long career. The difficulties of mapping the terrestrial ellipsoid on a sphere and plane led him, in 1816, to formulate and solve in outline the general problem of mapping one surface on another so that the two were "similar in their smallest parts." In 1822, the chance of winning a prize offered by the Copenhagen Academy motivated him to write these ideas in a paper that won for him first place and was published in 1825.

Surveying problems also inspired Gauss to develop his ideas on least squares and more general problems of what is now called mathematical statistics. His most significant contribution during this period, and his last breakthrough in a major new direction of mathematical research, was *Disquisitiones generales circa superficies curvas* (1828; *General Investigations of Curved Surfaces*, 1902), which was the result of three decades of geodesic investigations and that drew upon more than a century of work on differential geometry.

After the mid-1820's, Gauss, feeling harassed and overworked and suffering from asthma and heart disease, turned to investigations in physics. Gauss accepted an offer from Alexander von Humboldt to come to Berlin to work. An incentive was his meeting in Berlin with Wilhelm Eduard Weber, a young and brilliant experimental physicist with whom Gauss would eventually collaborate on many significant discoveries. They were also to organize a worldwide network of magnetic observatories and to publish extensively on magnetic force. From the early 1840's, the intensity of Gauss's activity gradually decreased. Increasingly bedridden as a result of heart disease, he died in his sleep in late February, 1855.

Significance

Carl Friedrich Gauss's impact as a scientist falls far short of his reputation. His inventions were usually minor improvements of temporary importance. In theoretical astronomy, he perfected classical methods in orbit calculation but otherwise made only fairly routine observations. His personal involvement in calculating orbits saved others work but was of little long-lasting scientific importance. His work in geodesy was influential only in its mathematical by-products. Furthermore, his collaboration with Weber led to only two achievements of significant impact: The use of absolute units set a pattern that became standard, and the worldwide network of magnetic observatories established a precedent for international scientific cooperation. Also, his work in physics may have been of the highest quality, but it seems to have had little influence.

In the area of mathematics, however, his influence was powerful. Carl Gustav Jacobi and Niels Henrik Abel testified that their work on elliptic functions was triggered by a hint in the *Arithmetical Inquisitions*. Évariste Galois, on the eve of his death, asked that his rough notes be sent to Gauss. Thus, in mathematics, in spite of delays, Gauss reached and inspired countless mathematicians. Although he was more of a systematizer and solver of old problems than a creator of new paths, the completeness of his results laid the basis for new departures—especially in number theory, differential geometry, and statistics.

Genevieve Slomski

Further Reading

Bell, Eric T. *Men of Mathematics*. Reprint. New York: Simon & Schuster, 1961. Historical account of the major figures in mathematics from the Greeks to Georg Cantor, written in an interesting, if at times exaggerated, style. In a lengthy chapter devoted to Gauss titled "The Prince of Mathematicians," Bell describes the life and work of Gauss, focusing almost exclusively on the mathematical contributions. No bibliography.

Boyer, Carl B. *A History of Mathematics*. New York: John Wiley & Sons, 1968. In "The Time of Gauss and Cauchy," chapter 23 of this standard history of mathematics, Boyer briefly discusses biographical details of Gauss's life before summarizing the proofs of Gauss's major theorems. Boyer also discusses Gauss's work in the context of the leading contemporary figures in mathematics of the day. Includes charts, an extensive bibliography, and student exercises.

Buhler, W. K. *Gauss: A Biographical Study*. New York: Springer-Verlag, 1981. The author's purpose is not to write a definitive life history but to select from Gauss's life and work those aspects that are interesting and comprehensible to a lay reader. Contains quotations from Gauss's writings, illustrations, a bibliography, lengthy footnotes, appendixes on his collected works, a useful survey of the secondary literature, and an index to Gauss's works.

Dunnington, Guy Waldo. *Carl Friedrich Gauss, Titan of Science*. With additional material by Jeremy Gray and Fritz-Egbert Dohse. Washington, D.C.: Mathematical Association of America, 2004. Originally published in 1955, this is an expanded version of a biography describing Gauss's life and times to reveal the man as well as the scientist. The new edition includes introductory remarks, an updated bibliography, and a commentary on Gauss's mathematical diary. It also reprints the features contained in the original edition, including appendixes on honors and diplomas, children, genealogy, a chronology, books borrowed from the college library, courses taught, and views and opinions.

Goldman, Jay. *The Queen of Mathematics: An Historically Motivated Guide to Number Theory*. Wellesley, Mass.: A. K. Peters, 1998. A history of number theory, described by Gauss as the "queen of mathematics," describing how number theory developed from the seventeenth through the nineteenth centuries. Includes a chapter on Gauss and his work, and another chapter on the ideas contained in *Disquisitiones arithmeticae*. Aimed at readers with a knowledge of mathematics.

Turnbull, H. W. *The Great Mathematicians*. New York: New York University Press, 1962. Useful as a quick reference guide to the lives and works of the major figures in mathematics from the Greeks to the twentieth century.

Sophie Germain

French mathematician

Germain overcame the limits of a haphazard education and a variety of social and institutional impediments to make fundamental advances in the proof of Fermat's last theorem and in the physics of elasticity. Those achievements represent the most original and significant contribution to mathematics by any woman before the end of the nineteenth century.

Born: April 1, 1776; Paris, France
Died: June 27, 1831; Paris, France
Also known as: Marie-Sophie Germain (full name); Sophia Germain

Early Life

Sophie Germain (zhayr-mehn) was born Marie-Sophie Germain, the daughter of Ambroise-François Germain, a prosperous French silk merchant. All that is known of her mother is the latter's name: Marie-Madeleine Gruguelin. Sophie also had two sisters—Marie-Madeleine, who was six years older, and Angélique-Ambroise, who was three years younger. That she shared the given name "Marie" with her mother and older sister probably explains her lifelong use of "Sophie" by itself. Destined to lead conventional lives of Parisian upper-middle-class women, both of Sophie's sisters married prominent professional men. In contrast to her sisters, Sophie never married, and her life was anything but conventional.

In 1789, the year of the French Revolution, Sophie's father was elected to the Estates General that King Louis XVI had been forced to convene. Over the next two years, the family shop and home was a center of political discussion. The earliest account of Sophie's life by her friend and fellow mathematician Guglielmo Libri-Carucci suggests that she disliked these discussions. Retreating to her father's study, she became absorbed in Jean-Etienne Montucla's *Histoire des mathématiques* (1799). In that book she discovered the story of the famous ancient Greek mathematician Archimedes, who, lost in the beauty of a geometrical demonstration and oblivious to the turmoil of the Roman conquest of Syracuse, was killed when he failed to respond to a soldier's order.

Ignoring the political drama of the revolution and the social expectations of her family, Sophie was similarly captivated by the study of mathematics. Libri-Carucci's account of her life depicts concerned parents depriving their daughter of candles and even heat, all in an effort to discourage her new interest, but all to no avail. Sophie soon taught herself Latin so she could read the works of Isaac Newton and Leonhard Euler. Eventually, her parents relented. As her father's foray into politics ended, the Germain house on the rue St. Denis increasingly played host to a company of scholars.

LIFE'S WORK

The years of the French Revolution constituted one of the most formative periods in the history of mathematics. The center of this ferment was the École Polytechnique, which opened in 1794. Its faculty included a true pantheon of late Enlightenment scientists. However, its classrooms excluded women. Nonetheless, an innovative pedagogy that made professors' lecture notes public and that invited student observations offered Sophie Germain an opportunity. She obtained the lecture notes from the college and, borrowing the name of a male student, Monsieur Le Blanc, she offered observations on Joseph-Louis Lagrange's mathematics lectures. It was not long before Lagrange uncovered Germain's deception, but he was so impressed with her ability he encouraged her to continue her work.

Pursuing an interest in number theory in 1798, Germain initiated a correspondence with Adrien-Marie Legendre, the author of an important recent treatise on the subject, and began to study the *Disquisitions arithmetical* that Carl Friedrich Gauss published in 1801. Again under the name M. Le Blanc, she began a correspondence with Gauss in 1804. Her identity was revealed only in the aftermath of the French-Prussian battle at Jena in October, 1806. Fearing that Gauss might suffer the same fate as Archimedes, Germain used her family connections in the French military to ensure the safety of the German mathematician.

Germain's correspondences with Legendre and Gauss are the source of the first of her two major contributions to mathematics. Around the year 1638, the French mathematician Pierre Fermat articulated a theorem while annotating a copy of the third century Greek mathematician Diophantus of Alexandria's *Arithmetica*. Fermat observed that, in contrast to the Pythagorean theorem, which holds that the square of the hypotenuse of a right triangle is equal to the sum of the squares of its other two legs, there were no solutions for the equation $xn + yn = zn$ when the factor n is greater than 2. When Fermat published his notes on Diophantus's *Arithmetica* in 1670, his observation stimulated important advances in number theory. However, proof of the tantalizing intuition that it was impossible to find a cube that was the sum of two other cubes or indeed to find any number raised to a power greater than 2 that was the sum of two other numbers raised to the same power—Fermat's so-called "last theorem"—turned out to be one of the most intractable problems in the history of mathematics.

Fermat himself had offered a proof of the case $n = 4$. In 1738, Euler proved the case $n = 3$. Germain demonstrated that the conditions for the case in which n is an odd prime number, such that $2p + 1$ was also prime, were so stringent as to be virtually impossible. In short, she moved beyond proofs of singular cases to a general strategy for many cases. Modern mathematics honors her achievement by referring to those prime numbers p in which $2p + 1$ is also prime as "Sophie Germain primes."

During 1807-1808, Germain's interests turned to applied mathematics. In 1809, the Institute of France announced a two-year prize competition for a mathematical theory of elasticity. Two years later, Germain submitted the only entry in the competition. The institute acknowledged that her "experiments presented ingenious results" but judged the rigor of her analysis inadequate and renewed the competition. Working with Lagrange, Germain again submitted the only entry in 1813. This time the institute awarded her an honorable mention but questioned the derivation of her equations from established principles of physics. Only after a second extension did the institute finally award Germain the prize, making her the first women ever to achieve such a public recognition in mathematics.

Germain did not attend the award ceremony on January 8, 1816. Some historians speculate that she wished to avoid the attendant notoriety that the event would bring to her. Other historians suggest that she was angry that the institute had made its award to her with reservations and that one of her judges, Siméon-Denis Poisson, was using her work without attribution for his own alternative theory of elasticity.

During the 1820's, Germain's friendship with the mathematician Joseph Fourier, another rival of Poisson, played a role in allowing Germain to be the first woman who was not a wife of a member to attend sessions of the French Academy of Sciences. By this time, she was increasingly writing on the philosophy of science. In her essay, "General Considerations on the Condition of the Arts and Sciences at Different Stages of their Cultivation," she—much as her younger contemporary Auguste Comte—argued that just as the natural sciences proceeded from observation and classification of phenomena through generalization and mathematical systemization, so similar progress was possible in the human sciences. Germain herself, however, was unable to contribute to this progress. On June 27, 1831, she died in Paris after a two-year battle with breast cancer.

Significance

Shortly before Sophie Germain died, Gauss recommended her to the University of Göttingen for an honorary degree, but his request was refused. Germain never received a university degree, she never attended university classes, and, indeed, never seems to have had much formal education at all. Nevertheless, in number theory, she took what at that time was the single most significant step forward in the proof of Fermat's last theorem—a theorem that was finally proved only in 1994. Likewise, in applied mathematics, it was not the model of the impeccably rigorous scientific insider Poisson that provided the foundation for the physics of elasticity but the model of the self-taught scientific outsider Sophie Germain.

The life of Sophie Germain is eloquent testimony to the difficulties women have historically faced in pursuing scientific careers. Through sheer determination, Germain was able to transcend social expectations. In the end, her parents' generosity provided her enough income to live independently. She was also able to challenge some of the institutional barriers to women in science. Through what Libri-Carucci aptly called her courage, Germain secured the support of many of the most eminent professional mathematicians of her time. However, subtle pressures always weighed on her while she worked. For example, when she was preparing her theory of elasticity and had to visit the École Polytechnique or Institute of France, she needed formal invitations and escorts. When she submitted a paper to the Academy of Sciences extending her theory in 1825, her submission was simply ignored.

Posthumous recognition has come to Germain almost as reluctantly. In 1889, when the opening of Paris's Eiffel Tower commemorated seventy-two people whose contributions to the mathematics of elasticity made possible the tower's construction, there was no mention of Sophie Germain. Thanks to the increase in interest in women's history in late twentieth century Europe, Germain's name now appears on Parisian schools and streets. However, in many standard histories of mathematics and science, the achievements of Sophie Germain are still overlooked.

Charles R. Sullivan

Further Reading

Bucciarelli, Louis L., and Nancy Dworsky. *Sophie Germain: An Essay in the History of the Theory of Elasticity.* Dordrecht, Netherlands: D. Reidel, 1980. This is the only extended study in English of Germain's achievements, but Bucciarelli and Dworsky present a lifeless Germain, and their presentation of the theory of elasticity is too technical for average readers.

Dahan Dalmédico Amy. "Sophie Germain." *Scientific American* 265, no. 6 (December, 1991): 116-122. A clear, if brief, summary of Germain's career by a leading historian of mathematics.

Petrovich, Vesna Crnjanski. "Women and the Paris Academy of Sciences." *Eighteenth-Century Studies* 22, no. 3 (1999): 383-390. Places Germain in the context of the strategies that eighteenth and early nineteenth century women used to achieve recognition for their scientific work.

Singh, Simon. *Fermat's Enigma.* New York: Doubleday-Anchor, 1997. This lucid introduction to the history of Fermat's last theorem includes an appreciative discussion of Germain's contribution.

Kurt Gödel

Austrian-born American mathematician

Gödel did fundamental work in many areas of mathematical logic and contributed to philosophy and physics. His most famous achievement was the enunciation and proof of results about the incompleteness of arithmetic. The consequences of this discovery cut across all branches of mathematics and gave rise to results in computer science as well. Mathematical logic assumed a more central position in mathematics following his career.

Born: April 28, 1906; Brünn, Moravia, Austro-Hungarian Empire (now Brno, Czech Republic)
Died: January 14, 1978; Princeton, New Jersey
Also known as: Kurt Friedrich Gödel (full name)

Early Life

Kurt Gödel (GURD-ehl) was born in Brünn, Moravia, to Rudolf Gödel and Marianne (née Handschuh) Gödel five years after their marriage. The Gödels were part of the German-speaking minority in an area that was to become Czechoslovakia after World War I, and the couple retained their linguistic and cultural identity.

The recollections of an elder son, Rudolf, who was born in 1902, serve as the best source of the early years of his brother. Gödel's parents came from different religious traditions, and neither son became part of any formal religious community. Gödel's father was a factory owner, and the family's financial circumstances were good enough to place no limits on the opportunities for schooling. Gödel had a distinguished academic career, even as a child, but suffered from rheumatic fever and felt that he bore the effects for the rest of his life. This led to hypochondria, which often interfered with his participating in certain events.

In 1916, Gödel enrolled in the closest gymnasium (high school) and spent eight years there. He was never particularly outgoing but early on displayed interests in mathematics and physics. From the gymnasium he proceeded to the University of Vienna, an academic center for the German-speaking community of the new Czechoslovakia. Gödel, in fact, renounced his Czech citizenship and took up Austrian citizenship during his years at the university. While there he pursued studies in physics, philosophy, and mathematics, subjects for which the university had distinguished faculty. In addition to the classroom setting, he took part in the meetings of the Vienna Circle, a group of philosophers and scientists dedicated to reexamining basic issues from what they believed to be a scientific point of view. Although he did not agree with all members of the circle, participation in its discussions put him at the forefront of current developments in the application of logic to science and other areas.

Life's Work

Gödel's doctoral dissertation, which was submitted in 1929, demonstrated the completeness of first-order logic. First-order logic examines statements in which one could talk about collections of objects (called sets) but not collections of sets. Completeness refers to the match between the true statements within an area and the statements that are provable from the axioms or starting points. Demonstrating completeness involves coming up with a technique for showing how any true statement can be proved, and the result was not especially surprising at the time. Far different was the result associated with Gödel's name.

Quite a controversy erupted around 1930 that had to do with the foundations of mathematics. One popular

Kurt Gödel. (Photographer unknown. From The Shelby White and Leon Levy Archives Center, Institute for Advanced Study, Princeton, NJ, USA.)

point of view was that mathematics was really about what was provable from a given set of axioms, equating truth and provability. Much to the surprise of the advocates of that point of view, Gödel was able to demonstrate the incompleteness of a subject as fundamental as arithmetic. He accomplished this by translating general statements into arithmetic and then combining facts about arithmetic with a variant on what was known as the Liar paradox (which points out that there are difficulties with deciding whether the statement "This statement is false" is true or false). Gödel translated the statement "This statement is not provable" into his arithmetic and from that concluded that there was indeed a true statement that was not provable. He was able to extend his result to general conclusions about the inability of a system of axioms to demonstrate its own freedom from contradiction.

It took a while for the mathematics community at large to appreciate the importance of Gödel's work, but the growing group of mathematical logicians around the world came rapidly to recognize the significance of his theorems. In particular, Oswald Veblen, an American logician who was involved in the creation of the Institute for Advanced Study in Princeton, New Jersey, invited Gödel to come to the United States as a member of the institute. There he had the chance to do research without any teaching obligations. This served Gödel well, because he never had been either a dedicated or effective instructor. For much of the decade of the 1930's, Gödel commuted between Vienna and Princeton, although various bouts of ill health (especially psychological) interfered with some of his travels. Princeton gave Gödel an audience of logicians and computer scientists, like John von Neumann and Alan Mathison Turing, who were able to appreciate his work and build on it.

Despite the increasing political tension in Austria, Gödel continued to return to Vienna. Although he had no Jewish ancestry, the Vienna Circle had been associated with the political Left and so Gödel was not deemed politically trustworthy. In 1938, Gödel married Adele Porkert Nimbursky, a dancer whom he had known for a number of years, although his family was not enthusiastic about the match. Two years later, finding himself in Vienna after the outbreak of World War II, he and his wife made their way to the United States through the Soviet Union. Gödel never returned to Europe.

Through the 1930's, Gödel's attention had moved somewhat in the direction of set theory. One of the statements of greatest interest to logicians had been formulated by Georg Cantor, the founder of set theory in the nineteenth century. Cantor's continuum hypothesis states that there are no infinite sets whose size is between that of the whole numbers and that of the real numbers. What Gödel succeeded in doing was showing that this hypothesis (and a generalized version of it) could be added to the standard axioms of set theory without contradiction. Although many mathematicians were not convinced that the continuum hypothesis was true, the techniques that Gödel introduced to demonstrate his result were taken up to look at other questions about sets.

Once Gödel came permanently to the Institute for Advanced Study, his attention shifted somewhat to questions about philosophy and physics. His proof of the incompleteness theorem had not settled issues about the foundations of mathematics, and Gödel's point of view was not in the mainstream. He felt that one could have a special kind of "perception" of mathematical objects, as one could "see" physical objects. This philosophical position would not have received so much respect from his contemporaries had it not come from so distinguished a mathematician.

In addition to his philosophical pursuits, Gödel also spent time chatting with his colleague, Albert Einstein, about questions of cosmology. Gödel came up with a solution to Einstein's general relativistic equations that would make time travel an apparent possibility. The idea of time travel had all kinds of associated problems, though, suggesting that there were problems with the role that time played in physical theory.

Much of Gödel's work from his later years was not published during his lifetime. He would work on an article and produce numerous drafts, finally concluding that none of those articles was satisfactory. He also grew increasingly reclusive and suspicious of the outside world in general. Although he received honorary degrees from Harvard and Yale, there were other degrees he did not accept for fear of traveling or fear of being in public.

After retiring from the institute in 1976, Gödel and his wife were faced with mounting medical problems. It may have been his concern for her that contributed to his self-starvation and death in 1978. There is some irony in the possibility that so great a logician may have starved himself to death out of fear of being poisoned.

Significance

No other mathematician had so much influence on the mathematics and philosophy of the twentieth century as did Gödel. The techniques he introduced to establish his incompleteness theorem became standard for logicians. His discovery of an unprovable statement encouraged those in other areas to look for similar results. His ideas in set theory laid a foundation for subsequent work in the area, while his claims in the philosophy of mathematics and in physics were hotly debated in technical and nontechnical settings. Popularization of the ideas of Gödel extended to connections with art, music, and literature.

Within mathematics, Paul J. Cohen investigated the problem of the continuum hypothesis and demonstrated in 1963 that the negation of that hypothesis could also be added to the standard axioms of set theory without contradiction. When this was combined with Gödel's earlier result, the question of how to decide the "truth" of axioms of set theory was rendered more puzzling. Many questions about the philosophy of mathematics had been raised before Gödel without expectation of a definite answer. What he demonstrated was that mathematics had the technical resources to answer some questions about itself. Mathematical logic had been moved closer to the center of the stage of intellectual inquiry.

Thomas Drucker

Further Reading

Davis, Martin. *Engines of Logic: Mathematicians and the Origin of the Computer*. New York: W. W. Norton, 2000. View of Gödel's work from the standpoint of the development of computer science.

Dawson, John W., Jr. *Logical Dilemmas: The Life and Work of Kurt Gödel*. Wellesley, Mass.: A. K. Peters, 1997. The definitive biography by a mathematician whose acquaintance with Gödel's work is unparalleled and a good antidote to some popular treatments.

Franzén, Torkel. *Gödel's Theorem: An Incomplete Guide to Its Use and Abuse*. Wellesley, Mass.: A. K. Peters, 2005. Attempt to clarify the technical content of the incompleteness theorem and its misapplications in philosophy.

Gödel, Kurt. *Collected Works*. New York: Oxford University Press, 1986-2003. Five volumes of papers, including letters and paper previously unpublished, with introductions helpful for all readers.

Hofstadter, Douglas R. *Gödel, Escher, Bach: An Eternal Golden Braid*. New York: Basic Books, 1979. Wide-ranging and imaginative connection of mathematical ideas with art and music by a computer scientist. Considered a classic work.

Nagel, Ernest, and James R. Newman. *Gödel's Proof*. New York: New York University Press, 2002. Revised edition of the classic exposition of the ideas in Gödel's paper for those with limited mathematical experience.

Penrose, Roger. *The Emperor's New Mind: Concerning Computers, Minds, and the Laws of Physics*. New York: Oxford University Press, 1989. Most ambitious attempt to use the incompleteness theorem to show that the human mind is more than a machine.

Smullyan, Raymond. *Forever Undecided: A Puzzle Guide to Gödel*. New York: Alfred A. Knopf, 1987. Explanation of incompleteness by means of a sequence of logical puzzles.

Yourgrau, Palle. *Gödel Meets Einstein: Time Travel in the Gödel Universe*. Chicago: Open Court, 1999. Careful philosophical explanation of Gödel's work in physics.

James Gregory

Scottish astronomer and mathematician

Gregory designed the first practical reflecting telescope and proposed utilizing light intensity to estimate stellar distances. He formulated methods anticipating the discovery of calculus, developed infinite series representations for various trigonometric functions, and was the first person to propose and prove the rudimentary theoretical proposition today known as the fundamental theorem of calculus.

Born: November, 1638; Drumoak, near Aberdeen, Scotland
Died: October, 1675; Edinburgh, Scotland
Also known as: James Gregorie

Early Life

James Gregory was born in the Manse of Drumoak, 9 miles (15 kilometers) west of Aberdeen, to the Rev. John Gregory, an Episcopalian clergyman, and Janet Anderson. The youngest of three children, James was often sick as a child. Perhaps for this reason, his mother, an intelligent, educated woman, taught him mathematics and geometry. Following their father's death in 1650, James's twenty-three-year-old brother, David Gregory (an amateur mathematician), gave James a copy of Greek mathematician Euclid's *Elementa* (c. 300 B.C.E.; *Elements*, 1570) to encourage his latent talent. Easily mastering the material, James was sent on to the Aberdeen Grammar School. He then proceeded to Marischal College, Aberdeen, where he focused his studies on astronomy and mathematical optics.

After graduating in 1657, Gregory devoted his energy to studying optics and telescope construction. Encouraged by his brother David, he wrote a treatise summarizing five years of original research. Titled *Optica Promota* (1663, the advance of optics), this work proved theorems on the reflection and refraction of light, presented propositions on mathematical astronomy, and discussed photometric methods to estimate stellar distances.

The book's greatest contribution, however, was an exposition of the first practical reflecting telescope utilizing a concave mirror to focus light. Gregory's innovation employed a small concave mirror near the top of the telescope to reflect light from the focusing mirror back down the telescope tube through a small aperture in the center of the primary mirror, where it formed an image that could be examined with an eyepiece. Gregory's design had two advantages over refracting telescopes: The tube was more compact and the color distortion (chromatic aberration) introduced by an objective lens was nonexistent. Because Gregory did not possess the skill to grind and polish a mirror to the correct shape, he abandoned his concept in 1664 and journeyed to Padua, Italy, devoting himself exclusively to mathematical studies.

Life's Work

Enrolling at the University of Padua in 1664, Gregory spent the next four years studying geometry, mechanics, and mathematical astronomy under the tutelage of Stefano degli Angeli. While studying at Padua, Gregory produced his first mathematical treatise, *Vera Circuli et Hyperbolae Quadratura* (1667; the true squaring of the circle and of the hyperbola). This text was the first to distinguish between an infinite series of summed terms that converge and those that diverge. (When all the terms of a converging series are added, a finite limit is approached, unlike a diverging series, whose sum approaches infinity.) Gregory used convergent infinite series to calculate, respectively, the areas of circles and hyperbolas.

The succeeding year saw the publication of a second insightful treatise, even more general and abstract than the first. Titled *Geometriae Pars Universalis* (1668; the universal part of geometry), this opus presented rules for finding the areas of curves and the volumes of their solids of revolution (that is, the volume generated when a two-dimensional curve is rotated about an axis). In the process of producing this work, Gregory formulated two key aspects of calculus, differentiation and integration, in a consistent systematic manner. Although Sir Isaac Newton has been given priority for inventing the calculus, Gregory and several other mathematicians were working out the ideas independently at about the same time.

Gregory returned to England in the spring of 1668. Based on his books, Gregory had acquired a sufficient commendatory reputation in mathematics that he was elected a fellow of London's prestigious Royal Society soon after his return. Through Gregory's connections in the Royal Society, King Charles II was persuaded to create an endowed chair of mathematics at the University of Saint Andrews, Scotland, to provide Gregory a professorship from which he could continue his distinguished mathematical research. His reputation now secured, Gregory took up residence at Saint Andrews late in 1668. The succeeding year, he married a young widow, Mary Jamesome, who would bear him two daughters and a son.

During his tenure at Saint Andrews, Gregory carried out much important mathematical and astronomical work. In his *Exercitationes geometricae* (1669; geometrical exercises), Gregory developed an analytical method of drawing tangents to curves (in the parlance of calculus, this would become known as differentiation). He kept in touch with current research by corresponding with other members of the Royal Society, including Newton, whom Gregory greatly admired. Based on this correspondence, Gregory incorporated many of Newton's ideas into his own teaching, even though these concepts were considered quite controversial at the time.

Due to an earlier controversy with Christiaan Huygens, who falsely claimed authorship of sections of *Vera circuli et Hyperbolae Quadratura*, Gregory was reluctant to publish much of his work or disclose the methods by which he made discoveries. Consequently, it was not until the James Gregory Tercentenary in 1938, when his papers were exhumed from the archives of Saint Andrews's library, that the full extent of his brilliance was realized. For example, he had discovered the principle now known as Taylor's theorem in February, 1671; it was not published by Brook Taylor until 1715.

Gregory made another important scientific discovery when he utilized the feather of a sea bird to observe the diffraction of light, a phenomenon explicable only if light consisted of waves. Because Newton believed light had a corpuscular nature and Gregory had an enormous respect for Newton, he pursued this concept no further. Consequently, Gregory received only a fraction of the credit he deserved during his lifetime and even less during the ensuing centuries. The true magnitude of his achievements was not acknowledged until the 1930's, when, in the wake of the tercentenary, his notes and correspondence were examined and published by H. W. Turnbull as *James Gregory: Tercentenary Memorial Volume* (1939).

Increasing prejudice against the brilliant mathematician by Saint Andrews's classically oriented faculty and administration caused Gregory to resign his position at the end of the 1674 spring term. Eager to acquire this young, productive scholar, Edinburgh University created a new position for him, their first chair of mathematics. Unfortunately, this was a post he was to occupy for only one year. In October, 1675, while observing the moons of Jupiter through a telescope, Gregory suffered a blinding stroke, dying several days later at the age of thirty-six.

Significance
Gregory was one of the most important mathematicians of the seventeenth century, significant especially in the steps that led to the calculus. Unfortunately, like so many other scientific luminaries of the seventeenth century, his brilliance was eclipsed by that of Isaac Newton. His lack of historical appreciation was further exacerbated by his reluctance to publish his methods and his relatively short life. Some of his remarkable contributions, only recently brought to light, include his discovery of the general binomial theorem several years before Newton and his exposition of so-called Taylor expansions forty years before Taylor. He studied infinite series and elucidated one of the earliest examples of a test for a series's convergence.

In calculus, Gregory's definition of the integral was well formulated in a completely general form, and he had acquired a profound understanding of the various solutions possible for differential equations. He was the first mathematician who attempted to prove that π and e are irrational numbers, and he knew how to express the sum of the nth powers of the roots of an algebraic equation in terms of their coefficients. His correspondence also suggests that at the time of his death he had begun to realize that algebraic equations of degree greater than four could not be solved by equations in closed form.

Although possessing an enormous talent for mathematics and exhibiting tremendous promise for outstanding future accomplishments, Gregory's relatively short life precluded him from realizing any major discoveries, publishing them, and receiving the critical acclaim he most definitely deserved. During his last years of life, Gregory was reluctant to publish important results, and he was reticent to engage in controversy or proprietary arguments with Newton once he heard of Newton's advances in calculus and infinite series. This reluctance posthumously exacted a heavy toll upon his place in history.

George R. Plitnik

Further Reading
Dehn, M., and E. Hellinger. "Certain Mathematical Achievements of James Gregory." *American Mathematical Monthly* 50 (1943): 149-163. This article discusses Gregory's anticipation of important mathematical discoveries in number theory, differential calculus, and infinite series.

Malet, A. "James Gregorie on Tangents and the 'Taylor' Rule for Series Expansions." *Archives for History of Exact Science* 46 (1993-1994): 97-137. Explains how Gregory's tangent rule is essentially equivalent to differentiation and how Gregory proposed, but never published, the famed Taylor expansion decades before Taylor.

Scriba, C. J. "Gregory's Converging Double Sequence: A New Look at the Controversy Between Huygens and Gregory Over the 'Analytical Quadrature of the Circle.'" *Historia Math* 10, no. 3 (1983): 274-285. A critical account of Huygens's attack on Gregory's *Vera circuli*, as well as Gregory's rebuttal, which proves that Huygen's aggressive assault was unfounded and unnecessary.

Simpson, A. D. C. "James Gregory and the Reflecting Telescope." *Journal of the History of Astronomy*. 23, no. 2 (1992): 77-92. A brief account of Gregory's design for a practical telescope and his futile search for a London optician who could correctly grind and polish the mirrors.

Turnbull, H. W. "James Gregory." In *James Gregory Tercentenary Memorial Volume*, edited by H. W. Turnbull. London: G. Bell & Sons, 1939. In addition to articles discussing Gregory's major works and the Gregory/Huygens controversy, this volume contains copies of Gregory's letters and posthumous manuscripts.

Sir William Rowan Hamilton
Irish mathematician

While questioning a commonly accepted three-dimensional concept of space on a plane, Hamilton discovered quaternions and, in doing so, drastically altered the study of algebra, forcing the abandonment of the commutative law of multiplication that was dominant in his day and leading the way to new methods of vector analysis.

Born: August 3/4, 1805; Dublin, Ireland
Died: September 2, 1865; near Dublin, Ireland

Early Life

William Rowan Hamilton was born exactly at midnight, a moment poised equally between August 3 and 4, 1805. His father, Archibald Hamilton, was away in the north at the time of his son's birth, carrying out his duties as agent to Archibald Rowan, a post he had held since 1800. Archibald Rowan, who was William Rowan Hamilton's godfather, had been in exile for eleven years. His agent, William's father, worked tirelessly to make possible Rowan's return to his estate at Killyleagh, an effort that resulted in Rowan's repatriation in 1806.

To help Rowan meet his expenses, Archibald Hamilton borrowed heavily at high interest rates. When these loans were called, Rowan failed to back Hamilton, who, in a year or two, had no alternative but to declare bankruptcy. By 1808, the family was sufficiently impoverished not to be able to provide for William, then three years old, and his sisters, Grace and Eliza, who had to be sent away to be cared for by relatives. The two girls presumably were sent to live with their father's sister, Sydney, and young William became the ward of his uncle, James Hamilton, a Church of England clergyman who ran the diocesan school at Trim, some forty miles to the northwest of Dublin in County Meath. William was to remain there until 1823, when he returned to Dublin as a student at Trinity College.

In retrospect, it appears to have been a stroke of good fortune that young William was forced by circumstance to live with his uncle, a man of considerable intellect. Before the boy was four years old, he was able to read English and showed a remarkable understanding of arithmetic. By the time he was five, William was able to translate from Latin, Greek, and Hebrew. He knew Greek and Latin authors well enough to recite from their works, and he was also able to recite passages from works by John Milton and John Dryden. He is said to have mastered fourteen languages by the time he was thirteen. Before he turned twelve, he had compiled a Syriac grammar, and two years later he was sufficiently fluent in Persian to compose a speech of welcome that was delivered to the Persian ambassador when he was a guest in Dublin.

Always advanced in mathematics, Hamilton was enormously exhilarated when he met the American mathematician Zerah Colburn in 1820. Colburn was able to perform complex mathematical computations quickly in his head, a skill that enticed the fifteen-year-old Hamilton. The youth had already read Sir Isaac Newton's *Philosophiae Naturalis Principia Mathematica* (1687) and Alexis-Claude Clairaut's *Elémens d'algèbre* (1746) by the time he met Colburn. The excitement generated by his meeting with Colburn led Hamilton in the following year to study the completed volumes of Pierre-Simon Laplace's five-volume *Traité de mécanique céleste* (1798-1825; *A Treatise of Celestial Mechanics*, 1829-1839).

Hamilton's detection of a flaw in Laplace's reasoning brought him to the attention of John Brinkley, a distinguished professor of astronomy at Trinity College who was then also president of the Royal Irish Academy. The following year, when he was seventeen, Hamilton sent a paper he had written on optics to Brinkley, who, upon reading the paper, declared to the Royal Academy that Hamilton was already the most important mathematician of his time.

Hamilton entered Trinity College in 1823. By 1825, he had completed his paper "On Caustics" and submitted it to the Royal Academy, only to be rebuffed because the members of the Academy could not follow his often convoluted reasoning. Hamilton was awarded the *optime* in both classics and mathematics, the first Trinity College student to achieve this dual honor. While still an undergraduate, in 1827, he submitted his paper "Theory of Systems of Rays" to the Royal Academy, establishing with that paper a uniform method of solving all problems in the field of geometrical optics. The paper was of sufficient significance that before he had finished his undergraduate studies at Trinity College, the school's faculty elected William Rowan Hamilton to the Andrews professorship in astronomy, a post that established him as royal astronomer of Ireland and an examiner of graduate students in mathematics at Trinity College. He assumed that post immediately upon graduation.

Life's Work

The post to which the Trinity College faculty elected Hamilton carried with it a residence at the Dunsink Observatory, some five miles from Trinity College. In October, 1827, Hamilton moved into that residence and remained there for the rest of his life. Although he did not have a distinguished career as an astronomer, Hamilton had a large following of people who attended his lectures on astronomy because the range of his literary as well as his mathematical knowledge was sufficient to enliven his presentations.

Hamilton read encyclopedically and regularly wrote poetry, although his friend, the poet William Wordsworth, advised him that his lasting contributions would lie in mathematics rather than in poetry. In

1832, Hamilton published an important supplement to his paper on the theory of rays. This supplement was purely speculative, postulating a new theory about the refraction of light by biaxial crystals. Augustin Fresnel had already developed the theory of double refraction, but Hamilton took the theory an important step beyond where Fresnel had left it. He contended that in certain circumstances, one ray of incident light could be refracted into an infinite number of rays in a biaxial crystal and would be formed in such a way that a cone would then result. Humphrey Lloyd, following Hamilton's speculative lead, proved this theory of conical refraction within two months.

In 1833, after six years of living alone in his official residence, Hamilton—a man of average height and ruddy complexion—married Maria Bayley, whose father had been an Anglican rector in County Tipperary. Maria bore three children, two sons and a daughter. Not renowned for her domestic abilities, Maria presided over a somewhat chaotic household. Hamilton considered liquor a more reliable source of nourishment than anything Maria's cook could provide, and, through the years, he became a heavy drinker.

Hamilton's "On a General Method in Dynamics," published in 1835, brought together his work in optics and dynamics. He proposed a theory that showed the duality that exists between the components of momentum in a dynamic system and the coordinates that determine its position. In many ways, this work was some of Hamilton's most significant, although it took nearly a century for the development of research in quantum mechanics to demonstrate the brilliance and importance of Hamilton's theory.

Hamilton served as the major local organizer of the British Association for the Advancement of Science meeting in Dublin in 1835, an activity that led to his being knighted in the closing ceremonies of that event. In 1837, he ascended to the presidency of the Royal Irish Academy. In 1843, the Crown awarded him an annual life pension of two hundred pounds. During his final illness, Hamilton received word that he had been ranked first on the list of foreign associates of the National Academy of the United States.

The contribution for which Hamilton is best remembered is his discovery of quaternions. This discovery has fundamentally changed the way in which mathematicians deal with three-dimensional space. Hamilton had begun his extensive investigation into ordered paired numbers more than ten years before he made his monumental discovery of quaternions on October 16, 1843, when, during a walk along Dublin's Royal Canal, the answer to a question that had been haunting him for nearly a decade flashed almost supernaturally into his mind. So excited was he by this flash of insight that he carved the formula for his discovery, $i^2 = j^2 = k^2 = ijk = -1$, into the Brougham Bridge.

Hamilton suddenly realized that in three-dimensional space, geometrical operations require not triplets, expressed as i, j, and k and representing space, as had been previously supposed, but rather that, because in three-dimensional space the orientation of the plane is variable, another element, a real term that represents time, must also be considered, resulting in quadruplets rather than triplets. One of the major consequences of this insight was its negation of the previously accepted commutative law of multiplication, which postulates $(a \times b) = (b \times a)$.

Hamilton's work with quaternions, to which he devoted the last two decades of his life, was essential to the development of vector analysis. More recently, further important applications of his theory of quaternions have been instrumental in the description of elementary particles. Hamilton published his *Lectures on Quaternions* in 1853, and his influential *The Elements of Quaternions* appeared posthumously in 1866. William Rowan Hamilton died of gout on September 2, 1865, after a lingering illness.

Significance

Sir William Rowan Hamilton's name lives in both the history of mathematics and the histories of physics and optics. His pioneering work in vector analysis forced specialists in that field to abandon the theory of double refraction and to replace it with Hamilton's expanded theory of conical refraction. The work that led to these changes began while Hamilton was still an undergraduate at Trinity College and reached its culmination in the supplement to his "Theory of Systems of Rays" in 1832.

Hamilton's next significant achievement posited a duality between the components of momentum in a dynamic system and the coordinates that determine its position, a theory that reduces the field of dynamics to a problem in the calculus of variations. This theory came to have considerable significance as the field of quantum mechanics developed.

Hamilton's most memorable contribution by far, however, was his discovery of quaternions, which forced mathematicians to break with the commutative law of multiplication. In its simplified form, termed vector analysis and adapted by J. Willard Gibbs from Hamilton's theory, Hamilton's theory of quaternions has been of great significance to modern mathematical physicists.

R. Baird Shuman

Further Reading

Bell, Eric Temple. *Development of Mathematics*. New York: McGraw-Hill, 1940. Bell relates Hamilton to some of the salient mathematical developments of his time. The coverage is sketchy and has been superseded by Thomas L. Hankins's biography (see below).

_____. *Men of Mathematics*. New York: Simon & Schuster, 1965. Bell puts Hamilton in historical perspective. The chapter "An Irish Tragedy" focuses on Hamilton, but, although interesting, it is not factually dependable in all respects.

Crilly, A. J. *Arthur Cayley: Mathematician Laureate of the Victorian Age*. Baltimore: Johns Hopkins University Press, 2005. Cayley (1821-1895) was a contemporary of Hamilton; the two men devised a matrix algebra theory that bears their names. Although focusing on Cayley, this biography also describes Hamilton and others who were part of a nineteenth century British mathematical vanguard.

Graves, R. P. *Life of Sir William Rowan Hamilton*. 3 vols. London: Longmans, Green, 1882. The three enormous volumes of this set include extensive selections from Hamilton's correspondence, poetry, and miscellaneous writings, as well as extensive commentary. The work, remarkable in its time for its thoroughness, is badly dated and suffers from lack of selectivity.

Hamilton, William Rowan. *The Mathematical Papers of Sir William Rowan Hamilton*. 4 vols. Cambridge, England: Cambridge University Press, 1931-2000. Volume 1, *Geometrical Optics* (1931), and volume 2, *Dynamics* (1940), are edited by A. W. Conway and J. L. Synge; volume 3, *Algebra* (1967), is edited by H. Halberstam and R. E. Ingram. Volume 4 (2000) is edited by Brendan Scaife and includes Hamilton's *Systems of Rays*, two lengthy letters regarding definite integrals and anharmonic coordinates, and reprints of numerous papers about geometry, astronomy, and other topics. Volumes 1 and 3 contain useful introductions. Despite some omissions, these volumes are superbly produced, and the highest standards of scholarship have been observed in their editing.

Hankins, Thomas L. *Sir William Rowan Hamilton*. Baltimore: Johns Hopkins University Press, 1980. Hankins's critical biography of Hamilton is the definitive work in the field. Meticulously documented, the book is written in such a lively style that it at times reads like a novel rather than like the eminently scholarly work that it is. The best book to date on Hamilton.

James, Ioan. *Remarkable Mathematicians: From Euler to von Neumann*. New York: Cambridge University Press, 2002. This collection of brief biographies of prominent mathematicians includes a seven-page biography of Hamilton.

Synge, J. L. *Geometrical Optics: An Introduction to Hamilton's Method*. Cambridge, England: Cambridge University Press, 1937. Highly technical in nature, this book contains a brief but valuable preface. This book is for the specialist rather than the beginner.

David Hilbert

German mathematician

Hilbert occupied a leading position in mathematics at the start of the twentieth century, advocating rigor and precise formalism in mathematics and mathematically rigorous formulations of physics. He is perhaps best known for his list of twenty-three fundamental questions that would guide mathematical work through the twentieth century.

Born: January 23, 1862; Königsburg, Prussia (now Kaliningrad, Russia)

Died: February 14, 1943; Göttingen, Germany

Early Life

David Hilbert was born into a prosperous family of jurists and physicians. His father, Otto, was a district judge. His mother, Maria Therese Erdtman, was the daughter of a merchant and interested in philosophy, astronomy, and mathematics, interests she would share with her young son. At the age of eight, Hilbert was enrolled in a gymnasium (preparatory school), the

Freidrichscolleg, which emphasized classical languages. In 1879, he transferred to the Wilhelm Gymnasium, which placed more emphasis on mathematics. Hilbert's school performance was satisfactory but not exceptional. In 1880 he entered the University of Königsburg as a mathematics student. At Königsburg, the spirit of the great German philosopher Immanuel Kant, who was born in the city and taught at the university, still was evident. Kant's view, that mathematics was a form of knowledge gained from pure reason alone, would influence Hilbert throughout his career.

The German universities of 1880 were havens of academic freedom, with professors free to lecture on topics of their own choosing and students free to select their own path of study. The universities also allowed for considerable student mobility. Hilbert chose to spend his second semester at the University of Heidelberg, but then returned to Königsburg. In the spring of 1882, he was joined there by the seventeen-year-old mathematical prodigy Hermann Minkowski, also a native of the city, who would become a close friend and collaborator.

Hilbert received his doctor of philosophy degree in 1885, submitting a thesis on algebraic invariants. He submitted a somewhat longer paper on the same subject for his habilitation exam, which made him a Privatdocent who could lecture without a salary to paying students. During the next few years Hilbert would lecture and travel to visit other mathematicians. In 1892 he married Käthe Jerosh. Their only child, Franz, was born a year later. In 1895, Hilbert was called to the University of Göttingen as a professor of mathematics.

Life's Work

Hilbert is known for important contributions in many areas of mathematics. At the time he received his professorship at Königsburg, the theory of numbers was considered the most important branch of mathematics. In 1893 the Association of German Mathematicians urged Hilbert and Minkowski to prepare a report on recent developments in number theory. The report, published in 1897 and known as the *Zahlbericht*, received high acclaim for its thoroughness and integrated presentation.

Hilbert then turned his attention to the foundations of geometry. Prior to the nineteenth century, geometry was considered the study of the properties of the physical space in which humans live. According to Kant, geometry could be conducted by reason alone. Geometry allowed for the deduction by logical means of many unclear statements from a small set of apparently self-evident statements systematized by the Greek geometer Euclid. One of these postulates, commonly known as the parallel postulate, asserted that given a straight line and a point not on the line, one could draw one and only one straight line through the point parallel to the line. By Hilbert's time mathematics was recovering from the finding that alternative postulates allowing no, or many, straight lines to be drawn through the point yielded different, but self-consistent, geometries and that there was no certainty that the space described by Euclidean geometry was in fact the space inhabited by humans.

Hilbert decided that the first step to certain knowledge in geometry was to treat the elementary notions of point, line, and plane as having meaning only in terms of their relation to each other. Hilbert devised twenty axioms, divided into five independent groups. One of the by-products of this work was finding that geometry, in its several forms, could be proved free of contradictions if arithmetic could be proved to be free of contradictions.

Hilbert was invited to make a keynote address at the Second International Congress of Mathematicians

David Hilbert. (Library of Congress)

in Paris in 1900. He chose to speak on those mathematical problems he thought would be the most important to solve in the new century. In the printed version of his talk he identified twenty-three urgent problems. He concluded his talk by saying, "We must know. We shall know." Hilbert's twenty-three item list ranges from the consistency of arithmetic—raised to greater importance by his work on the foundations of geometry—to the mathematical treatment of the laws of physics.

Physics has long provided a stimulus to the development of mathematics, and Hilbert found himself increasingly drawn to concerns of mathematical physics. He first turned to integral equations, useful in a number of areas of physics. Here he was able to cast the subject of linear integral equations as a case of geometry in an infinite dimensional space, now universally known as Hilbert space.

In 1910, Hilbert became only the second individual to win the Bolyai Prize of the Hungarian Academy of Sciences, an award that consisted of ten thousand gold crowns, solidifying his reputation as one of the two leading mathematicians of his time; the other was the great French mathematician Henri Poincaré.

By this time, however, Hilbert was devoting most of his attention to the mathematical problems of physics. He even took the unusual step of appointing an assistant to guide him in his study of physics. His last publications, though, from 1934 and 1939, returned to the issue of the consistency of arithmetic.

In 1914, World War I broke out and the German government pressured leading professors to add their signatures to a declaration asserting the justice of the German cause. Hilbert, like Albert Einstein, who was then working in Berlin, refused to sign the document. In refusing to sign, Hilbert retained his good standing in the international mathematics community; the remaining students and faculty at Göttingen, however, shunned him.

Significance

Hilbert was one of the leading figures in German mathematics at a time when the nature of mathematics and its role in physical science was receiving critical reevaluation. His twenty-three problems for future mathematicians would yield a number of surprising solutions. One problem prompted British mathematician Alan Mathison Turing to demonstrate the possibility of a universal automaton for solving problems of symbol manipulation, solutions that became the conceptual ancestor of the programmable digital computer. Perhaps most surprising was the attempt to demonstrate the consistency of arithmetic. Austrian logician Kurt Gödel demonstrated that any set of axioms consistent with ordinary multiplication would allow the formulation of assertions that could not be proved or disproved within the system.

Hilbert's work on the differential equations of physics challenged physicists to function at a new level of rigor. Physicists now routinely make use of the theory of Hilbert spaces, in which functions are represented in a geometrical space of an infinite number of dimensions. Over his long career Hilbert served as mentor to many distinguished mathematicians and physicists, including Nobel laureate Max Born and mathematical physicist Emma Noether, who would determine the fundamental connection between symmetry and conservation laws. Her work guided much elementary particle theory into the twenty-first century.

The University at Göttingen would come to play a pivotal role in the development of quantum physics, attracting students from the United States as well as Europe. After 1933, as the Nazi Party came to power, Hilbert and his colleagues did what they could to help former students and professors find academic positions outside Germany.

Donald R. Franceschetti

Further Reading

Hilbert, David. *Foundations of Geometry*. LaSalle, Ill.: Open Court, 2006. An English translation of the book in which Hilbert formulates the postulates of geometry in logically independent form. This work has been reprinted many times in English and German.

Klein, Morris. *Mathematics: The Loss of Certainty*. New York: Oxford University Press, 1980. Klein details the loss of confidence in mathematical intuition occasioned by the discovery of non-Euclidean geometries and the paradoxes of set theory. Hilbert's many contributions in this area are highlighted.

Reid, Constance. *Hilbert*. New York: Springer, 1996. A detailed biography by a nonmathematician. Includes many photographs of key mathematicians and the Göttingen locale in Hilbert's time.

Yandell, Ben H. *The Honors Class: Hilbert's Problems and Their Solvers*. Natick, Mass.: A. K. Peters, 2002. Following a brief biography of Hilbert, Yandell describes the twenty-three fundamental problems advanced by Hilbert at the 1900 international mathematics conference. Details the progress on each problem and includes biographical details on those who solved them.

Hipparchus

Greek astronomer

Hipparchus was the greatest astronomer of ancient times. He was the founder of trigonometry, which he used to determine the distances from Earth to the moon and sun, and the first to use consistently the idea of latitude and longitude to describe locations on Earth and in the sky.

Born: 190 B.C.E.; Nicaea, Bithynia, Asia Minor (now İznik, Turkey)
Died: After 127 B.C.E.; possibly Rhodes, Greece

Early Life

Very little is known about the life of Hipparchus (hih-PAHR-kuhs). He was born in Nicaea, a Greek-speaking city in Bithynia (modern İznik, Turkey), in the northwestern part of Asia Minor. Calculations in his works are based on the latitude of the city of Rhodes, on the island of the same name, so many historians believe that he spent a major portion of his life there. Rhodes was a merchant center, a convenient port from which to make voyages. At least one of Hipparchus's observations was made in Alexandria, so it seems that he visited and perhaps spent time as a student or research scholar at that great nucleus of scientific inquiry. Because he was intensely interested in geography, it is likely that he traveled to other places in the Mediterranean basin and the Near East. He seems to have been familiar with Babylonian astronomy, including eclipse records, but it is impossible to say how he came to know these.

Life's Work

Most of what is known of Hipparchus comes from the *Mathēmatikē syntaxis* (c. 150 C.E.; *Almagest*, 1948) of Ptolemy, whose work depends to a considerable extent on that of the earlier scientist, and from the *Geōgraphica* (c. 7 B.C.E.; *Geography*, 1917-1933) of Strabo. Of Hipparchus's own writings, only the *Tōn Araton kai Eudoxou* phainomenon exigesis (commentary on the phenomena of Eudoxus and Aratus) survives, in three books. It criticizes the less accurate placement of stars and constellations by two famous predecessors. It is certainly not one of his most important works, but it contains some information on his observations of star positions, which were the basis of his lost star catalog. Other lost works of Hipparchus include *Peri eviausiou megethous* (on the length of the year) and *Peri tes metabaseos tōn tropikon kai isemerinon semeion* (on the displacement of the solstitial and equinoctial points). He is also credited with a trigonometrical table of chords in a circle, a work on gravitational phenomena called *On Bodies Carried down by Their Weight*, an attack on the geographical work of Eratosthenes, a compilation of weather signs, and some aids to computational astrology.

A number of achievements are attributed to Hipparchus by Ptolemy and other ancient writers. A new star appeared in the constellation Scorpio in July, 133 B.C.E. Hipparchus realized that without an accurate star catalog, it was impossible to demonstrate that the star was indeed new, so he set about producing a complete sky map with a table of the positions of the stars, including the angle north or south of the celestial equator (latitude) and the angle east or west of the vernal equinox point (one of the two intersections between the celestial equator and the sun's path, or ecliptic).

In order to do this, he needed a means of measuring celestial angles, which led him to invent many of the sighting instruments, including the diopter and possibly the armillary astrolabe, used by astronomers before the invention of the telescope in the seventeenth century. He also knew how to calibrate water clocks. Hipparchus's star catalog included about 850 stars, along with estimates of their brightness. He divided the stars into six categories, from the brightest to the dimmest, thus originating a system of stellar magnitude. He also made a celestial globe, showing the locations of the fixed stars on its surface.

In comparing his own measurements of positions of stars with those of earlier astronomers, Hipparchus discovered that there had been a systematic shift in the same direction in all of them. He noticed the phenomenon first in the case of the bright star Spica. In 283 B.C.E. Timocharis had observed the star to be eight degrees west of the autumnal equinoctial point, but Hipparchus found the figure to be six degrees. He found a displacement for every other star that he was able to check. These discrepancies, he established, were the result of a shift in the position of the equinoxes—and therefore of the celestial equator and poles. In modern astronomy, this shift is called the precession of the equinoxes and is known to be caused by a slow "wobble" in the orientation of Earth's axis. The spot to which the north pole points in the sky (the north celestial pole) describes a circle in a period of more than twenty-six thousand years. Hipparchus was first to describe and to attempt to measure this phenomenon. He was, however, unable to explain its cause, since he held the geocentric theory, which postulates a motionless Earth at the center of a moving universe.

From the beginning of theoretical astronomy, the geocentric theory had been the accepted one. It assumed that the sun, moon, planets, and stars were carried on vast transparent spheres that revolved at different but constant speeds around Earth. Unfortunately, in order to explain the observed motions of the planets, which vary in speed and sometimes are retrograde relative to the stars, astronomers had to postulate the existence of additional spheres, invisible and bearing no celestial bodies but interconnected with the other spheres and affecting their motions. An Alexandrian astronomer, Aristarchus, had proposed the heliocentric theory, which holds that Earth, with its satellite, the moon, and all the other planets revolve around the central sun. The main appeal of this theory was its simplicity; it required fewer imaginary spheres to make it work.

Hipparchus rejected the heliocentric theory and instead adopted modifications of the geocentric theory to make it accord better with observations, perhaps following Apollonius of Perga. The main feature of the Hipparchan system is the epicycle, a smaller sphere bearing a planet, with its center on the surface of the larger, Earth-centered sphere and revolving at an independent speed. He also postulated eccentrics, that is, that the centers of the celestial spheres do not coincide with the center of Earth. The geocentric system with epicycles is often called "Ptolemaic," as Ptolemy made observations to support the theory developed by Hipparchus. Aristarchus's heliocentric theory is closer to the picture of the solar system provided by modern astronomy.

In developing his astronomical system, Hipparchus observed the period of revolution of the celestial objects that move against the background of the stars. That of the sun, which is the year, he found to be 365 1/4 days, less 1/300 of a day, a figure that was closer to the true one than that of any previous astronomer. He noticed the inequality in the lengths of the seasons, which he correctly attributed to the varying distance between Earth and the sun but incorrectly explained by assuming that the center of the sun's sphere of revolution was eccentric to the center of Earth. These conclusions were, perhaps, a step in the direction of recognizing that the relative motion of the two bodies describes an ellipse. He also achieved a measurement of the length of the lunar month, with an error of less than one second in comparison with the figure now accepted. The Roman scholar Pliny the Elder wrote that Hipparchus countered the popular fear of eclipses by publishing a list that demonstrated their regularity over the preceding six hundred years.

Hipparchus attempted to measure the distances of the moon and sun from Earth by observing eclipses and the phenomenon of parallax (the shift in the apparent position of the moon against the background of the stars under changing conditions). His figure for the distance of the moon (60.5 times the radius of Earth) was reasonably accurate, but his estimate of the sun's distance (2,550 times Earth's radius) was far too small. (The true ratio is about 23,452 to 1.) In fairness to Hipparchus, it should be noted that he regarded his solution to the problem of the sun's distance as open to question.

In order to make the mathematical computations required by these problems, it was necessary for Hipparchus to know the ratios of the sides of a right triangle for the various angles the sides make with the hypotenuse—in other words, the values of trigonometrical functions. He worked out tables of the sine function, thus becoming, in effect, the founder of trigonometry.

Geography also occupied Hipparchus's attention. He began the systematic use of longitude and latitude, which he had also employed in his star catalog, as a means of establishing locations on Earth's surface. Previous geographers show evidence of knowing such a method, but they did not employ it consistently. Hipparchus was able to calculate latitudes of various places on Earth's surface by learning the lengths of the days and nights recorded for different seasons of the year, although the figures given by him were often in error. As the base of longitude, he used the meridian passing through Alexandria. He was especially critical, probably too much so, of the descriptive and mathematical errors in the work of Eratosthenes. He even had some quibbles with the famous measurement of the spherical Earth, which is the latter's most brilliant achievement. It may be Hipparchus, rather than Eratosthenes, who first described climatic zones, bounded by parallels of latitude north and south of the equator.

SIGNIFICANCE

Hipparchus was a careful and original astronomer whose discoveries, particularly that of precession, were of the greatest importance in the early history of the science. He was a meticulous observer who produced the first dependable star catalog and who determined the apparent periods of revolution of the moon and the sun with an exactitude never before achieved. As a mathematician, he originated the study of trigonometry, compiling a sine table and using it in an attempt to measure distances in space beyond Earth that was, at least in the case of the moon, successful. Both as astronomer and as geographer, he pioneered the systematic use of the coordinates of latitude and longitude. He devised instruments for use in these observations and measurements.

Unfortunately, almost all Hipparchus's writings have disappeared, so modern assessments of his work must depend on ancient writers who happened to mention him. His influence was important enough to cause several later scientists whose works survive to refer to and summarize him. Most notable among these were Ptolemy and Strabo. It is sometimes hard to tell when these authors, particularly Ptolemy, are following Hipparchus and when they are going beyond him to present their own conclusions. Ptolemy's work became the standard textbook on astronomy until the time of Nicolaus Copernicus in the sixteenth century; thus Hipparchus's name was deservedly remembered. One of Hipparchus's most important mathematical successors was Menelaus of Alexandria (fl. c. 100 C.E.), who developed the study of spherical trigonometry.

J. Donald Hughes

FURTHER READING

Dicks, D. R. *Early Greek Astronomy to Aristotle.* Ithaca, N.Y.: Cornell University Press, 1985. Hipparchus is not given major treatment, although he does appear as an important figure in the history of astronomy. The discussion of his criticisms of Eudoxus and Aratus is particularly good.

Dreyer, John L. E. *A History of Astronomy from Thales to Kepler.* 2d ed. New York: Dover, 1953. This fine, accessible study places Hipparchus clearly in the context of the development of astronomy. Dreyer differs from common interpretation in crediting Ptolemy, not Hipparchus, with the theory of epicycles.

Heath, Thomas. *A History of Greek Mathematics.* 2 vols. Reprint. New York: Dover, 1981. Includes a section on Hipparchus in the second volume, emphasizing his probable contributions to the origin of trigonometry and establishing his place in the history of mathematics.

Lloyd, G. E. R. *Greek Science After Aristotle.* New York: W. W. Norton, 1973. Rather than giving a separate treatment to the subject, this work discusses the contributions of Hipparchus as they arise in a general study of ancient science from the fourth century B.C.E. to the end of the second century C.E. The attention given to Hipparchus is appropriate and appreciative.

Neugebauer, Otto. *A History of Ancient Mathematical Astronomy.* 3 vols. New York: Springer-Verlag, 1975. This work contains a section on Hipparchus in volume 1, briefly discussing what little is known about his life and chronology and devoting the rest of its space to a careful consideration of his astronomical work. There are some mathematical and astronomical symbols and formulas that the layperson may find difficult.

Ptolemy. *Ptolemy's "Almagest."* Translated by G. J. Toomer. Princeton, N.J.: Princeton University Press, 1998. Much of what is known about Hipparchus is based on Ptolemy's words. This fine translation has complete notes and a useful bibliography.

SOFYA KOVALEVSKAYA

Russian mathematician

Kovalevskaya was the first woman in the world to receive a doctorate in mathematics from a European university and was also the first woman to teach at a European university during the nineteenth century. Her achievements in mathematics provided evidence of the ability of women to conduct research at the highest level and were recognized throughout Europe.

Born: January 15, 1850; Moscow, Russia
Died: February 10, 1891; Stockholm, Sweden

Also known as: Sofya Vasilyevna Korvin-Krukovskaya (birth name); Sonya Kovalevskaya; Sophia Kovalevskya; Sofia Kovalevskya

EARLY LIFE

Sofya Kovalevskaya (kah-vah-LYAYF-skah-yah) was born Sofya Korvin-Krukovskaya. Her family had an estate located near the borders of what are now Russia, Lithuania, and Belarus; was affluent; and had a tradition of education. Her mother had German roots; her mother's maiden name was Schubert, and her mother's grandfather had been a German mathematician and astronomer.

Sofya's mother had inherited her grandfather's concern for education, and Sofya herself enjoyed the attention of a sequence of governesses of various nationalities, none of them Russian. This was in accord with the attitude of the educated Russian class of the time, who saw the benefit in exposing children to cultures and languages other than their own. As a result, Sofya was able to read in English and French from an early age and made good use of her language skills.

Sofya also displayed an early interest in mathematics, supposedly linked to the fact that copies of lectures given by a distinguished Russian mathematician were used to paper the wall of a bedroom. One of her uncles stimulated her interest in the subject, and she taught herself trigonometry. Her interests also extended to other sciences, and she used a microscope she purchased herself to study biology. Within her home there was no limit to the extent to which she could pursue intellectual interests. However, in the wider world, it was a different story.

The University of St. Petersburg had opened its lecture halls to women in 1861, when Sofya was eleven, but subsequent agitation led to government crackdowns and the withdrawal of the privilege for women to attend. As a result, it was clear that Sofya would have to go abroad if she wanted to pursue the various studies to which she was attracted. The difficulty was that unmarried Russian women were not allowed to travel abroad on their own. The standard solution was for women who were interested in studying elsewhere to enter into marriages of convenience with men willing to support their endeavors without necessarily expecting anything more from their marriages.

The man who played the appropriate role in Sofya's life was Vladimir Kovalevskii, an individual with scientific interests as well as political ones. There was a definite political slant to the intellectual circles in which the two spent their time, usually characterized by the term "nihilist." That term, which came from the attitudes of a character in Ivan Turgenev's novel *Fathers and Sons* (1862), summarized the view of a generation of intellectuals who found unpersuasive the extensive rules governing life and political action in the old Russian Empire. Instead, they wanted to start over again without regard for rank and wealth. Not surprisingly, this attitude was not encouraged by the government, and ongoing battles between nihilists and government agents provided another reason for going abroad.

After Sofya married Kovalevskii in September, 1868, she went abroad to study mathematics. She and her husband were also eager to find an interesting political circle, so they went first to Vienna, Austria, and then to England. Neither place fully met their needs, so by the end of the summer of 1869 they had gone to Heidelberg, Germany. Kovalevskii studied paleontology there but decided to finish his degree at the University of Jena, another German institution. Sofya, meanwhile, had decided to go the fountainhead of mathematics to study at the University of Berlin under the eminent mathematician Karl Weierstrass. Sofya's separation from her husband was to be typical of much of the couple's marriage, which ended when Kovalevskii committed suicide in 1883.

LIFE'S WORK

The prestigious position that Weierstrass held in the world of German mathematics gave him opportunities to do things that lesser scholars would not have been permitted to do; however, even he could not manage to arrange for Kovalevskaya to get credit for the work that she did with him. It was difficult for him simply to get her permission to use the university library. While the general political atmosphere in Berlin might have been more liberal than it was in Russia, that did not mean that Berlin provided a comfortable environment for a woman to make academic progress. There existed a tendency to discount the mathematical achievements of women—especially women as physically attractive as Kovalevskaya—and to attribute any success they demonstrated to their borrowing the work of their male colleagues. One indication of the stature of both Weierstrass and Kovalevskaya is the fact that no whisper of such claims was ever made about Kovalevskaya's work.

The work for which Kovalevskaya was to become best known pertained to the study of partial differential equations, which govern the behavior of most physical processes and describe how certain quantities change with changes in other quantities, such as time and position. Solutions to such equations raised great difficulties, but Kovalevskaya was able to apply Weierstrass's ideas to build a theoretical foundation for the subject, while looking for solutions. Meanwhile, she earned a doctoral degree from the University of Göttingen; she received it *in absentia*, which was a means of avoiding the issue of her being a woman. Although women had previously earned reputations in Europe as mathematicians, Kovalevskaya was the first woman to get a doctorate, a degree whose importance in the research arena was of relatively recent creation.

When Kovalevskaya returned to Russia and her husband, she looked for a suitable job but found opportunities at the university level and academies of science limited—partly because of her unpopular political views. In 1878, she gave birth to the only child she had by her husband. Two years later, she returned to Berlin. From there she hoped to be able to find a position in mathematics in some other country, but even Weierstrass's recommendation was not enough for some institutions. For example, the University of Helsinki did not offer her a position for which she was well qualified, but it is unclear whether being a Russian or being a woman was considered the greater disqualification in Finland.

In 1884, Kovalevskaya finally received a position in Stockholm, thanks to the efforts of the mathematician Gosta Mittag-Leffler, who had been so impressed by her work that he had traveled to St. Petersburg to hear her speak before she returned to Berlin. Mittag-Leffler arranged for Kovalevskaya to get a salaried five-year professorship after she had proven her mathematical skills during her first year in Stockholm. Her responsibilities in Stockholm included lecturing and tutoring, as well as carrying out research in collaboration with Mittag-Leffler.

Even at that stage in Kovalevskaya's career, she was not allowed to attend lectures in Berlin, a tribute to the continuing difficulties with which women mathematicians had to contend in Europe. It is also true that the distinguished Swedish playwright August Strindberg was openly negative about Kovalevskaya's presence on the faculty in Stockholm, although he was scarcely qualified to criticize her mathematical work.

Despite the efforts of Mittag-Leffler to make her feel at home, Kovalevskaya became bored with Stockholm. Her stock in the mathematical world rose considerably when she received the Borodin Prize of the French Academy of Sciences in 1888 for her solution of a problem about the motion of a rigid body rotating about a fixed point. The subject had been under investigation for some time, and Kovalevskaya was far from the only entrant in the competition for the prize. It was a measure of her success that the academy doubled its cash award to her in recognition of the elegance of her solution. On the strength of this award, Kovalevskaya hoped that she could find an academic post in Paris or Russia.

Kovalevskaya's quest for alternative employment proved unavailing, but the Russian mathematician Pafnuty Chebyshev succeeded in having her named the first woman corresponding member of Russia's Imperial Academy of Sciences. In 1891, she taught the first classes of the spring semester in Stockholm but died shortly afterward.

Significance

At the time of her death, Sofya Kovalevskaya was best known to the general public in Russia for her writings, especially an autobiographical account of her childhood. The mathematical community recognized her for her ability to carry on the research program of Weierstrass as applied to a variety of particular problems.

On one hand, Kovalevskaya was a mathematician whose work need not fear comparison with that of any of her contemporaries. She was educated in the best style of Weierstrassian analysis, and she made contributions to the study of partial differential equations in the form of textbooks and research articles. Those attacking more general problems involving rotations were able to build on her efforts.

On the other hand, there is no doubt that Kovalevskaya's life has received a great deal of attention because she was the first woman to do much of what she accomplished. She certainly did not enjoy having to overcome the difficulties placed in the way of women trying to do research in mathematics, and the intervals in her life when she stopped doing mathematics are representative of her own ambivalence. Nevertheless, the success that she achieved despite the handicaps enabled women following after to point to her distinguished precedent. During the late twentieth century a Kovalevskaya Fund was set up in her honor to help support the educational efforts of women in the sciences in underdeveloped countries. Her political and scientific testament could not have been better expressed.

Thomas Drucker

Further Reading

Cooke, Roger. *The Mathematics of Sonya Kovalevskaya*. New York: Springer-Verlag, 1984. The most extensive analysis of Kovaleskaya's mathematics, tracing its historical roots from the work of Weierstrass and his predecessors.

James, Ioan. *Remarkable Mathematicians*. Cambridge, England: Cambridge University Press, 2002. The sketch of Kovalevskaya does not go into detail about her mathematics but spells out some of the mathematical connections that she built during her career.

Kennedy, Don H. *Little Sparrow: A Portrait of Sophia Kovalevsky*. Athens: Ohio University Press, 1983. A political and literary biography that steers away from mathematics.

Koblitz, Ann Hibner. *A Convergence of Lives*. 2d ed. New Brunswick, N.J.: Rutgers University Press, 1993. One of the books that started a revival of interest in Kovalevskaya's work in English, paying attention to her literary, political, and scientific careers.

Kovalevskaya, Sofya. *A Russian Childhood*. Translated and introduced by Beatrice Stillman. New York: Springer-Verlag, 1978. The autobiography that won Kovalevskaya the most recognition in her lifetime, translated into modern English. Contains an analysis of her mathematics by P. Y. Kochina.

Spicci, Joan. *Beyond the Limit: The Dream of Sofya Kovalevskaya*. New York: Forge Press, 2002. Although well researched, this biography borders on fiction and traces Kovalevskaya's career only up to the moment that she earned her doctorate in 1874.

Joseph-Louis Lagrange
French mathematician

One of the most brilliant mathematicians of the mid- to late eighteenth century, Lagrange accomplished astonishing syntheses of the mathematical innovations of his predecessors, especially in the systems underlying classic physics. Almost as remarkable for his winning personality as for his incisive intellect, Lagrange created the mathematical basis of modern mechanics.

Born: January 25, 1736; Turin, Sardinia (now in Italy)
Died: April 10, 1813; Paris, France
Also known as: Giuseppe Luigi Lagrangia (birth name); Comte de l'Empire

Early Life

Born in what was then the kingdom of Sardinia of mixed French and Italian though predominantly French descent, Joseph-Louis Lagrange was the first son in an influential and wealthy family. His father, however, once a highly placed cabinet official, burned with the speculative fevers of the early eighteenth century and ended by losing everything. Typically, Lagrange took that in stride, remarking later that losing his inheritance forced him to find a profession; he chose wisely. Although early in his formal education he found mathematics boring, probably because it began with geometry, at age fourteen he chanced on an essay by the astronomer Edmond Halley, which changed his mind, and his life. In this essay, Halley, one of Isaac Newton's disciples, proclaimed the superiority of the new analytical methods of calculus to the old synthetic geometry. From that moment, Lagrange devoted as much time as he could to the new science, becoming a professor of mathematics at the Royal Artillery School in Turin before the age of eighteen.

From the beginning, Lagrange specialized in analysis, starting the trend toward specialization that has since characterized the study of mathematics. His concentration on analytical methods also liberated the discipline for the first time from its dependence on Greek geometry. In fact, of his major work, *Mécanique analytique* (analytical mechanics), first conceived when he was nineteen but not published until 1788, he boasted that it contained not a single diagram. He then stated offhandedly that in the future the physics of mechanics might be approached as a geometry of four dimensions, the three familar Cartesian coordinates combined with a time coordinate; in such a system, a moving particle could be defined in time and space simultaneously. This system of analyzing mechanics reemerged in 1916, when Albert Einstein employed it to explain his general theory of relativity.

From the ages of nineteen to twenty-three, Lagrange continued as a professor at Turin, producing a number of revolutionary studies in the calculus of variations, analysis of mechanics, theory of sound, celestial mechanics, and probability theory, for which he won a number of international prizes and honors. In 1766, he succeeded Leonhard Euler as court mathematician to Frederick the Great in the Berlin Academy, the most prestigious position of the time. There, freed from lecturing duties, he continued to produce epochal studies in celestial mechanics, number theory, Diophantine analysis, and numerical and literal equations. He also found it possible to marry a younger cousin; the marriage was successful, and Lagrange was later devastated when his wife died of a wasting disease. Characteristically, he tried to overcome his grief by losing himself in his work.

Life's Work

For most of Lagrange's life, overwork was a habit. Yet it enabled him to achieve much at an early age. At twenty-three, Lagrange wrote an article on the calculus of variations, in which he foreshadowed his later unifying theory on the whole of mechanics, both solids and fluids. This integrated general mechanics in much the same way that Newton's law of gravitation unified celestial motion. Lagrange's theory proceeds from the disarmingly simple observation that all physical force is identical, whether operating in the solid or liquid state, whether aural, visual, or mechanical. It thus integrates a diverse array of physical phenomena, simplifying their study. In the same work, Lagrange applied differential calculus to the theory of probability. He also surpassed Newton by absorbing the mathematical theory of sound into the theory of elastic physical particles, becoming the first to understand sound transmission as straight-line projection through adjacent particles. Furthermore, he put to rest a controversy over the proper mathematical description of a vibrating string, laying the basis of the more general theory of vibrations as a whole. At this early age, Lagrange already ranked with the giants of his age, Euler and the Bernoulli family.

The next problems Lagrange attacked at Turin were those involved in the libration of the Moon in celestial mechanics: Why does the Moon present the same surface to the Earth at every point in its revolution? He deduced the answer to this special instance of the three-body problem, a classic in mechanics, from Newton's law of universal gravitation. For solving this problem, Lagrange won the Grand Prix of the French Academy in 1764. The academy followed by proposing a four-body problem; Lagrange solved this, winning the prize again in 1766. The academy then proposed a six-body problem involving calculating the relative position of the Sun, Jupiter, and its four then-known satellites. This problem was not completely solvable by modern methods before the development of computers. Nevertheless, Lagrange developed methods of approximation that were superseded only in the twentieth century. After his move to Berlin, for further work on similar problems—the general three-body problem, the motion of the Moon, and cometary disturbances—Lagrange won further awards.

His career in Berlin lasted twenty years; during this career, he distinguished himself by unfailing courtesy, generosity to other mathematicians, and diplomacy in difficult situations—he was a stranger in a strange court, but he thrived. In addition to working on celestial mechanics there, he diverted himself by investigations

Joseph-Louis Lagrange. (Library of Congress)

into number theory, the humble matter of what his age considered higher arithmetic. Quadratic forms and Diophantine analysis—exponential equations—particularly interested him: He first solved the problem of determining for which square numbers x^2, $nx^2 + 1$ is also a square, when n is a nonsquare, for example, $n = 3$, $x = 4$. This problem was an ancient one; Lagrange's paper is a classic, couched in his elegant language and supported by his equally elegant reasoning. He followed this by offering the first successful proofs of some of Pierre de Fermat's theorems and the one of John Wilson that states that only prime numbers are factors of the sum of the factorial series of the next lowest number plus one—that is, p divides $(p-1)(p-2)\cdots 3 \cdot 2 \cdot 1 + 1$ only if it is prime. His most famous proof in number theory shows that every positive integer can be represented as a sum of four integral squares—a theorem that has had extensive applications in many scientific fields. He later did great work—which proved preliminary—on quadratic equations in two unknowns.

Perhaps the most important work of the Berlin period, however, relates to Lagrange's work in modern algebra. In a memoir of 1767 and in later sequels, he investigated the theoretical bases for solving various algebraic equations. Though once again he fell short of providing definitive answers, his work became an invaluable source for the nineteenth century algebraists who succeeded in finding them. The essential principles—that both necessary and sufficient conditions be established before solution—eluded him, but his work contained the clue.

Eventually, Lagrange's propensity for work broke both his body and his spirit. By 1783, he had sunk into a profound depression, in the grip of which he found further work in mathematics impossible. When Frederick died in 1786 and Lagrange fell out of favor in Berlin, he willingly accepted a position with the French Academy. Still, a change of scene brought no renewal of his interest in mathematics. When his monumental *Mécanique analytique* was published in 1788, Lagrange took no notice of it, leaving a copy unopened on his desk for more than two years. Instead, he turned his attention to various other sciences and the humanities.

It took the French Revolution to reawaken Lagrange's interest in mathematics. Although he could have fled, as many aristocratic scholars did, he did not. The atrocities of the Terror appalled Lagrange, and he had little sympathy with the destructive practices of revolutionary zealots. Yet when appointed to the faculties of the new schools—the École Normale and the École Polytechnique—intended to replace the abolished universities and academies, Lagrange took up his professional duties enthusiastically. Because he became aware of the difficulties his basically unprepared students had with the theoretical bases of calculus, he reformulated the theory to make it independent of concepts of infinitesimals and limits. His attempt was unsuccessful, but he prepared the foundation on which modern theories are built.

Part of his duties at the École Polytechnique required Lagrange to supervise the development of the metric system of weights and measures. Fortunately, he insisted that the base 10 be adopted. Radical reformers lobbied for base 12, alleging superior factorability; it is still occasionally proposed as more "rational," and for centuries it played an infernal role in the British monetary system. To suppress the reformers, Lagrange argued ironically for the advantages of a system with base 11, or any prime, since then all fractions would have the same denominator. A small amount of practice convinced the radicals that 10 was more functional.

Teaching and supervision alone, however, did not suffice to relieve Lagrange's besetting melancholy. He was saved from despair at the age of fifty-six, by the intervention of a young woman, the daughter of his friend the astronomer Pierre-Charles Lemonnier. She insisted on marrying him despite their disparity in age, and, contrary to all expectations, the marriage proved a brilliant success. For the following twenty years, Lagrange could not bear to have her out of his sight, and she proved to be a faithful companion, adept at drawing him out of his shell. At the end of his life, he worked on a second edition of his masterpiece, *Mécanique analytique*, adding many profound insights. He was still improving it when death came, gradually and almost imperceptibly, on April 10, 1813.

Significance

Joseph-Louis Lagrange ranks with the outstanding mathematicians of all time; in his prime, he was widely recognized as the greatest living mathematician, and he is certainly the most significant figure between Euler and Carl Friedrich Gauss. Beyond the quality of his work, he was noted equally for the brilliance of his demonstrations and for his accessibility and personal charm. He is particularly celebrated as one of the classic stylists of mathematical writing, almost the incarnation of mathematical elegance. His composition combines exceptional clarity of description and development with remarkable beauty of phrasing. His language is supple, never stilted or contorted; he somehow seems to ease the effort of strenuous thought. Lagrange once remarked that chemistry was as easy as algebra; in his writing, he is able to make things seem transparent, especially those which seemed particularly dense before reading him.

Perhaps because of this ease of expression, Lagrange is more important for the stimulus he provided for others than for his own original work. Time after time, his contemporaries and descendants found inspiration in him. He made his foundations so complete that others were able to apply them to other cases. In some instances, he was simply ahead of his time; his ideas have had to wait for the ground to be prepared. At any rate, Lagrange's work proved to be extraordinarily rich for those who labored after him.

Lagrange's most important contributions lie in mechanics and the calculus of variations. In fact, the latter is the centerpiece on which all of his achievements depend, the insight he used to integrate the theory of mechanics. This calculus derives from the ancient principle of least action or least time, which concerns the determination of the path a beam of light will follow when passing through or refracting off layers of varying densities. Hero of Alexandria began the inquiry by determining that a beam reflected from a series of mirrors reaches its object by following the shortest possible route; that is, it is the minimum of a function. René Descartes elaborated on the theory by experimenting with the effects of various lenses on a ray of light, showing that refraction also produced minima. Lagrange then proceeded to demonstrate that the general postulates for matter and motion established by Newton, which did not seem to harmonize, also fit this scheme of minima. Thus, he used a principle of economy in nature—that physical mechanics also tended to minimal extremes—to unify the principles of particles in motion. This not only was revolutionary in his time but also gave rise to the further integrating work of William Rowan Hamilton and James Clerk Maxwell, and eventually blossomed in Einstein's general theory of relativity.

James Livingston

Further Reading

Bell, Eric T. *Men of Mathematics*. New York: Simon & Schuster, 1937. Bell's work is famous for three features: readability, accessibility to the general reader, and general historical background. This is the preferred reference work, though Bell does not provide the technical detail of other sources.

Burton, David M. *The History of Mathematics: An Introduction*. Newton, Mass.: Allyn & Bacon, 1985. Burton's book has some very attractive features, especially the examples and practical exercises in real mathematics. However, readers should be aware that his focus is on major developments and broad concepts, so his treatment of Lagrange, while in one sense admirably concise, is also somewhat cursory.

Fraser, Craig G. *Calculus and Analytical Mechanics in the Age of Enlightenment*. Brookfield, Vt.: Variorum, 1997. This collection of essays written between 1981 and 1994 includes studies of Lagrange's early contributions to the principles and methods of mechanics, and his problems in the calculus of variations.

Grabiner, Judith V. *The Calculus of Algebra: J-L Lagrange, 1736-1813*. New York: Garland, 1990. Grabiner describes Lagrange's work in calculus.

Kline, Morris. *Mathematical Thought from Ancient to Modern Times*. New York: Oxford University Press, 1972. Kline offers a more thorough and more rigorously theoretical treatment than Burton (see above), but he requires considerable mathematical sophistication. Still, the book is not aimed at specialists, and Kline explains thoroughly, emphasizing the coherent evolution of mathematical thought. He highlights Lagrange's consistency admirably.

Porter, Thomas Isaac. "A History of the Classical Isoperimetric Problem." In *University of Chicago Contributions to the Calculus of Variations*. Vol. 3. Chicago: University of Chicago Press, 1933. Porter's article is a study for professionals and scholars, with much detail and requiring knowledge of advanced mathematics. It does, however, contain the most extensive account of Lagrange's most important work in the calculus of variations, with incidental reference to his other achievements.

Smith, David Eugene, comp. *A Source Book in Mathematics*. Reprint. Mineola, N.Y.: Dover, 1959. Smith's work is for historians of mathematics, but his selections of extracts from Lagrange's works are representative and reveal Lagrange's clarity of exposition, making them quite accessible.

Struik, D. J., ed. *A Source Book in Mathematics, 1200-1800*. Cambridge, Mass.: Harvard University Press, 1969. This is an anthology of extracts from the original works, such as David Eugene Smith's, but it is more extensive and representative of the entire body of Lagrange's work. The introductions and notes are useful and thorough, and particularly good in helping the reader reach an appreciation of Lagrange's accomplishments.

Pierre-Simon Laplace

French mathematician

Laplace made groundbreaking mathematical contributions to probability theory and statistical analysis. Using Isaac Newton's theory of gravitation, he also performed detailed mathematical analyses of the shape of the earth and the orbits of comets, planets, and their moons.

Born: March 23, 1749; Beaumont-en-Auge, Normandy, France
Died: March 5, 1827; Paris, France
Also known as: Marquis de Laplace

Early Life

Pierre-Simon Laplace (lah-plahs) was born into a well-established and prosperous family of farmers and merchants in southern Normandy. An ecclesiastical career in the Church was originally planned for Laplace by his father, and he attended the Benedictine secondary school in Beaumont-en-Auge between the ages of seven and sixteen. His interest in mathematics blossomed during two years at the University of Caen, beginning in 1766.

In 1768, Laplace went to Paris to pursue a career in mathematics; he remained a permanent resident of Paris or its immediate vicinity for the rest of his life. Soon after his arrival in Paris, he sought and won the patronage of Jean Le Rond d'Alembert, a mathematician, physicist, and philosopher with great influence among French intellectuals. D'Alembert found Laplace employment teaching mathematics to military cadets at the École Militaire, and it was in this position that Laplace wrote his first memoirs in mathematics and astronomy.

In 1773, Laplace was elected to the Academy of Sciences as a mathematician. This achievement, at the relatively young age of twenty-four, was based upon the merits of thirteen memoirs he had presented to academy committees for review. Some of Laplace's earliest mathematical interests involved the calculation of odds in games of chance. At a time when there was not yet a field of mathematics devoted to the systematic study of probability, Laplace played a major role in carrying the early development of this topic beyond the rules of thumb of gambling and the preliminary conclusions of earlier mathematicians. In addition, Laplace emphasized the relevance of probability to the analysis of statistics. He believed that, because all experimental data are imprecise to some extent, it is important to be able to calculate an appropriate average or mean value from a collection of observations. Furthermore, this mean value should be calculated in such a way as to minimize its difference from the actual value of the quantity being measured.

Statistical problems of this type inspired Laplace's initial interest in astronomy. He became intrigued by the process through which new astronomical data should be incorporated into calculations of probabilities for future observations. In particular, he concentrated on the application of Sir Isaac Newton's law of gravitation to the motions of the comets and planets. Laplace's interest in physics thus had a strong mathematical orientation. Throughout his career, he retained his early concentration on the solution of problems suggested by the mathematical implications of physical laws; he never devoted himself to extensive experimental investigation of new phenomena. Laplace's primary motivation was a deep conviction that, even if human limitations prevent an exact knowledge of natural laws and experimental

Pierre-Simon Laplace. (Library of Congress)

conditions, it is still possible progressively to eliminate error through increasingly accurate approximations.

Very little is known about Laplace's personal life during these early years. He does not seem to have stimulated strong friendship or animosity. In 1788, he married Marie-Charlotte de Courty de Romanges, who was twenty years younger than himself, and they had two children. Laplace established and maintained comfortable but disciplined living habits, and he retained an undiminished mental clarity to the moment of his death.

LIFE'S WORK

Although a brief summary of Laplace's life's work requires some classification by topics and an emphasis on final results rather than chronology, the highly integrated and developmental nature of his research should not be forgotten. For example, mathematical techniques that he invented for the solution of problems in probability theory often were immediately applied to similar problems in physics or astronomy. Because Laplace was particularly interested in approximate or probable solutions and the analysis of error, he repeatedly revised his mathematical techniques to accommodate new data.

Laplace's contributions to probability theory were both technical and philosophical. This twofold concern is expressed in the titles of the influential volumes in which he summarized his work, *Théorie analytique des probabilités* (1812; analytic theory of probability) and *Essai philosophique sur les probabilités* (1814; *A Philosophical Essay on Probabilities*, 1902).

The *Théorie analytique des probabilités* was the first comprehensive treatise devoted entirely to the subject of probability. Laplace provided a groundbreaking, although necessarily imperfect, characterization of the techniques, subject matter, and practical applications of the new field. He relied on the traditional problems generated by games of chance, such as lotteries, to motivate his mathematical innovations, but he pointed toward the future by generalizing these methods and applying them to many other topics. For example, because the calculation of odds in games of chance so often requires the summation of long series of fractions in which each term in the series differs from the others according to a regular pattern, Laplace began by reviewing some of the methods he had discovered to approximate the sums of such series, particularly when very large numbers are involved.

Laplace then proceeded to state what has since come to be called Bayes's theorem, after an early predecessor of Laplace. This theorem states how to use partial or incomplete information to calculate the conditional probability of an event in terms of its absolute or unconditional probability and the conditional probability of its cause. Laplace was one of the first to make extensive use of this theorem; it was particularly important to him because of its relevance to how calculations of probability should change in response to new knowledge.

The *Théorie analytique des probabilités* includes Laplace's applications of his mathematical techniques to problems generated by the analysis of data from such diverse topics as census figures, insurance rates, instrumentation error, astronomy, geodesy, election prognostication, and jury selection. In particular, he gave an important statement of what has since been called the least square law for the calculation of a mean value for a set of data in such a way that the resulting error from the true value is minimized.

A Philosophical Essay on Probabilities has been one of Laplace's most widely read works; it includes the conceptual basis upon which Laplace constructed his mathematical techniques. Most important, Laplace stated and relied upon a definition of probability that has been a source of considerable philosophical debate. Given a situation in which specific equally possible cases are the results of various processes (such as rolling dice) and correspond to favorable or unfavorable events, Laplace defined the probability of an event as the fraction formed by dividing the number of cases that correspond to or cause that event by the total number of possible cases. When the cases in question are not equally possible (as when dice are loaded), the calculation must be altered in an attempt to include this information. Laplace's definition thus calls attention to his treatment of probability as an application of mathematics made necessary only by human ignorance.

In one of the most famous passages in *A Philosophical Essay on Probabilities*, Laplace expresses this view by describing a supreme intelligence with a complete knowledge of the universe and its laws at any specific moment; for such an intelligence, Laplace believed that probability calculations would be unnecessary because the future and past could be calculated simply through an application of the laws of nature to the given perfectly stipulated set of conditions. Because knowledge of natural laws and the state of the world is always limited, probability is an essential feature of all human affairs. Nevertheless, Laplace's emphasis was not on the negative aspect of this conclusion but on the mathematical regularities to which even seemingly arbitrary sequences of events conform.

The domain in which Laplace saw the closest human approach to the knowledge of his hypothetical supreme intelligence was the application of Newton's theory of gravitation to the solar system. Since Newton's publication of his theory in 1686, mathematicians and physicists had reformulated his results using increasingly sophisticated mathematics. By Laplace's time, Newton's theory could be stated in a type of mathematics known as partial differential equations. Laplace made major contributions to the solution of equations of this type, including the famous technique of "Laplace transforms" and the use of a "potential" function to characterize a field of force.

Laplace made remarkably detailed applications of Newton's results to the orbits of the planets, moons, and comets. Some of his most famous calculations involve his demonstration of the very long-term periodic variations in the orbits of Jupiter and Saturn. Laplace thus contributed to an increasing knowledge of the stability and internal motions of the solar system. He also applied gravitation theory to the tides, the shape of Earth, and the rings of Saturn. His hypothesis that the solar system was formed through the condensation of a diffuse solar atmosphere became a starting point for more detailed subsequent theories.

Newtonian gravitation theory became Laplace's model for precision and clarity in all other branches of physics. He encouraged his colleagues to attempt similar analyses in optics, heat, electricity, and magnetism. His influence was particularly strong among French physicists between 1805 and 1815. By his death in 1827, however, this attempt to base all physics upon short-range forces had achieved only limited success; aside from the mathematical methods he developed, Laplace's conceptual contributions to physics were not as long-lasting as his more fundamental insights in probability theory.

Significance

Pierre-Simon Laplace's cultural influence extended far beyond the relatively small circle of mathematicians who could appreciate the brilliant technical detail in his work. In several ways he has become a symbol of some important aspects of the rapid scientific progress that took place during his career as a result of his role in institutional changes in the scientific profession and the implications that have been drawn from his conclusions and methods.

Laplace was very active within the highly centralized French scientific community. As a member of the French Academy of Sciences, he served on numerous research or evaluative committees that were commissioned by the French government. For example, following the French Revolution in 1789, he was an influential designer and advocate of the metric system, which has become the most widely used international system of scientific units. The academy was disbanded during the radical phase of the Revolution in 1793, but in 1796 Laplace became the president of the scientific class of the new Institute of France.

Highly publicized institute prizes were regularly offered for essays in physics and mathematics, and Laplace exerted a powerful influence on French physics through the attention he devoted to choice of topic and support for his preferred candidates. He also played an important part in the early organization of the École Polytechnique, the prestigious school of engineering founded in 1795. Although Laplace lived through turbulent political changes, he remained in positions of high scientific status through the Napoleonic era and into the Bourbon Restoration, when he was raised to the nobility as a marquis. Laplace seems to have held few strong political views, and he thus is sometimes cited as an example of a powerful scientist indifferent to social or political conditions.

Aside from his work in probability and statistics, which has quite direct impact on modern societies, other aspects of Laplace's work have contributed to general perceptions of the goals, limitations, and methods of science. With Newton's theory of gravitation as his model, Laplace was convinced that, although human knowledge of nature is always limited, there are inevitable regularities that can be expressed approximately with ever-increasing accuracy. Laplace thus has become a symbol of nineteenth century scientific determinism, the view that the uncertainty of the future is only the result of human ignorance of the natural laws that determine it in every detail.

When Napoleon I asked Laplace why God did not play a role in Laplace's analysis of the stability of the solar system, Laplace replied that he had had no need for such a hypothesis. Laplace thus contributed to a growing association of the scientific tradition with atheism and materialism. Finally, Laplace's style of mathematical physics has become a primary example of a reductionistic research strategy. Just as the gravitational effect of a large mass is determined by the sum of the forces exerted by all of its parts, Laplace expected all phenomena to reduce to collections of individual interactions. His success in implementing this method contributed to widespread perceptions that this is a necessary component of scientific investigation.

James R. Hofmann

FURTHER READING

Arago, François. "Laplace." In *Biographies of Distinguished Scientific Men*. New York: Ticknor & Fields, 1859. Arago was a student and colleague of Laplace for many years. His essay discusses only Laplace's work in astronomy and concentrates on his study of the stability of the solar system.

Brush, Stephen G. *The Origin of the Solar System and the Core of the Earth from Laplace to Jeffreys: Nebulous Earth*. Vol. 1 in *A History of Modern Planetary Physics*. New York: Cambridge University Press, 1996. Traces the evolution of Laplace's nebular hypotheses, the most popular nineteenth century explanation for the origin of the solar system.

Fox, Robert. "The Rise and Fall of Laplacian Physics." *Historical Studies in the Physical Sciences* 4 (1974): 89-136. This is an excellent summary of Laplace's efforts to direct French physics according to a research program based upon short-range forces.

Gillespie, Charles Coulston, Robert Fox, and Ivor Grattan-Guiness. *Pierre-Simon Laplace, 1749-1827: A Life in Exact Science*. Princeton, N.J.: Princeton University Press, 1997. Focuses on Laplace's research program and his work with the Academy of Science. Includes biographical information from a scientific point of view, a description of Laplace's efforts to gather young physicists who would work with the Newtonian model in physics, and an overview of the Laplace transform.

_____. "Pierre-Simon Marquis de Laplace." In *Dictionary of Scientific Biography*. Vol. 15. New York: Charles Scribner's Sons, 1978. This chronological survey of Laplace's scientific career combines discussion of significant concepts with summaries of important mathematical derivations.

Hahn, Roger. *Laplace as a Newtonian Scientist*. Los Angeles: Williams Andrew Clark Memorial Library, 1967. This short essay describes the philosophical debate concerning the status of laws of nature that occurred during Laplace's formative period at the University of Caen and his early years in Paris. Laplace's convictions about the law-governed structure of the universe are traced to his reading of d'Alembert and Marquis de Condorcet.

_____. *Pierre Simon LaPlace, 1749-1827: A Determined Scientist*. Cambridge, Mass.: Harvard University Press, 2005. Full biography of Laplace by a scholar who has studied him for decades.

Todhunter, Isaac. *A History of the Mathematical Theory of Probability from the Time of Pascal to That of Laplace*. New York: Chelsea House, 1965. Chapter 10 provides a technical and chronological account of the chief results and some of the derivations found in Laplace's publications on probability theory.

GOTTFRIED WILHELM LEIBNIZ

German philosopher and mathematician

Leibniz contributed to the development of rationalist metaphysics and a dynamic theory of motion. Contrary to Descartes, he believed that activity, and not extension, is essential to substance. He coined the term "monad" to refer to these fundamental units of existence, metaphysical entities that are not extended and are not of a material nature but are dynamic units of psychic activity.

Born: July 1, 1646; Leipzig, Saxony (now in Germany)
Died: November 14, 1716; Hanover (now in Germany)

EARLY LIFE
Gottfried Wilhelm Leibniz (GAWT-freet VIHL-hehlm LIP-nihts) was born into an academic family; his mother's father was a professor, as was his own father (who died when Leibniz was six). Leibniz was intellectually gifted; he taught himself Latin and read profusely in the classics at an early age. When he was an adolescent, Leibniz began to entertain the notion of constructing an alphabet of human thought from which he could generate a universal, logically precise language. He regarded this language as consisting of primitive simple words expressing primitive simple concepts that are then combined into larger language complexes expressing complex thoughts. His obsession with this project played an important role throughout his life.

Leibniz was formally educated at the University of Leipzig, where he received his bachelor's and master's degrees for theses on jurisprudence, and at the

University of Altdorf, where he received a doctorate in law in 1666. He declined a professorship at Altdorf and entered employment as secretary of the Rosicrucian Society. Eventually he was employed as a legal counsel by Johann Philipp von Schönborn, a governing official of Mainz.

Life's Work

Leibniz's philosophy was rationalist. According to this theory, human knowledge has its origins in the fundamental laws of thought instead of in human experience of the world as in the doctrine of empiricism. In fact, Leibniz argued that the laws of science could be deduced from fundamental metaphysical principles and that observation and empirical work were not necessary for arriving at knowledge of the world. What was needed instead was a proper method of calculating or demonstrating everything contained in certain fundamental tenets. For example, he believed that he could deduce the fundamental laws of motion from more basic metaphysical principles. In this general conception, he followed in the intellectual footsteps of René Descartes. The great problem with interpreting Leibniz's contribution to this tradition of thought is that he published only one major book during his lifetime, and it does not contain a systematic account of his full philosophy. Accordingly, it is necessary to reconstruct his system from his short articles and his more than fifteen thousand letters.

Leibniz's youthful dreams of constructing a perfect language quickly evolved into a theory of necessary and contingent propositions. He claimed that in every true affirmation the predicate is contained in the subject. This idea evolved from his conception of a perfect language that (in all of its true, complex statements) would perfectly reflect the universe. The true propositions of this language are necessarily true, and all necessary propositions are, according to Leibniz, ultimately reducible to identity statements. Such a conception was more plausible in the case of purely mathematical statements since, for example, $4 = 2 + 2$ can be equated with $4 = 4$.

Yet this conception seemed impossible in the case of contingent statements such as "the house is blue." Leibniz avoided this problem by arguing that the necessity in what appears as contingent truths can be revealed (or resolved) only through an infinite analysis and therefore can be carried out in full only by God. It follows that, for humans, all contingent truths are only more or less probably true. Such truths are guaranteed by the principle of sufficient reason, which states that there must be some reason for whatever is the case. Necessary truths, or truths of reason, on the other hand, are guaranteed by the principle of contradiction, which states that the denial of such a truth is a contradiction (though this can be known only by God). A logical principle closely related to the principle of sufficient reason is the notion of the "identity of indiscernibles," now known as Leibniz's law. This principle states that it is impossible for two things to differ only numerically, that is, to be distinct yet have no properties that differ; if two things are distinct, there must be some reason for their distinctness.

Leibniz had elaborated the rudiments of his metaphysical system while at Mainz, but it was during his sojourn in Paris that his philosophy matured. In 1672, he was sent to Paris on a diplomatic mission for the German princes to persuade King Louis XIV to cease military activities in Europe and send forces to the Middle East. Leibniz remained in Paris for four years, and, though he failed to even gain an audience with the monarch, he met frequently with the greatest minds of the day, such as Christiaan Huygens, Nicolas de Malebranche, Antoine Arnauld, and Simon Foucher. He also carried out studies of the mathematics of Blaise Pascal and René Descartes, and built one of the first

Gottfried Wilhelm Leibniz. (Library of Congress)

computers—a calculating machine able to multiply very large numbers. While residing in Paris, he also made a brief trip to England, where he met with Irish chemist Robert Boyle and visited the Royal Society, to which he was elected.

When he returned to Hanover, he accepted a post as director of the library to John Frederick, the duke of Brunswick, where he remained for the next ten years. It was only after working with Huygens in Paris on the nature of motion that Leibniz finally came to grips with the problem of the continuum. On his return trip from Paris, during which he visited philosopher Baruch Spinoza in Holland, he composed "Pacidius Philalethi" (1676), an extended analysis of this subject. This issue is traced back to the ancient Greeks and concerns the problem of resolving the motion of an object into its motions over discrete parts of space. If the body must pass through each successive parcel of space between two points, then it can never get from one point to another, since there are an infinity of such discrete parcels between any two points. It was in the context of this problem of motion and the continuum that Leibniz developed, in 1676, the differential calculus, publishing his results in 1684. Sir Isaac Newton had already discovered the calculus but did not publish his results until 1693, several years after Leibniz published his discoveries. Priority of discovery is accorded to Newton though the consensus now is that they arrived at the calculus independently.

Leibniz argued that Cartesian physics renders motion ultimately inexplicable on the basis of fundamental concepts, since it is grounded in the notion of matter as extension and does not accommodate dynamic properties. For Leibniz, the fundamental tenet is that activity is essential to substance. Substantial being is what is simple—what can be conceived by itself and what causes itself. The term "monad" was adopted by Leibniz to refer to this fundamental unit of existence. Monads are metaphysical entities that are not extended and are not of a material nature but are units of psychic activity. All entities are monadic, from God, the supreme monad who has created all the other grades of monads, to the lowest grade of being. The universe of monads is divided into two realms on the scale of perfection, that of nature and that of grace.

Because monadic substances are psychic rather than material, Leibniz's philosophy has been labeled "panpsychistic idealism." On the level of phenomena, Leibniz retained a mechanical model: Matter in the phenomenal realm is "secondary matter," composed of monadic substances and having mass. Yet, according to Leibniz, substances and monads do not interact with each other. The universe consists of an infinity of such monadic substances, individuated by the principle of indiscernability and each of which undergoes changes. This change in the monad occurs entirely because of its own nature, according to a logically necessary law and not because of effects coming to it from outside. All these changes in the monads have been harmonized by God into what appears as a causal order. Leibniz referred to this as the "way of preestablished harmony" and likened it to the synchronized sounding of two clocks. Since each monad/substance is completely independent of all the others, Leibniz said (in his later writings) that they are "windowless"; that is, they do not look out on the world. Though this conception may appear to be rather unusual, it does account for the plurality of existents in the universe, since the substances are infinite and independent of one another.

The changes of a monad are changes in the degree to which it expresses the universe. This expression or "perception" occurs on all levels of being; all individuals express the rest of the universe through the changes that occur in it. Since each individual represents all individuals, metaphysical accommodation is made of the unity of the universe in the diversity of an infinity of monads. An exhaustive specification of the nature of one substance/monad would give an exhaustive specification of the natures of all other substances/monads (from a particular point of view). Since such a specification would be logically necessary (in any true assertion

Leibniz's Major Works

1666	*Dissertatio de arte combinatoria*
1686	*Discours de métaphysique* (pb. 1846; *Discourse on Metaphysics*, 1902)
1704	*Nouveaux essais sur l'entendement humain* (pb. 1765; *New Essays Concerning Human Understanding*, 1896)
1710	*Essais de théodicée sur la bonté de Dieu, la liberté de l'homme et l'origine du mal* (*Theodicy: Essays on the Goodness of God, the Freedom of Man, and the Origin of Evil*, 1951)
1714	*La Mondologie* (pb. 1840; also published as *Lehrsütze über die Monadologie*, 1720; *Monadology*, 1867)
1714	*Principes de la nature et de la grâce fondés en raison* (pb. 1768?; *The Principles of Nature and of Grace*, 1890)

the predicate is contained in the subject), the complete description of the universe is a tautology, though this could only be fully known by God.

The characteristics of the monad, activity and perception, are analogous to the features of the mental lives of human beings. In connection with the notion of perception, Leibniz later introduced the notion of "apperception." In *Principes de la nature et de la grâce, fondés en raison* (1714; *The Principles of Nature and of Grace*, 1890), he distinguished between perceiving the outer world and apperceiving the inner state of the monad (which is self-consciousness in a human being). In fact, differences between monads relate to their degree of clarity of perception and the presence of perception or apperception. At the bottom of the hierarchy of being are monads with confused perception and unselfconscious appetition. Leibniz's theory of human understanding is developed in his *Nouveaux essais sur l'entendement humain* (1765; *New Essays Concerning Human Understanding*, 1896), written in response to English philosopher John Locke but not published in his lifetime. The perceptions of the human soul are expressions of the perceptions occurring in the body and are confused and unclear. Since all changes occur according to internal principles, all the ideas of the human mind are innate.

Leibniz was the first thinker to employ explicitly the notion of the unconscious, which he did in connection with the distinction between apperception and perception—not all perceptions are apperceived. These perceptions he refers to as "petites perceptions" and gives as his favored example the sound of a wave crashing on the beach; the sound is composed of tiny perceptions of droplets hitting the beach; although one is *unaware* of the droplets hitting the beach, one still *perceives* them (hears the sounds they make).

During his years in Hanover, Leibniz grew very close to Sophia, the wife of his patron Ernest Augustus, first elector of Hanover, and to Sophia's daughter, Sophia Charlotte, who became the first queen of Prussia. Leibniz discussed many philosophical ideas with them, and from these conversations arose his only published book, *Essais de théodicée sur la bonté de Dieu, la liberté de l'homme, et l'origine du mal* (1710; *Theodicy*, 1951). In this text, Leibniz argued along Augustinian lines that evil exists in the world because the world could not be as good as it actually is without the evil that it contains. In fact, out of all the possible universes, Leibniz believed, this universe contains the greatest amount of good. This conception earned for Leibniz's theories the appellation a "philosophy of optimism."

Toward the end of his life, Leibniz became embroiled in an intellectual dispute with Samuel Clarke, a disciple of Newton. Leibniz claimed that Newtonian physics had contributed to a general decline of religion in England. Clarke defended Newtonian physics against this charge, while Leibniz attacked Newton's conceptions on philosophical grounds in a series of letters. Leibniz asserted that the notions of absolute space and absolute time violated the principle of sufficient reason and that the concept of gravity introduced the incomprehensible notion of action at a distance. Leibniz had earlier argued that space and time have no substantive

Leibniz's Method of Reason

Gottfried Wilhelm Leibniz believed foremost in the power and necessity of reason for solving all problems, not just mathematical or scientific ones. He wanted to find a way to calculate and note philosophical, moral, and ethical thoughts, beliefs, and concerns with precision and certainty by using signs and characters.

It is manifest that if we could find characters or signs appropriate to the expression of all thoughts as definitely and as exactly as numbers are expressed by arithmetic or lines by geometrical analysis, we could in all subjects, in so far as they are amenable to reasoning, accomplish what is done in Arithmetic and Geometry.

All inquiries which depend on reasoning would be performed by the transposition of characters and by a kind of calculus which would directly assist the discovery of elegant results. We should not have to puzzle our heads as much as we have to-day, and yet we should be sure of accomplishing everything the given facts allowed.

Moreover, we should be able to convince the world of what we had discovered or inferred, since it would be easy to verify the calculation either by doing it again or by trying tests similar to the of casting out nines in arithmetic. And if someone doubted my results, I should say to him "Let us calculate, Sir," and so by taking pen and ink we should soon settle the question.

Source: Leibniz, *On Method* (1677), excerpted in *The Age of Reason: The Culture of the Seventeenth Century*, edited by Leo Weinstein (New York: George Braziller, 1965), pp. 88-89.

existence and are only the ordered relations between coexistent entities and the ordering of successively existent entities, respectively. The death of Leibniz ended the debate with Clarke, who immediately published the correspondence. In spite of his extensive contacts with savants throughout Europe, Leibniz's death on November 14, 1716, was relatively unnoticed.

Significance

Leibniz remained in the humble employ of royal patrons his whole life, though at one point he was offered the position of head librarian at the Vatican, which he declined to accept. In 1700, the Berlin Society of Sciences was founded, and Leibniz was elected president for life. Throughout his life, Leibniz speculated about grandiose social-intellectual projects. He advocated the Christian conquest of the pagan lands, the compilation of a universal encyclopedia of human knowledge, the reuniting of the Protestant and Catholic churches, and the restoration of peace in Europe under the Holy Roman Empire. In a true Enlightenment spirit, Leibniz also advocated the establishment of scientific academies throughout the world and actually corresponded with Peter the Great concerning such an academy for Russia. In spite of such visionary plans, Leibniz was very conservative politically; he did not criticize existing institutions and was opposed to innovation in moral and religious matters. Yet he was friendly to all, avid of learning of the world from everyone he encountered.

Leibniz had a tremendous influence on his contemporaries. Virtually all philosophers in Germany were Leibnizian during the years after his death. One early Leibnizian who proved to be equally influential in Germany was Christian von Wolff. Wolff had corresponded with Leibniz from 1704 to 1716 on mathematical and philosophical topics. Wolff taught the Leibnizian system to Martin Knutzen, who in turn taught it to Immanuel Kant, who long remained a Leibnizian. One of Kant's early essays in metaphysics was on the principle of sufficient reason and its relation to the logical principles of identity and contradiction.

Writing in the light of the Lisbon earthquake of 1756, Voltaire bitterly satirized the philosophical optimism of Leibniz (along with Alexander Pope) in his work *Candide: Ou, L'Optimisme* (1759; *Candide: Or, All for the Best*, 1759). Leibniz's philosophical influence is still evident to this day. His law concerning the identity of indiscernibles is the starting point of much of the work done in the twenty-first century on semantics, and his notions of necessity and possibility are the ancestors of work by contemporary modal logicians on the nature of necessity.

Mark Pestana

Further Reading

Broad, C. D. *Leibniz: An Introduction*. New York: Cambridge University Press, 1975. A compilation of the lecture notes used by Broad, published after his death. Reconstructs and analyzes the whole of Leibniz's philosophy.

Hostler, John. *Leibniz's Moral Philosophy*. New York: Barnes & Noble Books, 1975. A full study of the metaethical dimensions of Leibniz's metaphysics. Argues that the metaphysics is worked out in the framework of his systematic moral ideas.

Jolley, Nicholas. *Leibniz and Locke: A Study of the New Essays on Human Understanding*. New York: Oxford University Press, 1984. A study of Leibniz's response to Locke. Attempts to substantiate the notion that the guiding motive of Leibniz in writing his study was to refute Locke's materialism.

_____, ed. *The Cambridge Companion to Leibniz*. New York: Cambridge University Press, 1995. Thirteen essays examine Leibniz's life, his theories of metaphysics, knowledge, and morality, and his reception in the eighteenth century.

Leibniz, Gottfried Wilhelm. *The Monadology and Other Philosophical Writings*. Translated with an introduction and notes by Robert Latta. Oxford, England: Clarendon Press, 1898. Contains a two-hundred-page introduction to the whole of Leibniz's philosophy. Extensive discussion of his influence on the development of psychology in late nineteenth century Germany.

McRae, Robert. *Leibniz: Perception, Apperception, and Thought*. Toronto, Canada: University of Toronto Press, 1976. Focuses on Leibniz's theory of knowledge and attempts to explain how perception and apperception combine to provide for thought.

Mates, Benson. *The Philosophy of Leibniz: Metaphysical Underpinnings*. New York: Oxford University Press, 1986. This is an excellent introductory work on Leibniz, written by a logician. Covers all aspects of his general metaphysics.

Rescher, Nicholas. *Leibniz: An Introduction to His Philosophy*. Totowa, N.J.: Rowman and Littlefield, 1979. Argues that Leibniz's unorthodox metaphysical system is ultimately aimed at providing a foundation for utterly orthodox views in ethics and religion.

_____. *On Leibniz*. Pittsburgh, Pa.: University of Pittsburgh Press, 2003. Eleven essays examine key aspects of Leibniz's philosophy, personality, and personal and scholarly development. The final chapter explores how current issues in the philosophy of science can be addressed in terms of Leibniz's principles.

Russell, Bertrand. *A Critical Exposition of the Philosophy of Leibniz*. Cambridge, England: Cambridge University Press, 1900. An important work, arguing that Leibniz's philosophy can be understood in terms of five fundamental principles that are ultimately inconsistent.

Rutherford, Donald, and J. A. Cover, eds. *Leibniz: Nature and Freedom*. New York: Oxford University Press, 2005. A collection of essays exploring Leibniz's ideas on free will, determinism, and the philosophy of nature.

LEONARDO OF PISA
Italian mathematician

Leonardo provided Western Europe with the earliest and most heralded Latin account of the Hindu-Arabic number system and its computational methods. He contributed substantially to the acceptance of the Arabic algebraic system and created a revolutionary mathematical technique known as the Fibonacci sequence.

Born: c. 1170; Pisa (now in Italy)
Died: c. 1240; Pisa
Also known as: Leonardo Fibonacci (given name); Leonardo Pisano

EARLY LIFE
Leonardo of Pisa was born Leonardo Fibonacci, the surname meaning "son of Bonaccio." Although very little is known about his life beyond the few facts gleaned from his mathematical writings, his father, Guglielmo, was a successful merchant who was the chief magistrate of the community of Pisan merchants in the North African port of Bugia (now Bejaïa, Algeria). As a young boy, he joined his father there and began the study of mathematics in this culturally diverse environment.

Desiring his son to be a successful merchant or commercial agent, Guglielmo sent Leonardo to study with a Muslim master who introduced him to the intricacies of Arabic mathematics, especially al-Khwārizmī's *Kītab al-jabr wa al-muqābalah* (algebra; c. 820) and the practical value of the Hindu-Arabic numeral system represented by the nine Indian figures (1, 11, 111, 4, 5, 6, 7, 8, 9), the fourth through the ninth symbols representing the first letters of the Hindu names for these integers. As he grew older, he traveled around the Mediterranean area, especially Egypt, Syria, Greece, Sicily, and Provence; visited the dominant commercial centers; acquired knowledge of the arithmetical systems used by hundreds of merchants; and mastered the theoretical subtleties of Greek and Arabic mathematics, chiefly those of Plato of Tivoli, Savasorda, Euclid, Archimedes, Hero of Alexandria, and Diophantus. He even resided for a time at the court of Frederick II, the Holy Roman Emperor, where he engaged in scientific speculations with Frederick and his court philosophers, the most notable being Michael Scot, to whom Leonardo dedicated one of his works.

In his later years, Leonardo probably served Pisa administratively as an examiner of municipal accounts. This commercial expertise, however, was always secondary to his lifelong passion for mathematics.

LIFE'S WORK
The rapid improvements that marked the history of Western mathematics in the thirteenth century, particularly in the fields of arithmetic and algebra, were largely a result of the genius of Leonardo, although a second mathematician of originality, Jordanus Nemorarius, made significant contributions, especially to the theory of numbers and to mechanics. Yet Jordanus showed no trace of Arabic influence. Developing the Greco-Roman arithmetical tradition of Nicomachus and Boethius, he habitually used letters for generalizing proofs in arithmetical problems, an awkward method that Leonardo could avoid because of his employment of Arabic numbers.

Leonardo's pioneering achievements in mathematics began in 1202, when he wrote his first work, *Liber abaci* (English translation, 2002). Even though the title, which means "book of the abacus," is a misnomer because Leonardo eschewed Roman numerals and the methods of the abacus, the work became the earliest in the West to extol the superiority of the nine-numeral Arabic system of numbers when used in conjunction with the zero. When the *Liber abaci* first appeared, Arabic numerals were known to only a few European philosophers through the Latin translations of al-Khwārizmī's ninth century treatise. Leonardo understood fully the advantages of this system for mathematical operations. He displayed the system brilliantly in this edition and in a second, revised edition of 1228 dedicated to Michael Scot, the emperor's chief scholar, and provided more rigorous demonstrations than in any previous or contemporary work. Leonardo realized that the great merit of this system was that it contained the symbol for zero and that any number could be represented simply by arranging digits in order, the value of a digit being shown by its distance from zero or from the first digit on the left.

Predominantly theoretical in nature, the *Liber abaci* was also valuable for its commercial arithmetic, which covered such operations as profit margins, barter, money changing, conversions of weights and measures, partnerships, and interest. After his death, Italian merchants generally adopted Leonardo's Arabic system of numeration, and his book remained the standard in Europe for more than two centuries.

Besides popularizing a new system of numerals throughout the West, the *Liber abaci* was revolutionary for two reasons. First, it introduced Arabic algebra to European civilization. Leonardo's algebra was rhetorical, but it was unique because of its employment of geometrical methods in its descriptions. He dealt primarily with the extraction of square and cube roots, progressions, indeterminate analysis (an equation with two or more unknowns for which the solution must be in rational numbers, whole numbers or common fractions), false assumptions (when a problem is worked out by incorrect data, then corrected by proportion), the rules of three and five (methods of finding proportions), the solution of equations of the third degree, and other algebraic and geometrical operations.

Of even greater importance was Leonardo's famous sequence of numbers known before the nineteenth century as the Series of Lamé but now called correctly the Fibonacci sequence. In answer to the problem of how many pairs of rabbits could be produced from a single pair if each pair produced a new pair each month and every new pair became productive from the second month onward (supposing that no pair died), he devised the recurrent, or recursive, series of 1, 1, 2, 3, 5, 8, 13, 21, 34, 55, 89, 144, 233, and so on. In this number sequence, in which the relation between two or more successive terms can be expressed by a formula, each term is equal to the sum of the two preceding ones. In the nineteenth century, the series proved of immense value in the study of divisibility, prime numbers, and Mersenne numbers. In the modern world, the Fibonacci sequence is used in botany for determining the patterns of natural growth.

In addition to the *Liber abaci*, Leonardo wrote three other significant works. In 1220, his *Practica geometriae* (practice of geometry) presented theorems based principally on two of Euclid's works. It applied algebra to the solution of geometrical problems, a radically innovative technique for thirteenth century Europe. In 1225, two smaller works appeared, the *Flos* (prime) and the *Liber quadratorum* (*The Book of Squares*, 1987). More original than the *Liber abaci*, they were devoted to questions involving quadratic and cubic equations, in addition to several refinements to his earlier algebraic discourses. *The Book of Squares* may be considered Leonardo's masterpiece. Although the *Liber abaci* made his reputation, *The Book of Squares* made him the most important contributor to number theory until Pierre de Fermat, the celebrated seventeenth century French mathematician who was instrumental in early experimentation aimed at determining the exact length of a quadrant of Earth's meridian, the scientific basis of the metric system of weights and measures.

Significance

The impact of Leonardo on future generations was enormous. His pioneering achievements helped spread Arabic numeration and Arabic algebra throughout the West. Popular diffusion followed in the form of almanacs, calendars, and literary and poetic productions. Merchants accepted his new system—the Italians first and other Europeans by the end of the sixteenth century. Even as early as the second half of the thirteenth century, lectures in the universities incorporated the new numbering system. His use of geometry in algebraic problems and, conversely, his use of algebra in solving geometric problems ushered in a new era in these disciplines. In time, the Fibonacci sequence revolutionized many divergent scientific fields. Last, aside from their scientific merit, his works were of tremendous cultural influence, particularly as they relate to metrology and to the major economic conditions of his time. In short, Leonardo was the greatest Christian mathematician of the Middle Ages. The mathematical renaissance in the West dates from him.

Ronald Edward Zupko

Further Reading

Crombie, A. C. *Augustine to Galileo: The History of Science, A.D. 400-1650*. 1953. Reprint. Cambridge, Mass.: Harvard University Press, 1979. Includes much discussion of Leonardo's influence on mathematics and number theory prior to the scientific revolution.

———. *Medieval and Early Modern Science*. 2 vols. 1959. Reprint. Cambridge, Mass.: Harvard University Press, 1961. Excellent descriptive bibliographies in both volumes, with coverage of Leonardo's precursors, his impact on popularizing Arabic numerals, and his contributions to later medieval mathematics.

Fibonacci, Leonardo. *Fibonacci's "Liber abaci": A Translation into Modern English of Leonardo Pisano's Book of Calculation*. New York: Springer, 2002. Notes on the translation provide information on Leonardo and his work. Bibliography.

Gies, Joseph, and Frances Gies. *Leonard of Pisa and the New Mathematics of the Middle Ages*. Gainesville, Ga.: New Classics Library, 1969. Contains a summary of Leonardo's life, a general survey of his works, and a brief overview of the history of numerical notation.

Hardy, G. H., and E. M. Wright. *An Introduction to the Theory of Numbers*. 1960. Reprint. Oxford, England: Oxford at the Clarendon Press, 1983. This volume provides a detailed account of the Fibonacci numbers and sequences and is meant for the mathematician or serious student of science.

Kibre, Pearl. *Studies in Medieval Science: Alchemy, Astrology, Mathematics, and Medicine*. London: Hambledon Press, 1984. Leonardo's standing in the quadrivium (arithmetic, geometry, astronomy, and music) of later thirteenth century universities appears in the first of these republished articles.

Vorobiev, Nicolai N. *Fibonacci Numbers*. Boston: Birkhäuser Verlag, 2002. A mathematical work examining the Fibonacci numbers.

Nikolay Ivanovich Lobachevsky
Russian mathematician

Lobachevsky was the boldest and most consistent founder of a post-Euclidean theory of real space. His persistence in holding open his revolutionary line of inquiry into the reality of geometry helped to set the stage for the radical discoveries of twentieth century theoretical physics.

Born: December 1, 1792; Nizhny Novgorod, Russia
Died: February 24, 1856; Kazan, Russia

Early Life
Nikolay Ivanovich Lobachevsky (luh-buh-CHAYF-skuh-ih), whose parents were a minor government clerk and an energetic woman of apparently no education, was a member of what most nearly corresponded to a middle class in preemancipation Russia. As a government-supported student, he was recorded in school as a *raznochinets* (person of miscellaneous rank—not a noble, a peasant, or a merchant). Despite the early marriage of his mother, Praskovya, to the collegiate registrar Ivan Maksimovich Lobachevsky, the evidence strongly suggests that Nikolay and his two brothers were the illegitimate sons of an army officer and land surveyor, S. S. Shebarshin. Ivan Maksimovich Lobachevsky died when Nikolay, the middle son, was only five years old. The widowed Praskovya left Nizhny Novgorod and moved eastward along the Volga to the provincial center of Kazan. She enrolled all three boys in the local gymnasium (preparatory school). Nikolay attended the school between 1802 and 1807.

Lobachevsky's student years at Kazan University (1807 to 1811) were a time when Russia was eager to learn from the West and to give more than it had received. Lobachevsky was awarded Kazan's first master's degree for his thesis on elliptic movement of the heavenly bodies. He worked closely with Johann Martin Bartels, who had earlier discovered and taught Carl Friedrich Gauss, a great mathematician of the day.

Lobachevsky taught at Kazan University from 1811 until his mandated but most unwilling retirement in 1846. The tenure of Mikhail Magnitskii as curator from 1819 to 1826 was the school's most difficult period. A religious fanatic who attempted to give this particularly science-oriented university the atmosphere of a medieval monastery, Magnitskii was imprisoned in 1826 for his gross incompetence. He was particularly suspicious of the philosophy of Immanuel Kant. All the distinguished German professors left; for a time, the young Lobachevsky carried the burden of providing all the advanced lectures in mathematics, physics, and astronomy alone. His own development and integrity were only strengthened during this phase. It did Lobachevsky no harm that he too was anti-Kantian; he completely disagreed with Kant's view that Euclidean geometry was proof of the human mind's inborn sense of lines, planes, and space.

Life's Work

As a young professor in 1817, Lobachevsky was intrigued by the problem of Euclid's fifth postulate, which implies the possibility of infinitely parallel lines. More technically, one may draw a single line through a given point on a given plane that will never intersect another given line on the same plane. On one hand, this is not a simple axiom that has no need of proof. On the other, it cannot be proved. Two thousand years of general satisfaction with Euclidean geometry had seen many vain attempts to prove the fifth postulate and thereby give this geometry its final perfection. Such attempts became particularly frenzied in the eighteenth century. A rare few thinkers began to entertain the idea that the postulate was wrong, but they denied it even to themselves.

From 1817 to 1822, Lobachevsky made repeated attempts to prove the fifth postulate, already resorting to non-Euclidean concepts such as an axiom of directionality. Once he perceived the hidden tautology of even his best attempts, he concluded that the postulate must be wrong and that geometry must be put on a new foundation.

In addition to the resistance of intellectual tradition, Lobachevsky could expect little support in a country whose ruling house saw itself as the very embodiment of stability and conservatism. The unsettling implications of losing true parallelism and rocking the foundations of classical geometry were as unwelcome in czarist Russia as they could possibly have been anywhere in the world. This resistance makes Lobachevsky's boldness all the more impressive. Simultaneously and independently, two leading mathematicians of the day—Gauss in Germany and János Bolyai in Hungary—were facing the same conclusion as Lobachevsky. Despite their secure reputations, both refrained from pursuing the implications of negating the fifth postulate, correctly assessing that the world was not ready for it.

To exacerbate the radicalism of his approach in a highly religious country, Lobachevsky, though not an atheist, was a materialist in a most fundamental and original sense of the word. In his mathematical syllabus for 1822, he made the extraordinary statement:

> We apprehend in Nature only bodies alone; consequently, concepts of lines and planes are derived and not directly acquired concepts, and therefore should not be taken as the basis of mathematical science.

In 1823, Lobachevsky's full-length geometry textbook *Geometriya* was submitted to school district curator Magnitskii, who sent it to the St. Petersburg Academy of Sciences for review. The text was emphatically rejected, and Lobachevsky's difficulties with the academy began. A subsequent manuscript, "O nachalakh geometrii" (1829-1830; on the elements of geometry), was also submitted to the academy. Not only did the academy reject the manuscript but also an academician's flawed critique was fed to the popular press, which turned it into a lampoon of Lobachevsky.

The date February 7, 1826, marks the official debut of Lobachevskian geometry as an independent theory. On that day, Lobachevsky submitted to his department his paper entitled "Exposition succincte des principes de la géométrie avec une démonstration rigoureuse du théorème des parallèles." It was rejected for publication, as his colleagues ventured no opinion on it. Other major works continued to be largely ignored.

Nevertheless, the new school district curator who replaced Magnitskii, Count Mikhail Musin-Pushkin, was sufficiently impressed with Lobachevsky to make him rector of Kazan University in 1827. Thus began Lobachevsky's dual life as a brilliantly successful local administrator and a frustrated intellectual pioneer kept outside the pale of the St. Petersburg establishment. During his tenure as rector, Lobachevsky built Kazan University into an outstanding institution of high standards. He founded the scientific journal *Uchenye zapiski*, in which he published many of his works and that has flourished to the present day.

In 1846, Lobachevsky's life in the sphere of action fell apart, as he received a succession of blows: Musin-Pushkin was transferred to the St. Petersburg school district; the request to forestall Lobachevsky's mandatory retirement was denied; his eldest son, Aleksei, died of tuberculosis at the age of nineteen; his wife became seriously ill; his wife's half brother, dispatched to handle the sale of two distant estates, gambled away both the Lobachevskys' money and all of his own; and Lobachevsky's eyesight began to deteriorate. In the last year of his life, he was virtually blind, yet he dictated his best and strongest work, *Pangéométrie* (1855-1856). His views had evolved from rejection of Euclidean parallelism into a vision of reality that anticipated theories of the curvature of space and was validated by Albert Einstein's general theory of relativity.

Significance

When Nikolay Ivanovich Lobachevsky's ideas first caught the imagination of a wide audience during the late nineteenth century, he was dubbed "the Copernicus of geometry," partly because of his Slavic origin (Nicolaus Copernicus was Polish), but far more because of the profound reorientation of thought that he set in motion. Lobachevsky forced the scale of earthly dimension as the measure of the universe off its pedestal, as Copernicus had earlier shattered the illusory status of Earth as the center of the solar system. This upheaval, which initially met with great resistance, forced the mind to focus on awesome phenomena that were not so much abstract as invisible to the human eye. Lobachevsky promoted bold and fruitful speculation about the nature of reality and space.

Pangéométrie, Lobachevsky's crowning work, opens: "Instead of beginning geometry with the line and plane, as is usually done, I have preferred to begin with the sphere and the circle." For this geometry, there are no straight lines or flat planes, and all lines and planes must curve, however infinitesimally. Yet, while pointing to modern concepts of the curvature of space, Lobachevsky does not abolish Euclidean geometry. In some areas, his geometry and Euclid's coincide. However, the latter is a limited case, whose relative certainties hold true on a merely earthly scale. In the conclusion to *Pangéométrie*, Lobachevsky correctly predicted that interstellar space would be the proving ground for his theory, which he saw not as an abstruse logical exercise but as the real geometry of the universe.

— D. Gosselin Nakeeb

Further Reading

Bell, E. T. *Men of Mathematics*. New York: Simon & Schuster, 1937. Reprint. 1986. Chapter 16, "The Copernicus of Geometry," focuses on Lobachevsky's life and mathematical discoveries.

Bonola, Roberto. *Non-Euclidean Geometry: A Critical and Historical Study of Its Development*. Translated by H. S. Carslaw. Mineola, N.Y.: Dover, 1955. Up-to-date for its time, and still relevant to the general reader. Focuses on a basic exposition of Lobachevsky's theories without the highly sophisticated applications thereof. Contains several relevant appendixes.

Greenberg, Marvin Jay. *Euclidean and Non-Euclidean Geometries: Development and History*. 3d ed. New York: W. H. Freeman, 1993. Includes information about Lobachevsky's discovery of the hyperbolic geometric.

Kagan, Veniamin Fedorovich. *N. Lobachevsky and His Contributions to Science*. Moscow: Foreign Language Press, 1957. A solid, basic account by one of the chief Russian experts on Lobachevsky. Omits most of the human-interest material to be found in Kagan's 1944 biography. Includes a bibliography, necessarily of primarily Russian materials.

Kulczycki, Stefan. *Non-Euclidean Geometry*. Translated by Stanisuaw Knapowski. Elmsford, N.Y.: Pergamon Press, 1961. Another introduction for the general reader, which updates but does not supplant Bonola.

Mlodinow, Leonard. *Euclid's Window: The Story of Geometry from Parallel Lines to Hyperspace*. New York: Free Press, 2001. Lobachevsky's contributions to geometry are briefly discussed in this book focusing on the work of five other mathematicians.

Shirokov, Pëtr Alekseevich. *A Sketch of the Fundamentals of Lobachevskian Geometry*. Edited by I. N. Bronshtein. Translated by Leo F. Boron and Ward D. Bouwsma. Groningen, Netherlands: P. Noordhoff, 1964. Written in Russian during the 1940's, it appears to have been aimed at the secondary-school mathematics student.

Smogorzhevsky, A. S. *Lobachevskian Geometry*. Translated by V. Kisin. Moscow: Mir, 1982. Partly accessible to the general reader but of particular interest to the serious student of mathematics. Emphasizes specific mathematical applications of Lobachevsky's theories.

Vucinich, Alexander. "Nikolay Ivanovich Lobachevsky: The Man Behind the First Non-Euclidean Geometry." *Isis* 53 (December, 1962). A substantial, well-written article, abundantly annotated to point the reader in the direction of all the basic sources, which are primarily in Russian. Highlights some avenues not mentioned elsewhere, such as Lobachevsky's role in the mathematization of science. Includes a balanced account of Lobachevsky's life.

Colin Maclaurin
Scottish mathematician

Maclaurin, the greatest British mathematician of the eighteenth century, developed and extended Sir Isaac Newton's work in fluxions (calculus) and gravitation and made important new discoveries in geometry and mathematical analysis.

Born: February, 1698; Kilmodan, Argyllshire, Scotland
Died: January 14, 1746; Edinburgh, Scotland

Early Life

Colin Maclaurin was born in the western Scottish county of Argyll. He was the youngest of the three sons of John Maclaurin, a learned minister of Kilmodan parish. John, the eldest son, followed in his father's footsteps and became a minister. Daniel, the second son, manifested signs of extraordinary genius but died young. Colin, too, was a child prodigy, but he never knew his father, who died when Colin was six weeks old. Further tragedy struck when he was nine years old, with the death of his mother. His father's brother, also a minister, became guardian of the children.

In 1709, at the early age of eleven, Colin entered the University of Glasgow, where he studied theology for a year. During this time, he became friendly with Robert Simson, a professor of mathematics, from whom he acquired a passionate interest in Euclid's geometry and in other ancient Greek mathematics. He also became interested in Sir Isaac Newton's work, and this led to his thesis "On the Power of Gravity," which he publicly defended in 1715 and for which he was awarded a master of arts degree. He remained at Glasgow another year to study theology, after which he returned to live with his uncle in their Highland home beside Loch Fyne.

He enjoyed wandering over the hills as well as reading mathematics, philosophy, and the classics. Some of his notebook entries that have survived from this time reveal his sensitivity to the beauties of nature, which he deeply believed manifested God's perfections. He abandoned this Highland life in 1717, when, following a competitive examination, he was appointed to the chair of mathematics at Marischal College, Aberdeen, even though he was only nineteen years old. This first appointment marked the start of his brilliant mathematical career.

Life's Work

Colin Maclaurin's accomplishments grew out of Newton's. He first met Newton in 1719 on a visit to London. Newton was favorably impressed by the young Scottish mathematician, and they became friends. Maclaurin had already contributed papers to the *Philosophical Transactions of the Royal Society*, and he was soon elected a fellow of this society of which Newton was president.

During this time, Maclaurin was working on a book about geometry, which was published, with Newton's approval, in 1720. The book, whose full title was *Geometria organica: Sive, Descriptio linearum curvarum universalis* (organic geometry, with the description of universal linear curves), contained new and elegant methods for generating conics (circle, ellipse, hyperbola, and parabola). Maclaurin also devised an elaborate treatment of higher plane curves that was superior to Newton's earlier results. Maclaurin proved many important theorems that could be found, without proof, in Newton's work. He also discovered many new theorems; for example, he showed that many curves of the second and higher degrees could be described by the intersection of two movable angles.

In 1722, Maclaurin left Scotland to serve as companion and tutor to the eldest son of Lord Polwarth, British plenipotentiary at Cambrai in northern France. Maclaurin and his young charge visited Paris for a short time and then resided for a much longer period at Lorraine, where Maclaurin wrote a paper on the impact of bodies, for which he was awarded a prize by the Academy of Sciences. The sudden death of his pupil caused him to return to Aberdeen, but, because of problems connected with his three-year absence from the university without leave, he was unable to resume his position. Newton again stepped in to help, and it was largely through his strong recommendation that Maclaurin became, in November, 1724, deputy professor for the elderly James Gregory at Edinburgh University. (This James Gregory was the nephew of the famous Scottish mathematician of the same name, who had been appointed to the first chair of mathematics at Edinburgh in 1674, a year before his tragic death at the age of thirty-six.) Newton even wrote privately to the lord provost of Edinburgh, offering to contribute £20 a year toward Maclaurin's salary.

A short time later, in 1725, the Edinburgh position in mathematics became available, and on the recommendation of Newton, Maclaurin took up the position that he would occupy for the rest of his life. His outstanding success at Edinburgh fully vindicated Newton's trust in him. He lectured on a wide range of topics, including the *Elements* of Euclid, spherical trigonometry, astronomy, and Newton's *Philosophiae naturalis principia mathematica* (1687; *The Mathematical Principles of Natural Philosophy*, 1729; best known as *Principia*). His classes were well attended, and his lectures as well as his writings were models of lucid and logical construction.

Edinburgh provided Maclaurin with the opportunity to develop and share his many talents. He was a skilled experimenter who constructed clever and useful mechanical devices. He made valuable astronomical observations, and he advocated building an observatory in Scotland. He also made actuarial tables for the budding insurance companies of Edinburgh. He took an active part in improving the maps of the Orkney and Shetland Islands, and he was even eager to make a voyage to find a northeast polar passage by way of Greenland to the southern seas.

Maclaurin's growing fame also gave him the opportunity to play an important role in Edinburgh society. For example, he was influential in persuading the members of the newly formed Edinburgh Society for Improving Medical Knowledge to enlarge its scope. The new organization, named the Philosophical Society, reflected this change, and Maclaurin became one of its secretaries. This society later became the Royal Society of Edinburgh.

In 1733, Maclaurin married Anne Stewart, the daughter of the solicitor general for Scotland. They had seven children, of whom two sons and three daughters survived him. An engraving of Maclaurin from the Edinburgh period depicts him as a stocky man with heavy eyebrows, a strong nose, and a weak chin. His mien is determined and serious, as befits a distinguished Scottish professor and disciple of Newton.

Throughout his Edinburgh career Maclaurin sought to silence the criticism of Newton's work on differential calculus (which he and Maclaurin called fluxions). The most influential of these critics was George Berkeley, bishop of Cloyne. In 1734, Berkeley published *The Analyst: Or, A Discourse Addressed to an Infidel Mathematician*, in which he attacked Newton's ideas on fluxions. The infidel mathematician was Edmond Halley, the great astronomer and religious skeptic, who had convinced one of Berkeley's dying friends to refuse the last rites because of the untenability of Christian doctrines. Berkeley denied neither the utility of fluxions nor the validity of their results. He did confess, however, to confusion about the meaning of fluxions. According to their defenders, fluxions were neither finitely large nor infinitely small, and yet they were not nothing. Berkeley concluded acidly that they were the "ghosts of departed quantities."

Berkeley's criticism stung. Maclaurin felt that a reply was necessary. He thought that Berkeley had misrepresented the method of fluxions by depicting it as full of mysteries and based on false reasoning. Since fluxions were opaque to someone of Berkeley's intelligence, Maclaurin believed that Newton's method needed new and more vigorous arguments to support it.

A Treatise of Fluxions, Maclaurin's greatest work, was published in 1742. The book was an attempt to establish fluxions on as sound a basis as Greek geometry. Maclaurin began, like Newton, with the concepts of space, time, and motion, and then he systematically elaborated Newton's version of the calculus. Since his readers were more familiar with velocity than with strictly mathematical variables, Maclaurin approached Berkeley's difficulties by considering motion. Maclaurin agreed with Berkeley that the infinitely small was inconceivable, but he did not see any objection to bringing into geometry the idea of an instantaneous velocity. Indeed, for Maclaurin, mathematics included both the properties of motion and the properties of figures. Using this background analysis of motion, he went on to show how fluxions were measured by the quantities they would generate if they were to continue moving uniformly.

Maclaurin's search for a rigorous foundation for fluxions was commendable but in the end unsuccessful. Nevertheless, his analysis did leave hints that future mathematicians would fruitfully follow. Furthermore, his book was not only a defense of Newton's methods; it was also an investigation into a variety of other problems in geometry and physics. For example, he developed a test for the convergence of an infinite series, and, for the first time, he gave the correct method for deciding between a maximum and a minimum of a function by investigating the sign of a higher derivative.

In *A Treatise of Fluxions*, Maclaurin also built on many of the physical principles enunciated by Newton in the *Principia*. For example, he analyzed the tides as a problem in applied geometry. His interest in this subject had begun in 1740 when he submitted an essay on the cause of the tides for a prize offered by the French Academy of Sciences. He shared the award with Daniel Bernoulli and Leonhard Euler, both of whom also based their work on a proposition in the *Principia*. In his account, Maclaurin showed that a homogeneous fluid revolving uniformly about an axis under the action of gravity assumes at equilibrium the shape of an ellipsoid of revolution.

Maclaurin's concerns were not always centered on mathematics and physics. In 1745, when Charles Edward Stuart, the Young Pretender, landed in Scotland and proclaimed that his father James was rightful king, Maclaurin took an active role in opposing him. When the Young Pretender and his army of Highlanders marched against Edinburgh, Maclaurin helped prepare trenches and barricades for the defense of the city. Despite these efforts, the Jacobite rebels captured the city, and Maclaurin, to avoid submitting to the pretender, was forced to flee to York. Here he found refuge with Archbishop Thomas Herring. When it became clear that the Jacobites were not going to occupy Edinburgh, Maclaurin returned to Scotland, but the energy he had expended in the trench warfare and in the flight to York sapped his strength and severely undermined his health. He died soon after his return to Edinburgh, in 1746, shortly before his forty-eighth birthday. A few hours before his death, he dictated some passages of a work he had been writing on Newton's philosophy, in which he affirmed his unwavering belief in a future life.

Significance

Colin Maclaurin was the ablest and most spirited of the defenders and developers of Newton's methods. He was a strong advocate of Newton's geometrical techniques, and his success in using them influenced other, less able British mathematicians to try to follow. He has been best remembered for his defense of Newton against Berkeley, but he also extended Newton's work; for example, he applied Newton's analysis of the gravitation of a sphere to the problem of ellipsoids.

Like Newton, Maclaurin loved geometry. One of the ironies of Maclaurin's work is that, though he emphasized geometry over analysis, his name is commemorated for the discoveries he made in analysis. Continental mathematicians emphasized analysis over geometry, and some of them—Euler, for example—rejected geometry completely as a basis for the calculus and tried to work solely with algebraic (analytic) functions. Maclaurin, on the other hand, adopted a geometric style in his book on fluxions because of certain logical difficulties that seemed to him to be insurmountable unless one were to use geometry.

Newton and Maclaurin showed the power of the geometric approach to solve many mathematical and physical problems, but this success also had harmful consequences, for it led Britons to follow these geometric methods and to neglect the analytic methods that were being pursued so successfully in the rest of Europe. As a result, most British mathematicians came to think that many problems could be solved without using the calculus. This had the effect of retarding the more powerful analytic methods in Great Britain. Thus, after Maclaurin, British mathematics suffered an eclipse because he and Newton had inadvertently helped steer it into unproductive paths.

Despite the ultimate infertility of Maclaurin's geometric style, he still had admirers, such as Joseph-Louis Lagrange, the great French mathematician. Lagrange was proud that his famous book on analytic mechanics contained not a single geometric diagram. Maclaurin's work, on the other hand, dealt largely in lines and figures, and he saw the great book of the universe written in this geometric language. Lagrange, who pictured the universe in terms of numbers and equations, nevertheless appreciated the insight and integrity of Maclaurin, whose work, Lagrange once said, surpassed that of Archimedes—a compliment that would have deeply pleased Maclaurin.

Robert J. Paradowski

Further Reading

Boyer, Carl B. *A History of Mathematics*. New York: John Wiley and Sons, 1968. Written by a historian of mathematics, this book is intended to be a basic textbook for students of mathematics. Also appropriate, however, for general readers.

_____. *The History of the Calculus and Its Conceptual Development*. Mineola, N.Y.: Dover, 1959. An unabridged reprint of the work published in 1949 under the title *The Concepts of the Calculus*. Boyer traces the development of calculus from antiquity to the twentieth century, and offers a good account of Maclaurin's role as defender of Newton's theory of fluxions.

Grabiner, Judith V. "Was Newton's Calculus a Dead End? The Continental Influence of Maclaurin's Treatise of Fluxions." *American Mathematical Monthly* 104, no. 5 (May, 1997): 393. Explains how Maclaurin's treatise helped to further the development of Newtonian calculus.

Guicciardini, Niccoló. *The Development of Newtonian Calculus in Britain, 1700-1800*. New York: Cambridge University Press, 1989. A comprehensive survey of the research and teaching of Newtonian calculus, including information on Maclaurin's ideas about fluxions.

Hedman, Bruce. "Colin Maclaurin's Quaint Word Problems." *College Mathematics Journal* 31, no. 4 (September, 2000): 286. Discusses Maclaurin's solutions to several algebra word problems, including the Ptolemaic riddle and wage, age, and rate problems.

Kline, Morris. *Mathematical Thought from Ancient to Modern Times*. New York: Oxford University Press, 1972. Covers major mathematical developments from ancient times through the first few decades of the twentieth century. Kline aims to present the chief ideas that have shaped the history of mathematics rather than focus on individual mathematicians. Consequently, his treatment of Maclaurin emphasizes the principal themes of his work rather than the events of his life.

Maclaurin, Colin. *An Account of Sir Isaac Newton's Philosophical Discoveries*. Sources of Science 74. Introduction by L. L. Laundan. New York: Johnson Reprint, 1968. A reprint of a 1748 work on Newton by Maclaurin.

_____. *The Collected Letters of Colin Maclaurin*. Edited by Stella Mills. Nantwich, England: Shiva, 1982. Maclaurin's letters provide details of his life and ideas.

Mooney, John, and Ian Stewart. "Colin Maclaurin and Glendaruel." *Mathematical Intelligencer* 16, no. 1 (Winter, 1994): 48. Recounts Maclaurin's life and career by focusing on his hometown, Glendaruel, site of a memorial in his honor.

Turnbull, Herbert Westren. *The Great Mathematicians*. New York: New York University Press, 1961. This brief but excellent book, a biographical history of mathematics, attempts to show how mathematicians use both imagination and reason to make discoveries.

BENOIT B. MANDELBROT
Polish-born French-American mathematician

Mandelbrot's research in applied mathematics led him to discover a pattern appearing across the disciplines: the duplication of a pattern or shape at any scale (scale invariance), which he called "self-similarity." He then identified the mathematics underlying that pattern and gave it the name fractal, thereby creating a new branch of research in pure and applied mathematics.

Born: November 20, 1924; Warsaw, Poland

EARLY LIFE
Benoit B. Mandelbrot (behn-WAH MAHN-dehl-broht) was born in Poland shortly after that country regained its independence after more than a century of Russian, Prussian, and Austro-Hungarian rule. His father was in the clothing business but descended from a line of Jewish scholars, while his mother had medical and dental training. The name Mandelbrot was distinctively Jewish, if only because of the popularity of the pastry bearing that name in the Jewish communities of Eastern Europe (Mandelbrot means "almond bread"). Mandelbrot attributed some of his later intellectual independence to having been schooled at home for his first couple of years, after which he attended elementary school in Warsaw. When he was eleven years old, his family moved to France, in part because of the increasingly anti-Semitic atmosphere in Poland and because of his uncle's established position in the French mathematical community.

After the Nazi conquest of France in 1940, life became uncertain for the Jewish population of the Nazi-occupied and the Vichy parts of the country. Mandelbrot wandered between 1942 and 1944, taking up various jobs and keeping out of sight. After the end of the Nazi regime, he was able to return to Paris for college, and he started off at the École Normale, the leading university for mathematics in the country. He left after a few days and went to the École Polytechnique, which did not have quite the standing of the École Normale. His reason for making the change was that he found the mathematics at the École Normale too abstract, while the mathematics at the École Polytechnique was more applied.

Abstraction in mathematics was part of a general development in the field, spearheaded by a contingent of French mathematicians known collectively under the pseudonym Nicolas Bourbaki. They pushed French mathematics in the direction of abstraction, claiming that German mathematics had taken a lead in research in this area. The change was anathema to the young Mandelbrot, who found that his imagination and predilection for geometry was much stronger than his fondness for axiomatic deductions. After getting a degree from the École Polytechnique, he proceeded to the California Institute of Technology (Caltech) in Pasadena, California, and earned a master's degree in 1948. However, because the person with whom he wanted to work was no longer at Caltech, he returned to Paris for his doctoral study.

The combination of topics in Mandelbrot's dissertation (from word frequencies to statistical thermodynamics) was indicative of his reluctance to confine his research within too narrow a discipline. After a number of positions of short duration, he took a position as a member of the research staff at International Business Machines in the United States in 1958 and stayed with that company, with only a short break, until 1993. He married Aliette Kagan in 1955, and the couple had two children.

LIFE'S WORK

One of the problems Mandelbrot addressed early in his career was the daily and long-term fluctuation in stock prices. Economists tried to predict the *motion* of stock prices, but price *fluctuations* always seemed to get in the way of these predictions. One solution to the problem was to see how much time had to be taken into account before the effects of the fluctuations could be disregarded as insignificant. What Mandelbrot discovered was that a pattern of fluctuations remained, regardless of the time period he chose to examine. This indicated that fluctuations could not be ignored. Indeed, any model attempting to predict the motion of prices would have to take into account fluctuations in stock prices.

Mandelbrot also considered issues connected with what is called statistical noise in the transmission of information. The problem is that the transmission of a signal (information) is always accompanied by some distortion, and the receiver has to distinguish the signal from the distortion. The receiver needs to have a guarantee that there exists enough signal to eliminate the distortion. Mandelbrot discovered that, just as with stock prices, there is no way to be sure of telling distortion from signal, no matter how large a sample is taken. This duplication of a pattern at any scale (scale invariance) he called "self-similarity," and its appearance in widely different areas of study suggested that it was worth considering in its own right.

Mathematical constructions with the property of self-similarity existed prior to Mandelbrot's time. These constructions included, for example, the Cantor set. Named for nineteenth century German mathematician Georg Cantor, the Cantor set involved taking the set of points on the number line between 0 and 1 and deleting the middle third (leaving two disconnected pieces). The middle third of each of those pieces was then deleted, and the process continued. The result served as an important example for pure mathematics, but before Mandelbrot, no one had suspected that phenomena in the real world could be modeled by such constructions.

Perhaps the best-known example of this sort of self-similarity arose from considering the issue of measuring the length of a coastline. Mandelbrot himself devoted an article to the coastline of Great Britain. If one imagined measuring the coastline with a ruler of length one mile, he wrote, the coastline could be given a certain length. When one used a smaller ruler, however, then more of the little curves and inlets would have to be counted and the coast would end up with a slightly longer length. The smaller the size of the ruler, the longer the coastline. It was impossible to tell from looking at a drawing of the coastline what scale was used to measure it.

During the nineteenth century and into the twentieth century, scientists speculated that the atoms of the physical universe had a structure similar to those of the solar system, with the nucleus corresponding to the sun and the electrons to the planets. With the advent of quantum mechanics, however, the similarity between solar system and atomic structure waned. One of the reasons for the popularity of Mandelbrot's notion of self-similarity was that it revived the pre-quantum theory picture of identical structure at different scales. In searching for a name to describe the kind of curve that had this property of self-similarity.

It was important to establish that fractals were not just a curiosity, and Mandelbrot proceeded to do so through lectures and writings. In 1973 he lectured at the Collège de France, tying together his thoughts on the various applications of the idea of a fractal. In particular, he devised a notion of fractal dimension corresponding to the idea of dimension for ordinary geometrical shapes. What was new for mathematics was that the dimension of a fractal could end up as a fraction rather than a whole number. The dimension could be calculated by seeing what happened to the length of a fractal curve when a ruler of different size was used to measure it. The exponent that resulted was the fractal dimension. Mandelbrot wrote up his lecture and published it as *Les Objets fractals* (1975; *Fractals*, 1977). "Fractal" derives from the Latin adjectival root *fractus*, the same root as in fraction, fracture, and fragment. Mandelbrot claims to have consulted a Latin textbook used by his son to coin the term "fractal." The beauty of the pictures in *Fractals* helped guarantee an audience beyond that of professional mathematicians.

Another discipline that was coming into its own during the 1970's was chaos theory, which dealt with the complicated patterns that arose in the solution of problems in nonlinear dynamics. What made chaos theory appealing was that problems could be expressed in terms of the iteration of simple processes. As with fractals, a variety of simple processes could still lead to the same kind of result. What was even more striking was that the curves that resulted from such iterations looked like fractals. Chaos theory could be used to explain where fractals came from, a problem Mandelbrot was not able to solve. The presence of fractals in chaos theory ensured that Mandelbrot's ideas would survive. With the success of fractal research programs, Mandelbrot received a chair in mathematics at Yale University. A volume of papers in his honor was published by the American Mathematical Society in 2004.

Significance

Mathematics has its share of objects that lack any obvious connection to the real world. There are also problems in the real world that seem to have solutions only through extensive calculation and are not easily generalized. What Mandelbrot did was to find fractals across many areas of application, and not just within pure mathematics. He considered concrete examples of scaling, which led to his discovery of fractals and their wider use.

The conjunction of fractals research with chaos theory provided an explanation for how these self-similar curves arose and predicted where else they might be applicable. Many without mathematical training, or even interest, have been fascinated by pictures of fractals. A colored fractal can be the most accurate way of representing a repeated process, and it is certainly the most beautiful. Mandelbrot's work helped bring the geometrical back into the world of abstract mathematics.

Thomas Drucker

Further Reading

Albers, Donald J., and G. L. Alexanderson, eds. *Mathematical People*. Boston: Birkhäuser, 1985. Includes an interview with Mandelbrot, who discusses and explains the origins of some of his theories.

Gleick, James. *Chaos: Making a New Science*. New York: Viking, 1987. A popular account of chaos theory, with an important chapter on Mandelbrot.

Mandelbrot, Benoit B. *The Fractal Geometry of Nature*. New York: W. H. Freeman, 1983. Beautiful illustrations accompany an enumeration of those instances in which fractals appear.

_____. *Fractals: Form, Chance, and Dimension*. San Francisco, Calif.: W. H. Freeman, 1977. The first English-language general-audience monograph on the history and concepts of fractional dimension.

_____. *Fractals and Multifractals: Noise, Turbulence,* and Galaxies. New York: Springer, 1990. Mandelbrot's most comprehensive and rigorous treatment of fractals.

Peitgen, Heinz-Otto, and Peter H. Richter. *The Beauty of Fractals*. New York: Springer, 1986. Places Mandelbrot in perspective rather than in the foreground, allowing the book's illustrations to speak for themselves.

Smith, Peter. *Explaining Chaos*. New York: Cambridge University Press, 1998. An excellent exploration of the philosophical background for fractals.

Stewart, Ian. *Does God Play Dice? The New Mathematics of Chaos*. Malden, Mass.: Blackwell, 2002. Extensive tribute to Mandelbrot that also explains how the mathematics of chaos developed.

_____, ed. *The Colours of Infinity: The Beauty and Power of Fractals*. Bath, England: Clear Books, 2004. Perhaps the most straightforward, nontechnical outline and discussion of what Mandelbrot came to call fractals. Highly recommended for all readers at all levels. Includes a stunning DVD of the British television documentary that helped to popularize the subject. DVD introduced by writer Arthur C. Clarke.

Gaspard Monge
French mathematician

Monge founded modern descriptive geometry and revitalized analytic geometry. An enthusiastic supporter of the French Revolution, he helped establish the metric system and the École Polytechnique, an important engineering school.

Born: May 10, 1746; Beaune, France
Died: July 28, 1818; Paris, France

Early Life

Gaspard Monge (gah-spahr mohnzh) was born on May 10, 1746, in Beaune, a small town 166 miles southeast of Paris. He was the eldest son of Jacques Monge, an itinerant peddler and knife-grinder, and Jeanne Rousseaux, a woman of humble Burgundian origin. Jacques deeply respected education and sent his three sons to the local school run by the Oratorian religious order and to their Collège de la Trinité in Lyons. Although all three brothers eventually made successful careers in mathematics, Gaspard was clearly the genius. He was the golden boy of the Oratorians, and he regularly won academic prizes and became, at the age of sixteen, a physics teacher at Lyons.

In the summer of 1764, during a vacation to Beaune, Monge used surveying instruments of his own invention and construction to make a detailed map of the town. A military officer who later saw the map was so impressed by the boy's ability that he recommended Monge to the commandant of the military school at Mézières. Created in 1748, the École Royale du Génie had become a prestigious institution for the training of officers, who derived mostly from the nobility. Upon his arrival at Mézières in 1765, Monge learned that he would not study with the officers but would be trained as a draftsman to do the routine work of military surveying. Within a year, however, he had an opportunity to show that his mathematical skills were vastly superior to those of the officers. He was assigned the problem of computing the best places to locate guns in a proposed fortress at Metz. At the time, the calculation of positions shielded from enemy firepower in intricate fortifications was a long and arduous arithmetic procedure, but Monge developed a geometric method that obtained results so quickly that the commandant was initially skeptical. Upon detailed inspection by skilled officers, the advantages of Monge's invention, which formed the basis of what later came to be known as descriptive geometry, became evident. In fact, Monge's method was so highly valued that it was preserved as a military secret for twenty-five years.

Life's Work

Monge spent the first fifteen years of his career at the military academy of Mézières, where he was *répétiteur* (assistant) to the professor of mathematics, then a teacher of mathematics, and finally a royal professor of mathematics and physics. Through his excellent teaching, he was able to improve the French engineering corps and influence several students who went on to brilliant military careers. His lectures on descriptive geometry (then called stereotomy) allowed him to develop his ideas about perspective, the properties of surfaces, and the theory of machines. Descriptive geometry is basically a way to represent three-dimensional figures on a plane. Albrecht Dürer, the German painter and engraver, had used the idea of orthogonal projection of the human figure on mutually perpendicular planes in the early sixteenth century, and in 1738, A. F. Frézier had suggested a method of representing solid objects on plane

Gaspard Monge. (Library of Congress)

diagrams, but Monge developed descriptive geometry into a special branch of mathematics. He systematized its principles, developed its basic theorems, and applied this knowledge to problems of military engineering, mechanical drawing, and architecture.

Documents from his Mézières period reveal that Monge did extensive research in several areas of mathematics. He wrote memoirs on various curves and studied their radii of curvature. He analyzed evolutes (the loci of centers of curvature for a given curve) and systematically applied the calculus, in particular partial differential equations, to his investigations of the curvature of surfaces. In 1775, he presented to the French Academy of Sciences in Paris a paper on a developable surface, that is, a surface that can be flattened on a plane without distortion, a subject of great interest to mapmakers.

During the middle 1770's, Monge's interests began to switch from mathematics to physics and chemistry. In physics, he helped develop the material theory of heat (he called the heat substance "caloric"). This theory was useful to physicists and chemists in the eighteenth century (it was replaced by the kinetic theory of heat in the nineteenth century). Working alone at Mézières and with Antoine-Laurent Lavoisier on his trips to Paris, Monge carried out experiments on the expansion, solution, and reaction of various gases. To enable him to better carry out his research in chemistry and physics, Monge established a well-equipped laboratory in the late 1770's at the École Royale du Génie.

In 1777, Monge married a twenty-year-old widow, Catherine (Huart) Horbon, for whose honor he had earlier tried to fight a duel with one of her rejected suitors. The couple had three daughters. Since she owned a forge, Catherine had an indirect influence on her husband's interest in metallurgy. Supervising its operation led Monge to study the smelting and properties of metals. His outstanding work in mathematics and the physical sciences led to his election to the Academy of Sciences in 1780. This honor forced him to divide his time between Paris and Mézières. During his stays in Paris, he taught hydraulics (the science and technology of fluids) at the Louvre, and during his time at Mézières he taught engineering to the military officers and prepared memoirs on physics, chemistry, and mathematics for presentation at the Academy of Sciences.

Monge's researches in chemistry consumed so much of his time that he arranged for a substitute to deliver many of his lectures at Mézières. In the summer of 1783, he carried out his famous experiments on the synthesis of water from its component elements. Monge mixed hydrogen and oxygen gases (then called inflammable air and dephlogisticated air) in a closed glass vessel and ignited the explosive reaction between the gases by an electric spark from a voltaic battery. He found that the weight of the pure water he obtained was very nearly equal to the weights of the two gases. These studies became part of the so-called water controversy over the first discoverer of the compound nature of water. Monge deserves credit for showing quantitatively that water is composed of two elemental gases, but he did not formally publish until 1786. Henry Cavendish, the English physicist who published his results in 1784, showed that water is produced when inflammable air is burned in dephlogisticated air, but he interpreted his experiments in terms of the confusing phlogiston theory correctly interpreted the reaction as the oxidation of hydrogen.

In the fall of 1783, Monge accepted yet another responsibility in Paris—the examiner of naval candidates. For a while, he tried to continue his professorship at Mézières along with this new position, but this proved impossible; in December, 1784, he resigned from the school where he had spent nearly twenty years of his life. His post as examiner also required him to make tours of inspection of naval schools outside Paris, and this enabled him to reform the teaching of science and technology in the provinces. His time in Paris was spent participating in the activities of the academy and conducting research in chemistry, physics, and mathematics. During the 1780's, he did important work on the composition of nitrous acid, the liquefaction of sulfur dioxide, the nature of different types of iron and steel, and the action of electricity on carbon dioxide gas. In these chemical researches, he interpreted his results through Lavoisier's new oxygen theory rather than the outdated phlogiston theory. In physics, he did research on the double refraction of Iceland spar (a transparent calcite); he also studied capillary action. In mathematics, he continued his work on curved surfaces and partial differential equations.

When the French Revolution began in 1789, Monge was one of its most ardent supporters. His humble birth and his negative experiences with aristocrats gave him firsthand knowledge of the poverty of the masses and the corruption of the *ancien régime*. In 1791, he served on the committee that established the metric system. In 1792, he became minister of the navy and played a significant role in organizing the defense of France against the counterrevolutionary armies. In 1793, he voted in favor of the death of King Louis XVI. After all this, he was still bitterly attacked for not being revolutionary enough. These attacks forced him to resign from his ministerial post. Nevertheless, he continued to support the republic, and as a member of the committee on arms and munitions he worked hard to improve the extraction and purification of saltpeter and the construction and operation of powder-works in Paris and in the provinces. He also became involved in establishing a new system of scientific and technical education. It was during his teaching at the short-lived École Normale that his work on descriptive geometry was finally published.

Monge was very much concerned about preserving the nation's cultural and intellectual heritage during this time of revolutionary turmoil. He was convinced of the value of a national school for training civil and military engineers. As an influential member of the commission of public works, he helped institute the École Polytechnique in 1794. Monge became an important administrator and popular teacher at this school. His textbook on analytic geometry, which appeared in 1795, was used in the course he taught on the application of algebra to geometry. In pursuing the correspondence between algebraic analysis and geometry, Monge recognized that families of surfaces could be described both geometrically and analytically. He founded a school of geometers at the École Polytechnique, who would exert a powerful influence on the development of mathematics in the nineteenth century.

The last stage of Monge's career began in 1796 and was dominated by his fascination with—some have called it his mesmerization by—Napoleon I. Monge had actually met Napoleon earlier, when he cordially welcomed the young artillery officer from Corsica to the military school at Mézières. Though Monge had forgotten this meeting, Napoleon remembered and called Monge to Italy as a member of the committee supervising the selection of the paintings, sculptures, and other valuables that the victorious army was to bring back to France. Although this looting disturbed Monge's conscience, he accepted it as a way to finance Napoleon's military campaigns. Monge's duties took him to many cities throughout Italy and gave him the opportunity to become Napoleon's confidant and friend.

In the fall of 1797, Monge returned to Paris to begin his new post as director of the École Polytechnique, but his stay was brief, for Napoleon called him back to Rome to conduct a political inquiry. Monge also participated in the creation of the Republic of Rome and in the preparations for Napoleon's Egyptian adventure. Monge arrived in Cairo on July 21, 1798, the day after Napoleon's victory at the Battle of the Pyramids.

Napoleon made Monge president of the Institut d'Égypte (Egyptian Institute), modeled on the Institut de France, the revolutionary organization intended to replace the royal academies. Monge was heavily involved in many of the projects of the Institut d'Égypte. He was also a companion of Napoleon on a trip to the Suez region, on his disastrous Syrian expedition, and on his secret voyage back to France in 1799.

Upon his return to Paris, Napoleon rewarded Monge for his services. These favors continued throughout the period of the consulate as well as during Napoleon's reign as emperor. Monge was given more powerful administrative responsibilities along with extensive land grants. Napoleon created Monge count of Péluse, an honor he accepted gratefully, although he had once voted to abolish all titles. Napoleon also named Monge a senator, and by accepting the position Monge became publicly and irrevocably tied to Napoleon. A representation of Monge at this time depicts him in a powdered wig, looking slightly uncomfortable in the trappings of nobility. His strong and stocky build seems awkwardly confined by the expensive clothes, but his piercing eyes radiate intelligence and confidence, showing him to be a man ready to meet any challenge.

During these Napoleonic years, Monge divided his time among his duties in the senate, at the École Polytechnique, and in the Academy of Sciences. At the École Polytechnique, Monge influenced many young French mathematicians in various kinds of geometry—synthetic, analytic, and infinitesimal. Monge's great contribution to synthetic geometry was his *Géométrie descriptive* (1798; *An Elementary Treatise on Descriptive Geometry*, 1851), a summation of his life's work in descriptive geometry and a book that proved useful not only to mathematicians but also to artists, architects, military engineers, carpenters, and stonecutters. In 1801, Monge published a book on analytic geometry that revealed how useful geometry could be for algebra, and vice versa. In the same year, he published an expanded version of his lectures on infinitesimal geometry, his favorite subject, in which he used ordinary and partial differential equations to study complex surfaces and solids.

Monge's health began to decline in 1809, when he stopped teaching at the École Polytechnique. His health worsened during the autumn of 1812, when Napoleon's army suffered great losses on its retreat from Moscow. Monge deliberately fled Paris and did not participate in the senate session of 1814 that dethroned Napoleon. Seeing Napoleon as the standard-bearer of the revolutionary ideals of liberty, equality, and fraternity, Monge refused to condemn him, and during the so-called Hundred Days in 1815, when Napoleon tried to recover his throne, Monge pledged his allegiance to the emperor, remaining loyal even after Napoleon's defeat at Waterloo and his abdication. When the Bourbons were restored to the French monarchy in 1815, Monge, who refused to modify his anti-Royalism, was deprived of all of his honors and positions, even his membership in the Academy of Sciences. The last years of his life were filled with further humiliations and greater physical sufferings. Following a stroke, he died on July 28, 1818. Many students at the École Polytechnique asked to attend his funeral, but the king refused permission. Although they observed the king's refusal, the next day the students marched en masse to the cemetery and laid a wreath on the grave of their beloved teacher.

Significance

Gaspard Monge's reputation derives from his work in geometry, and there is no doubt that he was responsible for the revival of interest in geometry that occurred in the late eighteenth and early nineteenth centuries. As a result of his inspiration, a golden age of modern geometry began, and his methods flourished first in France and later throughout Europe, blazing the way for such nineteenth century mathematicians as Carl Friedrich Gauss and Georg Friedrich Bernhard Riemann. Though Monge was not strictly the inventor of descriptive geometry, he was the first to elaborate its principles and methods and to detail its applications in mathematics and technology. He also made valuable contributions to analytic and infinitesimal (or differential) geometry.

Despite his reputation as a geometer, Monge's accomplishments were actually much broader. Besides his exceptional sense of spatial relations, he was also an insightful analyst who could transform geometric problems into algebraic relations. For Monge, geometry and analysis supported each other, and in every problem he emphasized the close connection between the mathematical and practical aspects. His treatment of partial differential equations has a geometrical flavor, and he believed that problems involving differential equations could be solved more readily when visualized geometrically. On the other hand, some problems involving complex surfaces led to interesting differential equations. Many historians of mathematics ascribe to Monge the revival of the alliance between algebra and geometry. René Descartes may have created analytic geometry, but it was Monge and his students who made it a vital field.

Monge was a Renaissance man in the Age of the Enlightenment. He possessed a broad combination of talents: He was a creative mathematician, an excellent chemist, and a talented physicist and engineer. Furthermore, he was an adroit politician, a capable administrator, and an inspiring teacher. His skill as a teacher can be seen in his distinguished pupils, some who continued on paths he had opened and others who created new paths. Charles Dupin applied Monge's methods to the theory of surfaces. Victor Poncelet, the most original of Monge's students, became the founder of projective geometry. Jean Hachette and Jean Baptiste Biot developed the analytic geometry of conics and quadrics.

Throughout his career, Monge was interested in the practical consequences of his work in science and mathematics. At Mézières and at the École Polytechnique, he was interested in the structure and functioning of machines. He took his work on such practical problems as windmill vane design as seriously as the highly abstract problems of differential geometry. He believed that technical progress helped to augment human happiness, and, since technical progress depended on the development of science and mathematics, he supported France's efforts to improve education in these basic fields. In his unified view, science freed the intellect with the truth about the world, and this was the only valid way to social progress.

Robert J. Paradowski

Further Reading

Bell, Eric T. *Men of Mathematics*. New York: Simon & Schuster, 1937. Bell, who spent most of his teaching career at the California Institute of Technology, was skilled in unraveling the mysteries of mathematics for the general reader. In this book, he uses the lives of the men who created modern mathematics to explain some of the most important ideas animating mathematics today. Monge and Joseph Fourier are treated together in the chapter "Friends of an Emperor."

Boyer, Carl B. *History of Analytic Geometry*. New York: Scripta Mathematica, 1956. Reprint. Princeton Junction, N.J.: Scholar's Bookshelf, 1988. An important study of the development of analytic geometry from ancient times to the nineteenth century. Boyer's approach is conceptual rather than biographical, but Monge's work on analytic geometry is extensively analyzed. Boyer's emphasis is on the history of mathematical ideas, and some knowledge of algebra and geometry is assumed.

_____. *A History of Mathematics*. New York: John Wiley & Sons, 1968. A textbook for college students at the junior or senior level. Though he assumes an understanding of calculus and analytic geometry, much of the material is accessible to readers with weaker mathematical backgrounds. Boyer analyzes both Monge's contributions to mathematics and his involvement in politics. Extensive chapter bibliographies and a good general bibliography.

Crabbs, Robert Alan. "Gaspard Monge and the Monge Point of the Tetrahedron." *Mathematics Magazine* 76, no. 3 (June, 2003): 193. Explains the Monge Point of the Tetrahedron and discusses some of Monge's other geometric concepts. Includes a biographical profile of Monge, summarizing his career and his contributions to the development of applied and higher mathematics.

Kemp, Martin. "Monge's Maths, Hummel's Highlights." *Nature* 395, no. 6703 (October 15, 1998): 649. A brief, scientific discussion of the isometric perspective and how Monge developed its geometric properties.

Kline, Morris. *Mathematical Thought from Ancient to Modern Times*. New York: Oxford University Press, 1972. Monge's contributions to descriptive geometry and partial differential equations are extensively discussed. Kline's book is aimed at professional and prospective mathematicians, and a knowledge of advanced mathematics is necessary to understand his analysis of Monge's contributions.

Partington, J. R. *A History of Chemistry*. Vol. 3. London: Macmillan, 1962. This volume, dealing with the seventeenth, eighteenth, and early nineteenth centuries, contains an excellent analysis of chemistry in France during the time of Monge's career. Discusses Monge's contributions to chemistry in depth, with extensive references to original documents, some of which are translated into English. Accessible to the general reader.

Taton, René, ed. *The Beginnings of Modern Science*. Translated by A. J. Pomerans. New York: Basic Books, 1964. This work, the third volume in Taton's History of Science series, covers the period from 1450 to 1800. Monge's contributions to geometry and chemistry are discussed in their historical contexts, but this book is best used as a reference rather than for narrative reading.

JOHN NAPIER

Scottish mathematician, inventor, and theologian

Working alone, without the benefit of earlier work and the encouragement of mentors, Napier invented logarithms, which revolutionized arithmetic calculation and was the greatest boon to experimental science produced during the Renaissance.

Born: 1550; Merchiston Castle, near Edinburgh, Scotland
Died: April 4, 1617; Merchiston Castle
Also known as: John Neper

EARLY LIFE
John Napier (NAY-pyuhr), eighth lord of Merchiston, was born at Merchiston Castle, the son of Sir Archibald Napier by his first wife, Janet Bothwell. He was born into a family notable for several famous soldiers at a time when religious controversy was rife in Scotland.

Little is known of his childhood, but when Napier was thirteen, his mother died. He was subsequently sent to St. Salvator's College, St. Andrews University, a school not noted for its quiet academic environment. Although Napier remained at St. Andrews for only one year, he developed two intense interests that were to continue for the remainder of his life: theology and arithmetic. Because of the nonacademic environment, the bishop of Orkney advised that young John could better pursue an academic career at schools on the Continent. Although no direct evidence remains to confirm this, it is highly probable that he followed this course.

As young Napier traveled through a Europe divided into warring factions by the Protestant Reformation, he became a strong adherent of the Calvinist movement then sweeping Scotland. He was to remain a fervent and uncompromising believer, active in Protestant politics throughout his life, much of which was spent embroiled in bitter religious dissension aggravated by the embarrassing political activities of his papist father-in-law, Sir James Chisholm.

By 1571, he had returned to Scotland. The following year, he married Elizabeth Stirling and occupied a castle at Gartnes. In 1579, his wife died, leaving two children. Subsequently, Napier married Agnes Chisholm and fathered ten additional children. On the death of his father in 1608, Napier moved into Merchiston Castle, where he remained for the rest of his life.

As a member of the Scottish landed aristocracy, he had the time and resources to pursue his many interests. These included theology, agricultural improvements, and military science. In the latter field, he anticipated inventions three centuries before they were actually fabricated, and he invented an artillery so powerfully destructive that he refused, in horror, to develop or even to publicize it. Napier also experimented with fertilizers for crops and invented a mechanical device to pump water out of coal pits.

LIFE'S WORK

Napier's first literary work, *A Plaine Discovery of the Whole Revelation of St. John*, published in 1593 after five years of toil, was the first important work of biblical interpretation written in Scotland. In the book's introduction, the Scottish king James VI (the future James I of England) is entreated to safeguard the Scottish Protestant church and to purge and punish the Roman Catholic nobility. The body of this bitterly anti-Catholic exposition, among other things, identifies the pope as the anti-Christ described in the biblical Book of Revelations. Although from the perspective of history this enterprise may appear to be little more than fruitless theological supposition, it established Napier's reputation as both scholar and theologian. His theological interpretations followed the Greek form of mathematical argument, a form of theological reasoning that would not become popular for several centuries.

Although Napier's public life during these tumultuous times has been amply documented, the development of his mathematical work, conducted alone and almost in secret, is more difficult to trace. It seems as though mathematics was for Napier a solitary pursuit of leisure, while his highly visible public life focused on anti-Catholic proclamations meant to keep Catholicism out of Scotland.

An early treatise concerned with arithmetic and algebra was apparently assembled during his first marriage but remained unpublished until 1839. About 1590, he set out to make arithmetic easier; the task required twenty years of labor, but he succeeded by inventing logarithms, a system that simplified the computation of products, quotients, and roots. His fanatical dedication to Calvinist Protestantism shows the same obsessive persistence that enabled him to finish the grueling task of producing a usable set of logarithmic tables.

Logarithms, or "logs," are the exponents of a stated number, the "base," and are used to represent powers (exponents) of the base. Consider, for example, the powers of 2: 2^1, 2^2, 2^3, 2^4, 2^5, and so on. These correspond to 2, 4, 8, 16, 32, and so on. The exponents 1, 2, 3, 4, and 5 are the logs of these numbers to the base 2. To multiply any two numbers in the series, it is necessary only to add the exponents (or logs) of the numbers and then find to the antilog of the result, which corresponds to the sum desired. Thus, to multiply 32 by 4, take the log of 32, which is 5, and add it to the log of 4, which is 2, to get 7. The antilog of 7 (that is, the number with a log of 7) is 128, the desired result. Division is performed by subtracting logs.

By extension, numbers not found in the above series can be used if a noninteger number can be found such that when 2 is raised to this power the desired number is produced. For example, since $2 = 2^1$ and $4 = 2^2$, it follows that $3 = 2^x$, where x must be a number greater than 1 but less than 2. In fact, x is approximately equal to 1.585. Since any number can be expressed to a good approximation as a power of 2, any arithmetic operations can be performed provided a table of powers of 2 is provided.

Although Napier did not use 2 as the base of his logarithms, the principle is the same. Whereas logs make arithmetic computation considerably easier, Napier set himself the grueling task of computing, by various mathematical means, a complete set of log tables, that is, sufficient powers of the base to generate a complete set of numbers, including decimal fractions. The calculation of the tables occupied Napier for almost twenty years. While not entirely error-free, the calculations were basically accurate, forming the foundation for all subsequent log tables.

In 1614, Napier published the description of his logarithms together with a set of log tables, several uses for them, and rules for the solution of both plane and spherical triangles using the tables. This work, titled *Mirifici Logarithmorum Canonis Descriptio* (*Description of a Marvelous Canon of Logarithms*, 1857), omitted any explanation of his methods of calculation. Although the common folk who were Napier's neighbors had always suspected him of being a warlock who delved into the black arts behind his thick castle walls, his miraculous technique of logarithms, presented unexpectedly without explanation or rationale, seemed like black magic even to the relatively sophisticated people who had the occasion to use them.

A later work, published posthumously in 1619, *Mirifici Logarithmorum Canonis Constructio* (*Construction of a Marvelous Canon of Logarithms*, 1889), provides the explanation of his calculations, an outline of the steps leading to his invention, and the properties of his logarithmic function.

Napier sent a copy of his 1614 work to Henry Briggs, a professor at Gresham College. Briggs had the idea of making the base of the log tables 10, an innovation of which Napier approved because it simplified calculations. In 1624, Briggs published his tables of common logs (base 10 logarithms), but he gave full credit to Napier for the original idea.

Napier also invested considerable time in deriving complicated equations and exponential forms of trigonometric functions, since these played such important roles in astronomical computations. By mathematical manipulation, he was able to reduce the requisite number of spherical trigonometry equations from ten to just two general statements.

Napier's tables of logarithms were greeted with great enthusiasm by astronomers, since they simplified computations and removed some of the drudgery from analyzing data. Johannes Kepler (1571-1630), who inherited several decades of extremely accurate data on planetary motions from the great Danish astronomer Tycho Brahe, used Napier's logarithms to simplify the analysis. The results of his work led to Kepler's three laws of planetary motion, the first correct and accurate statement of planetary motion. Later, Isaac Newton (1642-1727) used Kepler's laws in formulating his theory of gravity.

In 1617, Napier published the results of his work on a mechanical system to simplify arithmetic computation, *Rabdologiae, seu numerationis per virgulas libri duo* (*Study of Divining Rods: Or, Two Books of Numbering by Means of Rods*, 1667). This involved manipulating a set of small counting rods (later termed "Napier's Bones") to multiply and divide numbers. This device could be considered the precursor of the slide rule (a set of sliding logarithmic scales that enabled rapid multiplication and division), a device widely used by scientists and students until the latter half of the twentieth century. Last but not least, Napier standardized and popularized the system now universally used for decimal notation, in which a decimal point is used to separate the integer from the fractional part of a number.

Significance

The 1614 publication of Napier's canon of logarithms is one of those extraordinary and exceptional events in the history of science whereby a new invention of great importance appears, seemingly out of thin air, with no obvious precursors foreshadowing its creation. Napier's invention removed much of the drudgery from reducing scientific data, particularly for astronomers attempting to use accurate measurements to predict planetary motions. When Johann Kepler used Tycho Brahe's accurate data to deduce his laws of planetary motion, Napier's logarithms helped make the arduous task possible.

In the centuries following their invention, log tables grew more detailed and more accurate, culminating in 1964 with the publication of a table of logarithms accurate to 110 decimal places. Until the 1970's, when inexpensive hand-held calculators and personal computers rendered them obsolete, log tables formed an essential component of college-preparatory secondary education, and no reputable engineer would be without his slide rule, a portable version of the log tables.

As a titled landowner, Napier, lord of Merchiston, devoted considerable energy to agricultural products to improve his crops and cattle. He tinkered with inventions and was granted a patent for a hydraulic screw to pump water from coal pits, and he outlined plans for (but never constructed) four new weapons of war, including an artillery piece that was designed to kill anything within a one-mile radius. Napier's first literary work, an interpretation of the Book of Revelation, secured his reputation as a scholar and as a theologian, although outside Scotland this work no longer commands high regard.

George R. Plitnik

Further Reading

Burton, David M. *The History of Mathematics: An Introduction*. 5th ed. Boston: McGraw-Hill, 2003. Survey of important developments in math and the people behind those developments. This edition adds broader coverage of important mathematicians, including women in math. Includes illustrations, bibliographic references, and index.

Gladstone-Millar, Lynne. *John Napier: Logarithm John*. Edinburgh: National Museums of Scotland, 2003. Biography of Napier emphasizes his importance, not just to mathematics, but to astronomy as well. Includes illustrations and bibliographic references.

Hobson, E. W. *John Napier and the Invention of Logarithms*. Cambridge, England: Cambridge University Press, 1914. This lecture is the most useful of the various reconstructions of Napier's invention of logarithms. Highly recommended.

Knott, C. G., ed. *Napier Tercentenary Memorial Volume*. London: Dawson's of Pall Mall, 1966. A reprint of a 1915 original. Contains a set of articles detailing different aspects of Napier's accomplishments by experts in various fields of mathematics, as well as some considerable detail on the historical background to his work. Also included is a complete bibliography of books exhibited at the July, 1914, Napier Tercentenary Celebration.

McLeish, John. *Number*. New York: Fawcett Columbine, 1991. Chapter 12, "John Napier: The Rationalization of Arithmetic," details his work on logarithms and Napier's Bones. Included are examples detailing the construction and use of both these inventions.

Napier, John. *Napier's Mathematical Works*. Translated by William F. Hawkins. 3 vols. Auckland, New Zealand: University of Auckland Press, 1982. A translation from Latin of all of Napier's writings on mathematics, including those published posthumously. Volumes 2 and 3 include a commentary on Napier's work and how it fits into the history of mathematics.

Napier, Mark. *Memoirs of John Napier of Merchiston: His Lineage, Life, and Times*. Edinburgh: W. Blackwood, 1834. Written by a direct descendant of John Napier with access to the family's private papers, this carefully researched work provides the original source material from which most later books were derived.

Neal, Katherine. *From Discrete to Continuous: The Broadening of Number Concepts in Early Modern England*. Boston: Kluwer Academic, 2002. Places Napier's invention of logarithms in the context of other changes in number theory in early modern England. Discusses Napier alongside such contemporaries and successors as Isaac Barrow and John Wallis. Includes illustrations, bibliographic references, and index.

JOHN VON NEUMANN

Hungarian-born American mathematician

A brilliant mathematician who laid the mathematical foundations of modern physics and computer science, von Neumann affirmed the importance of autonomous scientific research during the anticommunist McCarthy era.

Born: December 28, 1903; Budapest, Hungary
Died: February 8, 1957; Washington, D.C.
Also known as: Johnny von Newmann; Jancsi von Neumann

EARLY LIFE

The eldest of three boys, John von Neumann (NOY-mahn) was born in Budapest, Hungary. In the United States, he came to be known universally as Johnny, perhaps because he was already known by the Hungarian name Jancsi. Von Neumann belonged to the group of Hungarian mathematicians and physicists including Eugene Wigner, Edward Teller, Leo Szilard, and Dennis Gabor, who have substantially contributed to twentieth century science. In addition to knowing one another, some of them even attended the same high school, Budapest Lutheran.

Von Neumann's father, Max von Neumann, was a successful banker who had been elevated to the nobility. The Hungarian honorific "Margittai" was later Germanized to "von." The family was Jewish and bilingual in Hungarian and German. When John entered the gymnasium at age ten, he came into contact with László Rátz, a teacher who perceived his mathematical talents and arranged with his father for special tutoring. Von Neumann worked concurrently on a degree in chemical engineering, awarded by the Eidgenössische Technische Hochschule of Zurich in 1926, and a doctorate in mathematics, which was awarded by the University of Budapest, also in 1926.

Although von Neumann then held positions at the University of Berlin and at the University of Hamburg, he also visited Göttingen, where there was an amazing group of physicists and mathematicians, including David Hilbert, Werner Heisenberg, Max Born, and Erwin Schrödinger. Visitors included Albert Einstein, Wolfgang Pauli, Linus Pauling, J. Robert Oppenheimer, and Norbert Wiener.

After coming to the United States as a visiting professor at Princeton University in 1930, von Neumann accepted a permanent position in 1931. In 1933, he was invited to join the permanent faculty of the Institute for Advanced Study, also located at Princeton University, and became the youngest faculty member at the institute. He married Marietta Kövesi in 1930; she was Roman Catholic, and he at least nominally became a Catholic during his first marriage. His daughter, Marina, was born in Princeton in 1935. The marriage ended in divorce in 1937. In 1938, he visited Hungary and married Klára Dán, who joined him in Princeton.

Von Neumann was one of the rare individuals of extraordinary scientific genius who was as engaging personally as he was brilliant mentally. Colleagues relate anecdotes concerning his foibles, but all with a touch of nostalgia because his charm as well as his intelligence endeared him to those who knew him best. Von Neumann was of medium size, slender as a young man, plump as he grew older. His colleagues teased him about dressing like a banker—perhaps because he was the son of a banker; he habitually wore three-piece suits with a neatly buttoned coat and a handkerchief in his pocket. Cheerful and gregarious, he was a great raconteur. Not at all athletic, he had to watch his appetite for rich gravies, sauces, and desserts. He drove erratically, regularly acquiring speeding tickets and wearing out approximately a car a year. According to his friends, he was not mechanical enough to change a tire on his car.

Because of his powers of concentration, von Neumann could appear absentminded. He would sometimes start out on a trip and then have to call home to find out with whom he had an appointment and where it was to be. He loved off-color limericks and repeated them at parties. Especially fond of children, he enjoyed their toys so much that friends would give him toys as gifts on special occasions. Von Neumann's associates described him as sociable, witty, and party-loving.

History intrigued von Neumann. He had systematically read and learned most of the names and facts in the twenty-one volumes of the *Cambridge Ancient History* (1923-1939). During World War II, his colleagues were amazed at how frequently his forecasts were borne out by later events. Once asked by his colleague Herman Goldstine to recite Charles Dickens's *A Tale of Two Cities* (1859), von Neumann continued for so long that it was clear that he was prepared to recite from memory the entire book, even though he had read it twenty years before. While he had a photographic memory of books he had read decades earlier, he was quite capable of forgetting what his luncheon menu had been.

Life's Work

Von Neumann's first group of mathematical papers involved presenting an axiomatic treatment of set theory. Related to this concern with set theory was the problem of the freedom of contradiction of mathematics. Bertrand Russell and Alfred North Whitehead, in *Principia Mathematica* (1910-1913), contended that all mathematics derives from logic and is without contradiction. In 1927, following Hilbert, who wanted to separate number from experiential logic, where seven is related to seven objects, von Neumann argued that all analysis could be proved to be without contradiction. Three years later, the German mathematician Kurt Gödel upset these theories by showing that "in any sufficiently powerful logical system, statements can be formulated which are neither provable nor unprovable within that system unless the system is logically inconsistent." Von Neumann was entirely comfortable with theoretical issues of this kind.

In a series of important papers, culminating in his book entitled *The Mathematical Foundations of Quantum Mechanics* (1944), von Neumann showed that two different theories, Schrödinger's wave mechanics and Heisenberg's matrix mechanics, are equivalent. His work on the mathematical foundations of quantum theory had brought him into the small, closely knit circle of theoretical physicists, and so he shared from the beginning an awareness of the technological possibilities of the energy that could be generated by nuclear fission.

The first self-sustaining nuclear chain reaction was produced by a group of physicists headed by Enrico Fermi in Chicago on December 2, 1942. It was not until spring of 1943 that physicists and mathematicians were summoned to Los Alamos, New Mexico, where the Manhattan Project, charged with developing an atomic bomb, had established a research laboratory. Von Neumann was already engaged in scientific defense work, particularly in connection with the motions of compressible gases, but he did not arrive at Los Alamos until the fall of 1943. His contributions to the work at Los Alamos were substantial. He assisted in the development of a method of implosion and a means of calculating the characteristics of nuclear explosions.

Prior to and during the Manhattan Project, von Neumann became interested in problems of turbulence, general dynamics of continua, and meteorological calculation. Because these problems took too long to calculate even with the assistance of desk calculators, he became convinced that progress in developing electronic computing machines was essential.

The issue of who deserves the credit and patents for inventing the computer is complex and hotly debated. According to his colleague Stanislaw Ulam, von Neumann's contribution to the development of the computer was that he formulated the methods of translating mathematical procedures into a language of instructions for a computing machine. He developed the idea of a universal set of circuits in the machine, "a flow diagram," and a "code," or fixed set of connections that could solve a great variety of problems. Prior to that, each problem required a special and different set of wiring to perform operations in a given sequence. When von Neumann was awarded the Fermi Prize of the U.S. Atomic Energy Commission (AEC), he was especially cited for his work on using electronic computing machines.

After World War II, von Neumann was increasingly called on to act as an adviser to the government. From the perspective of the peace movement of the 1960's and 1970's, the arms race between the United States and the Soviet Union seemed irresponsible and inhumane. In the 1940's and 1950's, the very vivid memories of World War II contributed to the intense anti-Soviet atmosphere of the Cold War. After the Soviet Union tested its own atomic bomb, von Neumann supported work on the hydrogen bomb to maintain U.S. ascendancy in the arms race. Unlike Teller and Lewis Strauss, one of the commissioners of the AEC, von Neumann insisted on limiting his role to that of a technical expert; he resisted pressures to join scientists who publicly supported banning nuclear tests; he also engaged in political lobbying in support of development of the hydrogen bomb.

In 1954, when Oppenheimer, formerly the director of the Manhattan Project, was attacked as a security risk and hearings concerning his loyalty were conducted, von Neumann expressed complete confidence in Oppenheimer's integrity and loyalty. He acknowledged that their views concerning the importance of developing the hydrogen bomb differed, but he unequivocally opposed the political harassment of scientists. In October, 1954, the president offered von Neumann a position on the AEC. According to Ulam, von Neumann was flattered and proud that he, even though foreign-born, would be entrusted with such responsibility, but he was concerned about the Oppenheimer affair. Convinced that work on the commission was of great national importance, he accepted the post. In a written statement prepared for the hearing before the Special Senate Committee on Atomic Energy, which took place on January 31, 1946, von Neumann stated that "science has outgrown the age of independence from society." Observing that the combination of politics and physics could render the earth uninhabitable, he commented that regulation is necessary in both spheres. Restricting himself as a scientist to the scientific, he supported health and safety measures and protection by government police power, but unequivocally asserted the importance of freedom of information. He wrote, "There must, however, be no restriction in principle on research in any part of science, and none in nuclear physics in particular, and absolutely no secrecy or possibility of classification of the results of fundamental research."

A proponent of armament during the Cold War, von Neumann incorrectly predicted a massive war between the United States and the Soviet Union immediately after World War II; still, he accurately foresaw that the Soviet Union would take control of Eastern European countries, including his native Hungary, and crush all opposition to communism.

In the summer of 1955, von Neumann slipped in a corridor and injured his left shoulder. Diagnosis of the injury revealed that he had bone cancer. During this illness, he surprised his Jewish colleagues by consulting a Catholic priest, but it is likely that he had received instruction in Catholicism at the time of his first marriage. Until the last, he continued to function as a member of the AEC; he also worked on a number of projects, including the texts for the honorary Silliman lectures to

John von Neumann. (Library of Congress)

be given at Yale. The painful nature of his illness prevented him from concentrating with his accustomed intensity. He was forced to leave his lectures, entitled *The Computer and the Brain*, unfinished, but they were published posthumously in 1958.

Significance

When von Neumann answered a questionnaire in 1954 distributed by the National Academy of Science, which asked him to name his three most important contributions to mathematics, he identified his work on the rigorous formulation of quantum theory as one of those three. His papers on this topic represent one-third of his total work. As his most important contributions, von Neumann selected his work on the mathematical foundations of quantum theory and ergodic theorems and his theory of operators. In making this selection, von Neumann may have been motivated by a keen desire to maintain the importance of mathematics on a conceptual level in solving the problems of the physical sciences. Von Neumann's interest in the development of the electronic computing machine was prompted in part by the need for swift answers to problems in mathematical physics and engineering. Later, he pioneered in the use of computing machines to assist in weather prediction. In addition to working on the mathematics of weather prediction, he believed that control over climate might one day be possible.

Von Neumann was essentially the creator of game theory, a new branch of mathematics. He later coauthored a treatise with Oskar Morgenstern, *Theory of Games and Economic Behavior* (1944), which attempts to schematize mathematically the economic exchange of goods and to solve problems concerning monopoly, oligopoly, and free competition.

In his theory of automata, an area of study in which von Neumann was a pioneer, he effectively demonstrated that in principle it is possible to build machines that can reproduce themselves. His posthumously published lectures, *The Computer and the Brain*, return to these problems and draw on ideas and terminology from mathematics, electrical engineering, and neurology to outline a theory of representation of logical propositions by electrical networks or nervous systems.

Jeanie R. Brink

Further Reading

Goldstine, Herman. *The Computer from Pascal to von Neumann*. Princeton, N.J.: Princeton University Press, 1972. A history of the development of computing technology, written by a very close and loyal friend of von Neumann.

Hargittai, István. *The Martians of Science: Five Physicists Who Changed the Twentieth Century*. New York: Oxford University Press, 2006. Biography of von Neumann, Theodore von Karman, Leo Szilard, Edward Teller, and Eugene Wigner, five scientists who grew up in Hungary, studied in Germany, and emigrated to the United States to flee the Nazis. Recounts the events of their lives and describes their important scientific discoveries.

Macrae, Norman. *John von Neumann: The Scientific Genius Who Pioneered the Modern Computer, Game Theory, Nuclear Deterrence, and Much More*. 2d ed. Providence, R.I.: American Mathematical Society, 2000. An accessible account of von Neumann's personality, life, and work.

Neumann, John von. *Collected Works*. Edited by A. H. Taub. 6 vols. Elmsford, N.Y.: Pergamon Press, 1961-1963. A standard edition of the works of von Neumann.

Shurkin, Joel. "John von Neumann." In *Engines of the Mind: A History of the Computer*. New York: W. W. Norton, 1984. A survey of the difficulty in ascertaining who deserves credit for the development of the modern computer, inaccurate concerning von Neumann's life. Opposes Herman Goldstine's view.

Stern, Nancy. "John von Neumann's Influence on Electronic Digital Computing: 1944-1946." *Annals of the History of Computers* 2 (1980): 349-361. Discussion of von Neumann's contribution to the development of the computer.

Ulam, Stanislaw. "John von Neumann." *Bulletin of the American Mathematical Society* 64, no. 3, pt. 2 (1958): 1-49. Biography and highly mathematical assessment of von Neumann's career.

Wigner, Eugene. *Symmetries and Reflections*. Cambridge, Mass.: MIT Press, 1970. Autobiography of Wigner, a friend and colleague of von Neumann.

Simon Newcomb

Canadian-born American astronomer

As superintendent of The American Ephemeris and Nautical Almanac and later as director of the U.S. Naval Observatory, Newcomb undertook a complete revision of the data used for calculating the positions of the planets, and his work led to the adoption of a new set of international standards for astronomical calculations.

Born: March 12, 1835; Wallace, Nova Scotia, Canada
Died: July 11, 1909; Washington, D.C.

Early Life

Simon Newcomb was taught by his father, an itinerant Nova Scotia schoolmaster, but otherwise had no formal education as a child. Nevertheless, he developed an intense interest in learning and felt keenly deprived by the intellectual poverty of the primitive rural area in which he grew up. He later described his childhood as "one of sadness," likened its primitive conditions to growing up at the time of the American Revolution, and titled the chapter of his autobiography on his childhood "The World of Cold and Darkness."

At the age of sixteen, Newcomb entered into apprenticeship with an herbalist but soon became disenchanted with the man's unscientific practices and left Canada. In Maine, he signed aboard a ship that sailed for Salem, Massachusetts, where he rejoined his father, who had moved there previously. The two then moved to Maryland, where Simon taught in rural schools for two years. In 1856, he became a private tutor near Washington, D.C., and took advantage of the location by frequently visiting libraries in the capital city. In that setting, Newcomb finally began to experience the intellectual stimulation that he had craved. During his spare time, he studied such subjects as religion, economics, and astronomy.

In 1857, Newcomb took a job at *The American Ephemeris and Nautical Almanac*, which was then located in Cambridge, Massachusetts, and also began studying mathematics and conducting research at Harvard University. One of his first noteworthy achievements was a study of the orbits of the asteroids. His findings showed that it was unlikely that the asteroids had formed from the breakup of a planet, as was then widely supposed. In 1860, he participated in one of the most ill-fated eclipse expeditions in history. The journey required travel by rail, stagecoach, and canoe into remote central Canada and was repeatedly delayed by storms and flooded rivers. Members of the expedition finally reached a site for viewing the eclipse after twenty-four hours of nonstop canoe travel, only to have the eclipse itself hidden from their view by clouds.

Life's Work

When the U.S. Civil War broke out in 1861, a number of officers at the U.S. Naval Observatory resigned in order to join fighting units, and Newcomb was appointed to fill one of the vacancies. In 1864, he and other observatory staff members were called to active duty for a short time when Confederate forces briefly threatened to attack Washington, D.C.

As Newcomb gained experience in astronomical calculations, it became apparent to him that the best available tables for predicting the positions of the moon were yielding unacceptably large errors. In 1870, he took advantage of an eclipse expedition to Gibraltar to visit most of the major observatories of Europe and to make an extended visit to Paris.

Newcomb's visit to Paris in May, 1871, came at a remarkably unfortunate time. The Franco-Prussian War had only recently ended, and Paris was under siege and occupied by a revolutionary government, the Commune, which would be violently suppressed not long after Newcomb's departure. Although Newcomb described Paris in a letter as a "slumbering volcano," he was nevertheless able to spend six weeks at the Paris Observatory without mishap. During that time, he was delighted to discover many good records of the moon's position dating back to 1675—three-quarters of a century earlier than previously known historical data. He termed this discovery the greatest find he ever made.

Precise predictions of the motions of the sun, moon, and planets are necessary for accurate navigation and timekeeping. When Newcomb assumed the post of superintendent of *The American Ephemeris and Nautical Almanac* in 1877, the almanacs then being issued by various nations differed significantly in the fundamental quantities they used in calculating planetary predictions. To improve the accuracy of the predictions, Newcomb and his assistants had to recalculate the positions of the stars and planets from original observations. They also had to detect, assess, and correct errors in existing almanacs. Of this work, Newcomb said:

One might almost say it involved repeating, in a space of ten or fifteen years, an important part of the world's work in astronomy for more than a century past.

The work of Newcomb and his collaborators set such a high standard that an international conference in 1896 agreed to use their data compiled as the basis for all astronomical calculations. Despite this recognition of his work, Newcomb once wrote that he had never been able to confine his attention to astronomy "with that exclusiveness which is commonly considered necessary to the highest success in any profession." This was a remarkable admission from a scientist who achieved the highest success in his profession.

Another of Newcomb's interests quite separate from astronomy was economics. He first became recognized as an economist with a short book warning of the dangers of debased currency during the Civil War. In 1886, he published *Principles of Political Economy*, a 550-page book in which he approached economics in a way that might be expected from an astronomer: by stating simple, fundamental laws and then developing the consequences of those laws in greater and greater detail. However, Newcomb was quite aware of the limitations of economics as a science and its weaknesses in deciding policy.

One of Newcomb's less successful scientific efforts was his analysis of the possibility of powered heavier-than-air flight. He reasoned that the ability of a craft to fly depended on the area of its wings and therefore on the square of its dimensions, but concluded, incorrectly, that the aircraft's weight would increase as the cube of its dimensions, and therefore any machine large enough to support a human being would be impossibly heavy. What Newcomb failed to anticipate was that future construction methods would employ materials that maintained the necessary strength but reduce weight. More important, he failed to take into account the importance of wing cross-section in generating lift. Newcomb published an analysis of the problem in a news magazine, *The Independent*, in October, 1903. Less than two months later, two bicycle builders, Wilbur and Orville Wright, successfully flew a heavier-than-air craft at Kitty Hawk, North Carolina, after making careful studies of lift and wing design.

In 1908, Newcomb was diagnosed with cancer of the bladder. He died on July 11, 1909, in Washington, D.C.

The Possibility of Life on the Moon

The moon being much the nearest to us of all the heavenly bodies, we can pronounce more definitely in its case than in any other. We know that neither air nor water exists on the moon in quantities sufficient to be perceived by the most delicate tests at our command. It is certain that the moon's atmosphere, if any exists, is less than the thousandth part of the density of that around us. The vacuum is greater than any ordinary air-pump is capable of producing. We can hardly suppose that so small a quantity of air could be of any benefit whatever in sustaining life; an animal that could get along on so little could get along on none at all.

But the proof of the absence of life is yet stronger when we consider the results of actual telescopic observation. An object such as an ordinary city block could be detected on the moon. If anything like vegetation were present on its surface, we should see the changes which it would undergo in the course of a month, during one portion of which it would be exposed to the rays of the unclouded sun, and during another to the intense cold of space. If men built cities, or even separate buildings the size of the larger ones on our earth, we might see some signs of them.

In recent times we not only observe the moon with the telescope, but get still more definite information by photography. The whole visible surface has been repeatedly photographed under the best conditions. But no change has been established beyond question, nor does the photograph show the slightest difference of structure or shade which could be attributed to cities or other works of man. To all appearances the whole surface of our satellite is as completely devoid of life as the lava newly thrown from Vesuvius. We next pass to the planets. Mercury, the nearest to the sun, is in a position very unfavorable for observation from the earth, because when nearest to us it is between us and the sun, so that its dark hemisphere is presented to us. Nothing satisfactory has yet been made out as to its condition. We cannot say with certainty whether it has an atmosphere or not. What seems very probable is that the temperature on its surface is higher than any of our earthly animals could sustain. But this proves nothing.

Source: Simon Newcomb, *Side-Lights on Astronomy and Kindred Fields of Popular Science: Essays and Addresses* (New York: Harper & Brothers, 1906).

Significance

The awe that Newcomb felt as a young man when he finally came into contact with a world of learning never entirely left him. His autobiography suggests a man interested in almost everything he encountered. He wrote about historically obscure assistants with the same personal attention that he applied to the most illustrious scientists of his day. He spent far more effort describing the events around him and the people he met than he did on the technical details of his own work.

Precise projections of the motions of astronomical bodies are necessary for accurate navigation. After Newcomb assumed the post of superintendent of *The American Ephemeris and Nautical Almanac*, he developed measurements that served as the international standard for accuracy until 1984. His measurements were superseded only after the development of radar-ranging of planets and spacecraft, electronic computers, and improved mathematical techniques made it possible to achieve even greater accuracy. Nevertheless, in the twenty-first century, numerous astronomical computer programs were still using simplified versions of Newcomb's original methods.

Newcomb received a large number of scientific honors during his lifetime. In addition, Cape Newcomb, Greenland, was named after him, as was the World War II naval surveying ship the USS *Simon Newcomb*. Craters on the Moon and Mars bear his name, as does an asteroid named Newcombia. In 1978, the Royal Astronomical Society of Canada founded the Simon Newcomb Award to honor achievements in astronomical writing. When the the Astronomical Society of the Pacific awarded him a medal, it stated that Newcomb "has done more than any other American since [Benjamin] Franklin to make American science respected and honoured throughout the entire world."

Steven I. Dutch

Further Reading

Carter, Bill, and Merri Sue Carter. *Latitude: How American Astronomers Solved the Mystery of Variation*. Annapolis, Md.: Naval Institute Press, 2002. Careful measurements of the positions of stars show that the earth does not rotate smoothly on its axis but wobbles slightly. Newcomb's precise determination of star positions were central to the discovery and understanding of this phenomenon.

Dick, Steven J. *Sky and Ocean Joined: The U.S. Naval Observatory, 1830-2000*. Cambridge, England: Cambridge University Press, 2003. Newcomb's directorship of the U.S. Naval Observatory propelled the institution into the front ranks of world astronomical centers and occupies a prominent place in this history of the observatory.

Moyer, Albert E. *A Scientist's Voice in American Culture: Simon Newcomb and the Rhetoric of Scientific Method*. Berkeley: University of California Press, 1992. Biography, focusing on Newcomb's advocacy of the scientific method, and how his position spurred support for science and raised numerous social, culture, and intellectual issues.

Newcomb, Simon. *Principles of Political Economy*. New York: Augustus M. Kelley, 1966. A reprint of Newcomb's book of 1886. Newcomb the economist was strikingly like Newcomb the astronomer, viewing the world as the rational product of predictable forces. However, Newcomb was not a blind believer in economic forces, and he also stressed the limitations of economics in making societal decisions.

_____. *The Reminiscences of an Astronomer*. Boston: Houghton Mifflin, 1903. Newcomb's rambling autobiography is written in somewhat stuffy Victorian prose but beneath the formality, Newcomb's interest in people and events both great and small shows through.

Omar Khayyám

Persian mathematician and poet

Khayyám was a leading medieval mathematician and the author of Persian quatrains made famous through the English poet Edward FitzGerald's 1859 study, The Rubáiyát of Omar Khayyám.

Born: May 18, 1048?; Nishapur, Persia (now in Iran)
Died: December 4, 1123?; Nishapur

Early Life

Omar Khayyám (OH-mahr ki-YAHM) was born in all likelihood in Nishapur, then a major city in the northeastern corner of Iran. At his birth, a new Turkish dynasty from Central Asia called the Seljuks was in the process of establishing control over the whole Iranian plateau. In 1055, when their leader, Toghrïl Beg, entered Baghdad, the Seljuks became masters of the Muslim caliphate and empire. Of Omar's family and education, few specifics are known. His given name indicates that he was a Sunni Muslim, for his namesake was the famous second caliph under whose reign (634-644) the dramatic Islamic expansion throughout the Middle East and beyond had begun. The name Khayyám means "tentmaker," possibly designating the occupation of his forebears. Omar received a good education, including study of Arabic, the Qurʾān, the various religious sciences, mathematics, astronomy, astrology, and literature.

At Toghrïl Beg's death, his nephew Alp Arslan succeeded to the Seljuk throne, in part through the machinations of Niẓām al-Mulk (1019-1092), also from Nishapur, who was to serve the Seljuks for more than thirty years as a vizier (government administrator). Alp Arslan (r. 1063 to 1072 or 1073) was succeeded by his son Malik-Shāh, who ruled to 1092.

During this period of rule, Khayyám studied first in Nishapur, then in Balkh, a major eastern city in what is now Afghanistan. From there, he went farther northeast to Samarqand (now in Uzbekistan). There, under the patronage of the chief local magistrate, he wrote a treatise in Arabic on algebra, classifying types of cubic equations and presenting systematic solutions to them. Recognized by historians of science and mathematics as a significant study, it is the most important of Khayyám's extant works (which comprise about ten short treatises). None of them, however, offers glimpses into Khayyám's personality, except to affirm his importance as a mathematician and astronomer whose published views were politically and religiously orthodox.

From Samarqand, Khayyám proceeded to Bukhara and was probably still in the royal court there when peace was concluded between the Qarakhanids and the Seljuks in 1073 or 1074. At this time, he probably entered the service of Malik-Shāh, who had become Seljuk sultan in 1072.

Life's Work

Two of Malik-Shāh's projects on which Khayyám presumably worked were the construction of an astronomy observatory in the Seljuk capital at Eṣfahān in 1074 and the reform of the Persian solar calendar. Called *maleki* after the monarch, the new calendar proved more accurate than the Gregorian system centuries later.

Khayyám was one of Malik-Shāh's favorite courtiers, but after the latter's death Khayyám apparently never again held important positions under subsequent Seljuk rulers. In the mid-1090's, he made the *hajj* (pilgrimage) to Mecca and then returned to private life and teaching in Nishapur. It is known that Khayyám was in Balkh in 1112 or 1113. Several years later, he was in Marv, where a Seljuk ruler had summoned him to forecast the weather for a hunting expedition. After 1118, the year of Sanjar's accession, no record exists of any work by Khayyám. He died in his early eighties.

Some of the meager information available today regarding Khayyám was recorded by an acquaintance called Nizāmī ʿArūzī (fl. 1110-1161) in a book called *Chahár Maqála* (c. 1155; English translation, 1899). Nizāmī tells of visiting Khayyám's grave site in 1135 or 1136. Surprisingly, given Khayyám's reputation as a poet, the anecdotes regarding him appear in Nizāmī's "Third Discourse: On Astrologers," and no mention of him is made in the "Second Discourse: On Poets." In other words, though in the West Omar Khayyám is known for his poetry, no evidence in Persian suggests that he was a professional court poet or that he ever was more involved with poetry than through the occasional, perhaps extemporaneous, composition of quatrains (*rubai* or *robai*, plural *rubáiyát*). Because the quatrains first attributed to Khayyám are thematically of a piece and are distinct from panegyric, love, and Sufi quatrains, they can be usefully designated as "Khayyamic" even if authorship of many individual quatrains is impossible to determine definitively.

The following three quatrains are among the most typical and earliest to be attributed to the historical figure of Omar Khayyám:

> There was a drop of water, it merged with the sea.
> There was a speck of dirt, it merged with the earth.
> Your coming into the world is what?
> A fly appearing and disappearing.

> Drink wine: the universe means your demise,
> intends the death of your pure life and mine.
> Be seated on the grass and drink bright wine,
> for here will blooms bloom from your dust and mine.

> This ancient caravanserai called the world,
> home of the multicolored steed of night and day,
> is where a hundred Jamshids feasted and
> a hundred Bahrams ruled in splendor, and left.

In the centuries following Khayyám's death, increasing numbers of quatrains attributed to him appeared in manuscripts. Several of these manuscripts came to the attention of Edward FitzGerald (1809-1883), a serious student of Persian, who found them particularly appealing. His study of them inspired him to compose *The Rubáiyát of Omar Khayyám*, the first edition of which consisted of 75 quatrains and appeared in 1859. A second edition, expanded to 110 quatrains, appeared in 1868. The third edition in 1872 and the fourth in 1879 contained 101 quatrains, and the latter is the standard text. By FitzGerald's death, his work had begun to receive favorable critical attention, but its extraordinary fame, making it the single most popular poem of the Victorian Age, did not commence until later. A comparison of *The Rubáiyát of Omar Khayyám* with the Khayyamic Persian quatrains that FitzGerald had read and studied reveals that the themes, tone, and imagery of his poem are very close to those in the Persian quatrains, but that FitzGerald's poem is not a translation in any sense. It was the worldwide popularity of *The Rubáiyát of Omar Khayyám* that drew scholarly attention in Iran to Khayyám as a poet, so that he now is recognized as a leading figure in the Persian literary pantheon, along with Firdusi (between 932 and 941-between 1020 and 1025), Jalāl al-Dīn Rūmī (1207-1273), Saʿdi (1200-1291), and Hafiz (c. 1320-1389 or 1390).

Significance

The Persian quatrains attributed to Omar Khayyám express the point of view of a rationalist intellectual who sees no reason to believe in a human soul or an afterlife (as in the first quatrain quoted above). The speaker would like to live a springtime garden life, but his continuing awareness of his own mortality and his inability to find answers in either science or religion lead him to a modified *carpe diem* stance: In this far-from-perfect world, in which human beings do not have a decent chance at happiness, one should nevertheless endeavor to make the best of things (as in the second quatrain quoted above). Some slight consolation is offered in appreciating the fact that human beings have faced this situation from the beginning of time (as in the third quatrain quoted above).

In the orthodox Seljuk age, Khayyamic quatrains constituted a bold, individualistic voicing of skepticism. Because literary Iranians throughout history have admired individualists and free spirits, Omar Khayyám has been mythologized into a figure quite different from what the known facts about his biography imply. For example, he was a hero and inspiration to Sadegh Hedayat (1903-1951), Iran's most acclaimed twentieth century author, in whose novel *Buf-i kur* (1936; *The Blind Owl*, 1957) are palpable Khayyamic echoes.

Regardless of the historical facts, the view of Hedayat and many others is that Khayyám bucked the tide of religious orthodoxy and dared to say what many secular-minded people believe: that religion, science, and government fail to give an adequate explanation of the mystery of the individual lives of human beings.

Michael Craig Hillmann

Further Reading

Bloom, Harold, ed. Edward *FitzGerald's "The Rubáiyát of Omar Khayyám."* Philadelphia: Chelsea House, 2004. Presents an introduction to FitzGerald's infamous study and chapters that consider the "fin de siècle cult" of FitzGerald's work, comparisons with poets such as Tennyson, "forgetting" Fitzgerald's study, and more. Bibliography, index.

Boyle, J. A. "Omar Khayyám: Astronomer, Mathematician, and Poet." In *The Cambridge History of Iran*, edited by R. N. Frye. Vol. 4. Cambridge, England: Cambridge University Press, 1975. A succinct and careful review of the known facts about Khayyám's life, concluding with a brief review of the dispute over Khayyám's attitude toward Sufism, with which he presumably had little affinity.

Dashti, Ali. *In Search of Omar Khayyám*. Translated by L. P. Elwell-Sutton. London: Allen and Unwin, 1971. A very reliable study of Khayyám, which includes a review of his age and the known facts of his life, a collection of seventy-five quatrains that the author argues can be attributed with some confidence to Khayyám, and a sympathetic and sensitive identification of themes in the poems.

FitzGerald, Edward. *The Rubáiyát of Omar Khayyám*. 4th ed. London: Bernard Quaritch, 1879. This is the last edition the author saw to press and thus the official, final version of the poem.

Gray, Erik. "Forgetting FitzGerald's Rubáiyát." *Studies in English Literature, 1500-1900* 41, no. 4 (Autumn, 2001). Argues that the notion of "forgetting" or remembering "imperfectly" marks FitzGerald's poetic study as an important text in the context of Victorian poetry and in continuing literary work.

Heron-Allen, Edward. *Edward FitzGerald's "Rubáiyát of Omar Khayyám" with Their Original Persian Sources*. Boston: L. C. Page, 1899. A study of FitzGerald's stanzas paralleled with the Persian texts of possible sources, demonstrating that, although FitzGerald was inspired by Khayyamic and other Persian quatrains, The Rubáiyát of Omar

Khayyám is an original English poem and not a translation.
- Hillmann, Michael C. "Perennial Iranian Skepticism." In *Iranian Culture: A Persianist View*. Lanham, Md.: University Press of America, 1988. A treatment of the significance to Iranian culture today of the ideas expressed in Khayyamic quatrains, which are compared to FitzGerald's poem. Comprehensive bibliography.
- Kennedy, E. S. "The Exact Sciences in Iran Under the Saljuqs and Mongols." In *The Cambridge History of Iran*, edited by J. A. Boyle. Vol. 5. Cambridge, England: Cambridge University Press, 1968. Surveys the foundations of mathematics, algebra, trigonometry, planetary theory, observational astronomy, mathematical geography, specific gravity determination, and rainbow theory, with a discussion of Khayyám's contribution to polynomial equations and his possible contribution to observational astronomy.
- Khayyám, Omar. *The Algebra of Omar Khayyám*. Translated by Daoud S. Kasir. 1931. Reprint. New York: AMS Press, 1972. A great history of mathematics and Khayyám's most important extant work, prefaced with a discussion of the state of algebra before his time and Khayyám's methods and significance. Bibliography.
- Nasr, Seyyed Hossein. *The Islamic Intellectual Tradition in Persia*. Edited by Mehdi Amin Razavi. Richmond, Surrey, England: Curzon Press, 1996. Presents a chapter exploring Omar as a philosopher, poet, and scientist. Bibliography, index.
- Ozdural, Alpay. "A Mathematical Sonata for Architecture: Omar Khayyám and the Friday Mosque of Isfahan." *Technology and Culture* 39, no. 4 (October, 1998). Explores the possibility that Omar, with his theories on ornamental geometry and the triangle, was the designer of the North Dome Chamber (or Great Mosque), built in 1088-1089, in Eṣfahān, Iran. Includes technical language and geometrical drawings.
- Rashed, Rushdei, and Bijan Vahabzadeh. *Omar Khayyám: The Mathematician*. New York: Bibliotheca Persica Press, 2000. An exploration of Omar's work in mathematics. Part of the Persian Heritage series. Bibliography, index.

Pappus

Alexandrian mathematician

Pappus provided a valuable compilation of the contributions of earlier mathematicians and inspired later work on algebraic solutions to geometric problems.

Born: c. 300 C.E.; Alexandria, Egypt
Died: c. 350 C.E.; place unknown

Early Life

Almost nothing is known about Pappus's life, including the dates of his birth and death. A note written in the margin of a text by a later Alexandrian geometer states that Pappus wrote during the time of Diocletian (284-305 C.E.). The earliest biographical source is a tenth century Byzantine encyclopedia compiled by Suidas. This work lists the writings of Pappus and describes him as a "philosopher," which suggests that he may have held some official position as a teacher of philosophy. Nevertheless, this reference to philosophy may be no more than an indication of his interest in natural science. The geometer had at least one child, a son, since he dedicated one of his books to him. In addition, Pappus mentions two of his contemporaries in his texts: a philosopher, Hierius, although the connection between the two is not clear; and Pandrosian, a woman who taught mathematics. Pappus addressed one of his works to her, not as a tribute, but because he found several of her students deficient in their mathematical education.

Pappus lived at a time when the main course of Greek mathematics had been in decline for more than five hundred years; although geometry continued to be studied and taught, there were few original contributions to the subject. To alleviate this lack, he attempted to compile all available sources of earlier geometry and made several significant contributions to the subject. As the first author in this new tradition, sometimes called the silver age of mathematics, Pappus provides a valuable resource for all of ancient Greek geometry.

Life's Work

Throughout his life, Pappus maintained a lively interest in a number of areas dealing with mathematics and natural science. The bulk of his surviving works can be found in the *Synagogē* (c. 340 C.E.; partial translation *The Collection*, 1986). Other works either are in fragmentary form or else are no longer extant, although mentioned by other writers. There exists part of a commentary on the mechanics of Archimedes, which considers problems associated with mean proportions and constructions using straightedge and compass. There are two remaining books of a commentary on Ptolemy's *Mathāmatikā syntaxis* (c. 150 C.E.; *Almagest*, 1948) explaining some of the finer points of the text to the inexperienced reader. Pappus continued his interest in the popularization of difficult texts in a work, of which only a fragment survives, on Euclid's *Stoicheia* (c. 300 B.C.E.; *The Elements of Geometrie of the Most Aunciend Philosopher Euclide of Megara*, 1570, commonly known as the *Elements*), in which Pappus explains the nature of irrational magnitudes to the casual reader. The lost works include a geography of the inhabited world, a description of rivers in Libya, an interpretation of dreams, several texts on spherical geometry and stereographic projection, an astrological almanac, and a text on alchemical oaths and formulas. Pappus was more than a geometer; he was a person who lived in a world where the search for new knowledge was rapidly declining and where political instability was the order of the day. Yet he expressed a continuing interest in the education of those less fortunate than himself and showed a lively interest in affairs outside his city.

Pappus's claim to historical and mathematical significance is found in a compendium of eight books on geometry. This collection covers the entire range of Greek geometry and has been described as a handbook or guide to the subject. In several of the books, when the classical texts are available, Pappus shows how the original proof is accomplished as well as alternative methods to prove the theorem. In other books, where the classical sources are not easily accessible, Pappus provides a history of the problems as well as different attempts at finding a solution. An overall assessment of these books shows few moments of great originality; rather, a capable and independent mind sifts through the entire scope of Greek geometry while demonstrating fine technique and a clear understanding of his field of study.

A summary of the contents of the eight books shows that some are of only historical interest, providing information on or elucidation of classical texts. Other books, particularly book 7, have been a source of inspiration for later mathematicians. All of book 1 and the first part of book 2 are lost. The remainder of book 2 deals with the problems of multiplying all the numbers between 1 and 800 together and expressing the product in words using the myriad (10,000) as base. Pappus refers to a lost work by Apollonius of Perga which seems to be part of the problem of expressing large numbers in words that began with Archimedes' *Psiammites* (c. 230 B.C.E.; *The Sand-Reckoner*, 1897). Book 3 deals with construction problems using straightedge and compass: finding a mean proportion between two given straight lines, finding basic means between two magnitudes (arithmetic, geometric, and harmonic), constructing a triangle within another triangle, and constructing solids within a sphere. Book 4 consists of a collection of theorems, including several famous problems in Greek mathematics: a generalization of Pythagoras's theorem, the squaring of the circle, and the trisection of an angle. Book 5 begins with an extensive introduction on the hexagonal cells of honeycombs and suggests that bees could acquire geometric knowledge from some divine source. This discussion leads to the question of the maximum volume that can be enclosed by a superficial area and to a sequence of theorems that prove that the circle has the greatest area of figures of equal parameter. His proof appears to follow those formulated by an earlier Hellenistic geometer named Zenodorus, whose work is lost. In a later section of this book, Pappus introduces a section on solids with a Neoplatonist statement that God chose to make the universe in a sphere because it is the noblest of figures. It has been asserted but not proved that the sphere has the greatest surface of all equal surface figures. Pappus then proceeds to examine the sphere and regular solids. Book 6 is sometimes called "Little Astronomy"; it deals with misunderstandings in mathematical technique and corrects common misrepresentations.

Book 7 is by far the most important, both because it had a direct influence on modern mathematics and because it gives an account of works in the so-called *Treasury of Analysis* or *Domain of Analysis*, of which a large number are lost. These are works by Euclid, Apollonius, and others that set up a branch of mathematics that provides equipment for the analysis of theorems and problems. Classical geometry uses the term "analysis" to mean a reversal of the normal procedure called "synthesis." Instead of taking a series of steps through valid statements about abstract objects, analysis reverses the procedure by assuming the validity of the theorem and working back to valid statements. Through the preservation of Pappus's account of these works it is possible to reconstruct most of them.

His most original contribution to modern mathematics comes in a section dealing with Apollonius's

Cōnica (*Treatise on Conic Sections*, 1896; best known as *Conics*), where Pappus attempts to demonstrate that the product of three or four straight lines can be written as a series of compounded ratios and is equal to a constant. This came to be known as the "Pappus problem." Book 8 is the last of the surviving books of *Synagogē*, although there is internal evidence that four additional books existed. In this book, Pappus takes on the subject of mechanical problems, including weights on inclined planes, proportioning of gears, and the center of gravity.

There exist substantial references to various lost books of Pappus; among the lost works is a commentary on Euclid's *Elements*, although a two-part section does exist in Arabic. Several other works fit into this category, surviving only in commentary by later writers or in fragments of questionable authorship in Arabic. One of the more interesting Arabic manuscripts (discovered in 1860) shows that Pappus may have invented a volumeter similar to one invented by Joseph-Louis Gay-Lussac (1778-1850). Pappus was not merely a geometer; he was a conserver of classical tradition, a popularizer of Greek geometry, and an inventor as well.

Significance

The works of Pappus have provided later generations with a storehouse of ancient Greek geometry, both as an independent check against the authenticity of other known sources and as a valuable source of lost texts. For modern mathematics, Pappus offers more than merely historical interest. In 1631, Jacob Golius pointed out to René Descartes the "Pappus problem," and six years later this became the centerpiece of Descartes's *Des matières de la géométrie*, which was a section of his *Discours de la méthode* (1637; *Discourse on Method*, 1649). Descartes realized that his new algebraic symbols could easily replace Pappus's more difficult geometric methods and that the product of the locus of straight lines generated from conic sections could generate equations of second, third, and higher orders.

In 1687 Sir Isaac Newton found a similar inspiration in the "Pappus problem" using purely geometric methods. Nevertheless, it was Descartes's algebraic methods that would be utilized in the future. Pappus also anticipated the well-known "Guldin's theorem," dealing with figures generated by the revolution of plane figures about an axis. It can be argued that Pappus was the only geometer who possessed the ability to work out such a theorem during the silver age of Greek mathematics.

Victor W. Chen

Further Reading

Bulmer-Thomas, I. "Guldin's Theorem—or Pappus's?" *Isis* 75 (1984): 348-352. There exists some question whether the Pappus text is original or if the text was corrupted at a later date. A less significant issue here is the interpretation of the Pappus manuscript—a historical problem concerned with the extent to which Pappus anticipated Guldin.

Cuomo, Serafina. *Pappus of Alexandria and the Mathematics of Late Antiquity*. New York: Cambridge University Press, 2000. Sees Pappus's work as part of a wider context and relates it to other contemporary cultural practices, opening new avenues of research into the understanding of mathematics in antiquity.

Descartes, René. *The Geometry of René Descartes*. Translated by Davis E. Smith and Marcia L. Latham. Chicago: Open Court, 1952. It is possible to follow from Descartes's own text the relevant passages from Pappus's work, seeing how Descartes develops his new symbols and why this method would later become the preferred method.

Fried, Michael N. *Apollonius of Perga's "Conica": Text, Context, and Subtext*. Boston: E. J. Brill, 2001. An extensive discussion of this work, from which arose the "Pappus problem."

Heath, Sir Thomas. *From Aristarchus to Diophantus*. Vol. 2 in *A History of Greek Mathematics*. Reprint. New York: Dover Publications, 1981. This edition contains several long sections from the *Collection* as well as commentaries on the history and contents of these theorems.

Pappus. *Book 7 of the Collection*. Edited by Alexander Jones. 2 vols. New York: Springer-Verlag, 1986. These two volumes contain the most complete rendition of book 7; in addition, there are exhaustive commentaries and notes on every aspect of this text. Contains a detailed account of the history of various Pappus manuscripts and notes on the problems of translating ancient Greek text.

Blaise Pascal

French philosopher, mathematician, and physicist

Pascal was a genius in many areas who made important contributions to mathematics and physics and invented an early form of the calculator. His major contribution, however, is the record of his religious and philosophical struggle to reconcile human experience, God, and the quest for happiness and meaning.

Born: June 19, 1623; Clermont-Ferrand, France
Died: August 19, 1662; Paris, France

Early Life

Blaise Pascal (blehz pahs-kahl) was the third child of Étienne Pascal, a government financial bureaucrat, and Antoinette (Begon), who died when Pascal was about three. After his mother's death, Pascal and his family moved to Paris. Pascal's father decided to educate his children himself, rather than making use of either tutors or schools. Étienne Pascal was associated with the intellectual circles of Paris and thereby exposed Pascal to the best scientific and mathematical thought of his time.

While still a teenager, the precocious Pascal attracted the attention of the court and, in 1640, published his first mathematical treatise. In 1642, he began working on a mechanical calculator to help in his father's work. He continued improving the device for the next ten years and in 1652 sent a version of it to Queen Christina of Sweden. In 1646, Pascal and his two older sisters first came under the influence of Jansenism, a strict, pietistic movement within the Catholic Church that stressed a life of devotion, practical charity, and asceticism. Pascal experienced what is usually called his "first conversion," feeling the need for religious renewal but not wanting to give up his scientific and mathematical endeavors. His scientific work at this time included experiments with vacuums, an important area of exploration in seventeenth century physics.

Life's Work

By his mid-twenties, Pascal had assumed a pattern of life that he would continue until his death. In 1647, he entered into the first of the public religious controversies that would preoccupy him for the rest of his life. He also continued his scientific work on the vacuum, exchanging information with the great philosopher René Descartes and publishing his own findings. In 1648, he wrote a mathematical essay on conic sections. Throughout this period, Pascal was afflicted with serious illness, as he would be for the remainder of his life.

Pascal's sister Jacqueline continued to be influenced by Jansenism, and during this time she expressed her desire to enter the Jansenist religious community at Port-Royal. Both Pascal and his father objected, but after her father's death in 1651, Jacqueline entered the convent the following year. Pascal began a brief phase in which he indulged himself in the pleasures and pursuits of French society, finding the experience empty but also finding no other direction for his life at this time.

Pascal experienced a growing disillusionment with the skeptical worldliness of society life and greatly desired something more meaningful. During the middle of the night of November 23, 1654, he had an intense, mystical religious experience that lasted about two hours and changed the direction of his life. During this experience, Pascal felt powerfully and unmistakably the truth of God's existence and the blessing of His love and forgiveness. Pascal had been provided with the kind of

Blaise Pascal. (Library of Congress)

experiential certainty for which his scientific mind yearned and, consequently, saw everything thereafter in spiritual terms. In reaction to this experience, he went to Port-Royal, the center of Jansenism, for a two-week retreat in early 1655 to begin the reformation of his life that he now sought. He was particularly concerned with overcoming the willful pride that had marked his life since his spectacular intellectual accomplishments as a boy and the selfishness that showed itself in his resistance to his sister Jacqueline's entrance into the community at Port-Royal.

Jansenism was to dominate his life for the next few years. In 1653, Pope Innocent X had condemned the writings of Cornelius Otto Jansen, bishop of Ypres, upon which the Jansenist movement in the Catholic Church was based. The great enemies of the Jansenists were the rationalistic Jesuits, and in January of 1656, Pascal wrote the first of a series of anonymous letters now titled *Lettres provinciales* (1656-1657; *The Provincial Letters*, 1657). These letters, eighteen in all, came out in May, 1657, and are masterpieces of satire, wit, analytic logic, and French prose style. Especially in the early letters, the fictitious writer adopts a pose of objective, naïve curiosity about the controversy between the Jesuits and Jansenists, which he is purportedly trying to explain to his fellow provincial back home. In reality, the letters are an impassioned defense of the principles and principals of the Jansenist movement and a stinging attack on the Jesuits. The letters were enormously popular, and the local authorities went to great lengths to try to suppress them and discover their author. Pascal's letters have been admired ever since as masterpieces of French prose.

Pascal was not satisfied, however, merely to defend a particular movement within the Catholic Church. He desired to write a great defense of Christianity as a whole at a time when religious faith was increasingly under attack by skepticism, on one hand, and rationalism, on the other. Prompted in part by what he took to be the miraculous cure of his young niece, Pascal began in 1657 to take notes for this work, which he once said would take ten years of steady effort to complete. As it turned out, Pascal never completed the work or even a draft of it. Instead, he produced approximately one thousand notes, some only a few words, others pages long and substantially revised. The majority of these notes were written in 1657 and 1658, after which time he fell into an extremely painful and debilitating illness that would largely incapacitate him until his death. The notes were first published in abbreviated form as *Pensées* (1670; *Monsieur Pascal's Thoughts, Meditations, and Prayers*, 1688; best known as *Pensées*) and have become one of the classic documents of Western culture.

Pascal on Reconciling Reason's Possibilities and Limits

Blaise Pascal, a great thinker and innovator, was also deeply religious. The excerpt here shows Pascal "reasoning" the limits of reason, a seeming paradox. Pascal, however, concludes that in the face of overwhelming odds against both knowing absolutely and knowing nothing, one can still be sure that he or she can know something. It is this unique human ability to think that leads to moral, righteous behavior.

What is the rank man occupies in Nature? A nonentity, as contrasted with infinity; a universe, contrasted with nonentity; a middle something between every thing and nothing....

Such is our real state; our acquirements are confined within limits which we cannot pass, alike incapable of attaining universal knowledge or of remaining in total ignorance. We are in the middle of a vast expanse, always unfixed, fluctuating between ignorance and knowledge; if we think of advancing further, our object shifts its position and eludes our grasp; it steals away and takes eternal flight that nothing can arrest. This is our natural condition, altogether contrary, however, to our inclinations. We are inflamed with a desire of exploring every thing, and of building a tower that shall rise into infinity, but our edifice is shattered to pieces, and the ground beneath it discloses a profound abyss.

Man is the feeblest reed in existence, but he is a thinking reed. There is no need that the universe be armed for his destruction; a noxious vapour, a drop of water is enough to cause his death. But though the universe were to destroy him, man would be more noble than his destroyer, for he would know that he was dying, while the universe would know nothing of its own achievement. Thus all our dignity consists in the thinking principle. This and not space and duration, is what elevates us. Let us labour then to think aright; here is the foundation of morals.

Source: Pascal, *Pensées* (1670), excerpted in *The Age of Reason: The Culture of the Seventeenth Century*, edited by Leo Weinstein (New York: George Braziller, 1965), pp. 256-257.

Although Pascal never wrote his great apology for the Christian faith, he did organize many of his notes into groups, from which scholars have speculated as to his ultimate intentions. As enlightening as these speculations sometimes are, the timelessness of *Pensées* comes not from the tantalizing promise of some irrefutable defense of religious faith but from Pascal's compelling, often painful insights into the human condition and from the process of watching one of history's great minds struggle with eternal questions of faith, spirit, and transcendence.

Many of Pascal's most powerful entries poignantly explore the tragedy and folly of the human condition if there were no God. He depicts humankind as lost in an alien and inhospitable world, given over to the empty baubles and distractions of society. Pascal portrays the world as a psychologically frightening place. Men and women are caught between the infinitely large, on one hand, and the infinitely small, on the other. They are torn by a divided nature that is neither angel nor beast, to use one of his images, but is capable of acting like either. Human beings yearn for something sure and permanent but find only illusion and transience. Pascal finds the solution for the human dilemma in the grace of God as manifested in Jesus Christ. Only by knowing who created them, Pascal argues, can humans know who they are and how they can be happy. He does not, however, offer this solution as an effortless one. Part of Pascal's enduring appeal is his very modern awareness of the difficulty of religious faith in a scientific and skeptical world.

Pascal was seriously ill much of the last four years of his life, but that did not prevent him from at least sporadic efforts on a variety of projects. In 1658, he made further mathematical discoveries on the cycloid and publicly challenged the mathematicians of Europe to a contest in solving problems in this area. He was drawn briefly into the Jansenist controversy once again but then withdrew from it altogether. His concern for the poor led him to invent and launch a public transportation system in Paris in March of 1662. Additionally, when health permitted, he worked on his defense of Christianity that became *Pensées*. After much suffering patiently borne, Pascal died on August 19, 1662, at the age of thirty-nine.

Significance

Pascal is one of those handful of individuals in history whose wide range of accomplishments shows evidence of a fundamental genius that expressed itself wherever it was applied. Proof of his greatness is given by the number of different fields of intellectual effort that claim him. He is considered a mathematician of the first rank, an important physicist at the early stages of that science, an inventor, a literary master of French prose, and, most important, a philosopher and religious thinker who has written brilliantly about fundamental questions of the human condition.

Pascal stood at the beginning of the modern age, one who felt keenly the call of reason and science but who realized the price to be paid if one lost a sense of the spiritual and transcendent. He felt caught between two contrary forces: the rationalism of rising seventeenth century science and the skepticism about all human efforts, reason included, as epitomized by his French predecessor, Michel Eyquem de Montaigne. He sought an approach to life that avoided the arrogance and materialism of the former and the cynicism and moral passivity of the latter. In this sense, Pascal's situation anticipates the modern one. How does one find meaning, values, and faith in a rationalistic, skeptical world where most traditional guidelines are called into question? For more than three hundred years, men and women have found insight and inspiration in Pascal's answers.

Daniel Taylor

Further Reading

Adamson, Donald. *Blaise Pascal: Mathematician, Physicist, and Thinker About God*. New York: St. Martin's Press, 1995. A chronological survey of Pascal's work in mathematics, physics, religion, and philosophy.

Coleman, Francis X. J. *Neither Angel nor Beast: The Life and Work of Blaise Pascal*. New York: Routledge & Kegan Paul, 1986. A somewhat poorly organized but still-insightful overview of Pascal's life and work. Good at placing Pascal in the context of seventeenth century thought.

Davidson, Hugh M. *Blaise Pascal*. Boston: Twayne, 1983. A good introduction to Pascal. Short but adequate overview of his life and discussion of all of his major and most of his minor works, including detailed discussion of his mathematical contributions.

Groothius, Douglas. *On Pascal*. Belmont, Calif.: Thomson Learning/Wadsworth, 2003. Concise introduction to Pascal's most important ideas, placing these concepts within a historical context.

Hammond, Nicholas, ed. *The Cambridge Companion to Pascal*. New York: Cambridge University Press,

2003. A collection of essays, including discussions of Pascal's life and times; his work on probability, decision theory, and religion; and the reception of *Pensées* in the seventeenth and eighteenth centuries.

Krailsheimer, Alban. *Pascal*. New York: Hill & Wang, 1980. Brief but helpful summary of Pascal's mathematical and scientific accomplishments. Good on the cultural context and central concerns of the *Pensées*.

Moriarty, Michael. *Early Modern French Thought: The Age of Suspicion*. New York: Oxford University Press, 2003. Examines the philosophy of Pascal, René Descartes, and Nicolas Malebranche.

Nelson, Robert J. *Pascal: Adversary and Advocate*. Cambridge, Mass.: Harvard University Press, 1981. One of the more comprehensive and ambitious studies of Pascal. Takes a psychological approach to Pascal's biography and work, with extensive critical study of individual works.

Pascal, Blaise. *Pensées*. Translated by A. J. Krailsheimer. London: Penguin Books, 1966. One of the better of the many translations of Pascal's great work. Recommended.

ÉMILE PICARD

French mathematician

Picard's work with differential equations led to his having a theorem named after him. His theories advanced research into analysis, algebraic geometry, and mechanics.

Born: July 24, 1856; Paris, France
Died: December 11, 1941; Paris, France
Also known as: Charles Émile Picard (full name)

EARLY LIFE

The mother of Émile Picard (ay-meel pee-kahr) was the affluent daughter of a doctor from France's northern provinces. His father, from Burgundy, was a textile manufacturer who died during the Franco-Prussian War of 1870. Picard demonstrated brilliance early in his life. While in school, he developed interests in varied subjects such as literature, languages, and history. His accomplishments in these areas of scholarship were enhanced by his love for books and reading and by his exceptionally powerful memory. One theme that appears throughout Picard's life is his broad range of interests. He was an athlete as well as a scholar. Throughout his life, he maintained a love for such rigorous physical activities as gymnastics and mountain climbing.

Given the fact that Picard was a generalist, it was difficult for him to choose any one field in academics on which to focus. In fact, he only decided to study mathematics at the end of his secondary studies. The reason for this decision came from his having read an algebra book. After making the decision to study mathematics, Picard committed himself to this pursuit with a devotion that is rarely matched. Indeed, by 1877, at age twenty-one, he had already made a major contribution to the development of a portion of mathematics that focused on the theory of algebraic surfaces. He had also received, by this time, the degree of doctor of science.

Picard's scholarship was recognized by many important members of the academic community. One of the great French mathematicians of the time, Charles Hermite, became his mentor and lifelong friend. In fact, Picard's development as a mathematician was strongly guided by Hermite. In 1881, with support from Hermite, he was appointed to a professorship at the Sorbonne, and during the same period he married Hermite's daughter.

LIFE'S WORK

The diversity that marked his life also marked his professional development. His early career emphasized research and focused on algebra and geometry. Some of his major contributions came in the field of algebraic geometry. He soon, however, began to pursue other interests in mathematics. By 1885, he had begun to pursue work in the field of differential and integral calculus. Picard was elected to the chair of this subject at the Sorbonne during this period.

Picard's most famous work came from his investigations of differential equations. Indeed, one theorem for which he is still remembered in modern texts of all languages is Picard's theorem, a method for approximating the solution to a differential equation. More important, however, is the work that Picard did in trying to develop a general framework for finding solutions to differential equations.

In addition to his work in differential equations, Picard did work in complex analysis. In this area, he helped to extend the research of his colleague Henri Poincaré. The work on which he focused involved functions of two complex variables. Picard termed these functions hypergeometric and hyperfuschian. The work that he did here was collected in a two-volume set entitled *Théorie des fonctions algébriques de deux variables indépendantes* (1897, 1906). This work was coauthored by the mathematician Simart.

At the turn of the century, Picard was engaged in the study of algebraic surfaces. This series of investigations was inspired by his previous work on the nature of complex functions. One of the interesting side effects of Picard's career as a generalist is the fact that his investigations frequently led him into other areas of study.

Another interesting facet of Picard's career is the fact that it touched so many areas that one would not expect. For example, he was the chair of many government commissions, including the Bureau of Longitudes. Also, the quality and the variety of his scholarship led to his permanent election as the secretary to the Academy of Sciences. Picard's wide-ranging scholarship reached even into areas such as physics and engineering. His researches included the application of mathematical methods to physics problems of elasticity, heat, and electricity. One subject to which Picard added significantly was the way in which electrical impulses moved along wires. In engineering, Picard, who had originally begun as a theoretician in mathematics, developed into an excellent teacher and eventually became responsible for training ten thousand French engineers between 1894 and 1937.

It should be emphasized that the quality of Picard's scholarship did not suffer because of its variety. During the course of his career, he was responsible for the development of more than three hundred papers on various subjects. His *Traité d'analyse* (1891-1896) is considered a classic book on mathematics. At one time, this monumental work was considered required reading for obtaining a thorough background in mathematics.

Picard also published materials that were, strictly speaking, outside the realm of mathematics. He was, for example, responsible for collecting and editing the works of Charles Hermite, his mentor. In addition, he published a number of works on the philosophy of science and the scientific method, the majority of these after 1900.

Picard's career was long as well as productive. Many mathematicians find their most productive years early. Picard was a notable exception to this pattern. Again following his early path of intellectual diversity, he made significant contributions to the development of mathematical concepts such as similarity and homogeneity well after he was eighty years old. These concepts are important in algebra and engineering.

Picard's investigations were particularly significant in their effects on his fellows and successors in his field. Among those influenced by him were Henri-Léon Lebesgue, Émile Borel, and Otto von Blumenthal. Picard was one of the most honored scientists of his generation. In 1924, he was elected to the Academy of France. In 1932, he received the Gold Cross of the Legion of Honor. In 1937, he received the Mittag-Leffler Gold Medal from the Swedish Academy of Sciences. All told, he was awarded honorary doctorates by universities in five foreign countries.

In contrast to his almost unbroken string of professional successes, Picard had a personal life that was filled with tragedy. War was a common theme in the litany of misery that filled his personal existence. Besides the death of his father in the Franco-Prussian War, he lost a daughter and two sons during World War

Émile Picard. (Library of Congress)

I. During World War II, his grandsons were wounded in the invasion of France. The personal tragedy under which Picard lived was emphasized by the fact that he died while France was still under German occupation. Yet he had lived as one of the most productive and honored mathematicians in a period known for the brilliance of its mathematical researchers.

Significance

Émile Picard was, in all ways, a generalist. Many would have termed him a Renaissance man. This diversity of interests was reflected in his work both in and out of mathematics. In mathematics, his research involved such varied areas as geometry, algebra, differential equations, and complex analysis. It is extremely rare to encounter mathematicians in the modern world who make significant contributions in more than one specialty. Picard's most significant work was in differential equations. It is here that the mathematical world outside France most commonly remembers the great Parisian. Modern works still make reference to Picard's theorem for approximating solutions to differential equations, and these new efforts still mention Picard groups as a way of categorizing the transformations that can occur in linear differential equations.

In areas outside mathematics, Picard was known as a teacher, writer, editor, and administrator. He published an important survey of mathematics in France. He headed both the Academy of Sciences and the Society of Friends of Science, a group interested in helping needy scientists. Picard was not only a great mathematician and scientist but also a great man. In an age that is characterized by the specialist, it is good to reflect that men such as Picard have lived.

Lyndon Marshall

Further Reading

Bell, Eric T. *The Development of Mathematics*. New York: McGraw-Hill, 1940. This is a particularly good discussion of the history of mathematics from a developmental standpoint. Consequently, Picard gets fairly good treatment. This book also discusses Hermite fairly extensively.

_____. *Men of Mathematics*. New York: Simon & Schuster, 1937. Picard is mentioned only slightly in this text. Yet it provides an excellent look at one of his most famous colleagues, Henri Poincaré. It gives a view of the flavor of the times and the problems that were faced by mathematicians.

Considine, Douglass M., ed. "Picard's Theorem." In *Van Nostrand's Scientific Encyclopedia*. New York: Van Nostrand Reinhold, 1976. This includes a good discussion of Picard's theorem.

Griffiths, Phillip. "Œuvres de Émile Picard, Tome II." *Dialog Math-Sci Database*, February, 1989. A review of the collected works of Picard. It discusses the material addressed by Picard during his researches. It also includes a discussion of his life. The review is in English. Unfortunately, the book that it covers is not.

Hadamard, J. "Émile Picard." *Journal of the London Mathematical Society* 18 (1943). This biographical sketch, published not long after Picard's death, is the best description of his life.

Henri Poincaré

French mathematician

One of the most important mathematicians of the late nineteenth century, Poincaré developed the theory of automorphic functions, did extensive work in celestial mechanics and mathematical physics, and was a codiscoverer of the special theory of relativity. His writing style was so clear that his books about the philosophy of science were read widely by the general public and translated into many languages.

Born: April 29, 1854; Nancy, France
Died: July 17, 1912; Paris, France
Also known as: Jules Henri Poincaré (full name)

Early Life

Henri Poincaré (pwahn-kah-ray) was born into one of the most distinguished families of France's Lorraine region. His father, Leon, was a physician, and one of his cousins, Raymond, became president of the French Republic during World War I. Henri and his sister were adored by their mother, and she devoted herself to their education and rearing. When he was five, Henri contracted diphtheria, and the resulting weakness may have influenced his entire life. Because he was unable to join the other boys in their rough play, Henri was forced to entertain himself with intellectual pursuits. He developed a remarkable memory so that he could even cite page numbers for information in books that he had read many years earlier. In addition, because his eyesight was poor, he learned most of his classwork by listening, because he could not see the blackboard. Thus, he was forced to develop the ability to see spatial relationships in his mind at an early age.

Although he was a good student in his early years, there was no indication of his impending greatness until he was a teenager. He won first prize in a French national competition and in 1873 entered the École Polytechnique, where he exhibited his brilliance in mathematics. Upon his graduation, Poincaré entered the École des Mines in 1875 to study engineering. Although he was a careful student who did his work adequately, Poincaré spent much of his time pursuing mathematics as a recreation. He continued his practice of mathematics during his apprenticeship as a mining engineer.

Poincaré was not an extremely attractive man; he had thinning blond hair, wore glasses, and was short in stature; he was known for being absentminded and clumsy. Nevertheless, he maintained a happy personal life. He married at the age of twenty-seven, fathered four children, whom he adored, and never wanted for friends, because he was by nature humble and interested in other people.

Life's Work

In 1879, Poincaré submitted the doctoral thesis in mathematics that he had written during his work as an engineer, and he received his degree that same year. The subject was the first of his great achievements: the theory of differential equations. His first appointment was as a lecturer of mathematical analysis at the University of Caen in 1879, and in 1881 he was invited to join the faculty at the University of Paris. He continued this appointment until his death in 1912, although by then his responsibilities had expanded to include mechanics and physics.

During his tenure, he was elected to the Académie des Sciences in 1887 and the Académie Française in 1908. this second appointment is most unusual for a mathematician, for it is given to honor literary achievements and is thus a sure indication of his lucid writing style. He was named president of the Académie des Sciences in 1906. Other awards included a Fellowship in the Royal Society in 1894, the Prix Poncelet, Prix Reynaud, and Prix Bolyai, and gold medals from the Lobachevsky Fund.

Much of Poincaré's early work was in differential equations, a branch of calculus that is linked directly to the physical world. It was natural, then, for him to turn his attention from pure mathematics to physics and celestial mechanics. However, in his pursuit of solutions of physical and mechanical problems, he often created new tools of pure mathematics.

Poincaré was first drawn to celestial mechanics and astronomical physics by the classical three-body problem, which concerns the gravitational influence and distortions that three independent bodies in space would exert on one another; it held his interest throughout his life. Poincaré published partial results in his early years at the Sorbonne and later published work broadening the number of objects from three to any number. His results won for him a prize that had been offered by King Oscar II of Sweden.

In celestial mechanics, Poincaré was the first person to demand rigor in computations: He found the approximations used commonly at the time to be unacceptable, because they introduced obvious errors into the work. Consequently, more powerful mathematics had to be developed. This work was not centered on any one branch of mathematics but instead included calculus, algebra, number theory, non-Euclidean geometry,

Henri Poincaré. (Library of Congress)

and topology. In fact, the field of topology was begun in large part with Poincaré's study of orbits. He published much of this work in *Les Méthodes nouvelles de la mécanique céleste* (new methods in celestial mechanics), in three volumes between 1892 and 1899.

Poincaré's other early achievement was in the theory of automorphic functions—a method for expressing functions in terms of parameters in mathematical analysis. These are functions that remain relatively unchanged though they are acted on by a series of transformations. He found that one class of these, which he called Fuchsian (for German mathematician Immanuel Fuchs), was related to non-Euclidean geometry, and this became an important insight. Indeed, there was some argument over priority in this development between Poincaré and German mathematician Christian Felix Klein; however, scientific historians agree that Poincaré was the developer of these theories.

It seems that all branches of mathematics held Poincaré's interest. Poincaré was essential to the development of algebraic geometry. Of particular importance is his development of a parametric representation of functions. For example, the general equation of a circle $x^2 + y^2 = r^2$ can be rewritten as two equations that describe the variables x and y in terms of some angle A. The equations $x = r \sine A$ and $y = r \cosine A$ are the equivalent of the original equation since $x^2 + y^2 = r^2 \sine^2 A + r^2 \cosine^2 A = r^2 (\sine^2 A + \cosine^2 A)$, which in turn equals r^2 since $\sine^2 + \cosine^2 = 1$. Many problems can be solved using parameters that do not yield to any other methods.

Poincaré is equally important in physics. Although Albert Einstein is generally known for his theory of relativity, the special theory of relativity was discovered independently by Poincaré. He and Einstein arrived at the theory from completely different viewpoints, Einstein from light and Poincaré from electromagnetism, at about the same time (Einstein's first work was published in 1905, and Poincaré's was published in 1906). There can be no doubt that both men deserve a share of the credit. When Poincaré became aware of Einstein's work, he was quite enthusiastic and supportive of the Swiss physicist even though most scientists were skeptical. Max Planck, who developed quantum theory, was another physicist who was recognized by Poincaré while he was being scorned by others. In addition, Poincaré developed the mathematics required for countless physical discoveries during the early twentieth century. An example is the wireless telegraph. He also developed the theory of the equilibrium of fluid bodies rotating in space.

Poincaré had a rare gift for a mathematician: He was able to write clearly and to make mathematics and science exciting to people whose educations were directed toward other fields. One of his most widely known works in the philosophy of science, *Science et méthode* (1908; *Science and Method*, 1914), is devoted to a study of how scientists and mathematicians create. Poincaré believed that some things in mathematics are known intuitively rather than from observation or from classic logic. His articles and books in the philosophy of science were avidly read and translated into most of the European languages and even into Japanese.

Poincaré continued in relatively good health until 1908, and in 1912 he died of an embolism following minor surgery. The church Saint-Jacques-de-Haut-Pas, the site of his funeral several days later, was filled with eminent persons from all fields who had come to pay a last tribute to his greatness.

Significance

Henri Poincaré was clearly one of the great mathematicians of his time. In fact, some believe that he had no peer. He won virtually every mathematical prize available, and he also won several scientific awards. His work entered every field of mathematics at the time, and he created at least one new branch called algebraic topology. His discoveries inspired other mathematicians for years after his death. In addition, Poincaré did first-rate work in celestial mechanics and was a codiscoverer of the theory of relativity.

The more than thirty books and five hundred papers that Poincaré published are a testament to his prolific career, especially since he died during his productive years. In addition, his writings on the philosophy of science sparked public interest in mathematics and the physical sciences and foreshadowed the intuitionist school of philosophy. These works have helped define the way human beings think about mathematical and scientific creation and will continue to do so for years to come. The practical applications of Poincaré's work are numerous. Differential functions are the primary mathematics used in engineering and some of the physical sciences; his work in celestial mechanics was completely different from past works and altered the field's course. In addition, he offered many new ideas in pure mathematics.

Perhaps the most articulate tribute to Poincaré was given in the official report of the 1905 Bolyai Prize written by Gustave Rados: "Henri Poincaré is incontestably the first and most powerful investigator of the present time in the domain of mathematics and mathematical physics."

Celeste Williams Brockington

FURTHER READING

Barrow-Greene, June. *Poincaré and the Three-Body Problem*. Providence, R.I.: American Mathematics Society, 1996. Describes how Poincaré accidentally discovered chaos theory. Aimed at readers with an advanced level of mathematical knowledge.

Bell, E. T. "The Last Universalist." In *Men of Mathematics*. New York: Simon & Schuster, 1937. This book is a series of twenty-nine chapters, each introducing a different mathematician from the early Greeks to the early twentieth century. Its account of Poincaré focuses on three areas: the theory of automorphic functions, celestial mechanics and mathematical physics, and the philosophy of science. Biographical information is also included.

Galison, Peter. *Einstein's Clocks, Poincaré's Maps: Empires of Time*. New York: W. W. Norton, 2003. Examines how Poincaré and Einstein created the modern conception of time through their ideas about relativity.

Nordmann, Charles. "Henri Poincaré: His Scientific Work, His Philosophy." In *Annual Report of the Board of Regents of the Smithsonian Institution*. Washington, D.C.: Government Printing Office, 1913. Nordmann includes not only a summary of Poincaré's work and philosophy as the title indicates but also a considerable amount of biographical information.

Poincaré, Henri. *The Foundations of Science*. Translated by George Bruce Halsted. New York: Science Press, 1913. Contains a preface by Poincaré and an introduction by Josiah Royce. Argues Poincaré's philosophy of science.

_____. "The Future of Mathematics." In *Annual Report of the Board of Regents of the Smithsonian Institution*. Washington, D.C.: Government Printing Office, 1910. This article represents Poincaré at his best. After a brief introduction, he guides the reader through most of the prominent fields of mathematics and predicts what he believed was to come. His explanations are excellent.

_____. *Mathematics and Science: Last Essays*. Translated by John W. Balduc. Reprint. Mineola, N.Y.: Dover, 1963. Another work in the philosophy of science.

Slosson, Edwin E. "Henri Poincaré." In *Major Prophets of Today*. Freeport, N.Y.: Books for Libraries Press, 1968. Slosson chose several representatives from the modern era whom he viewed as having lasting prominence. His article on Poincaré includes biographical information as well as a discussion of Poincaré's work in mathematics and philosophy.

Zahar, Eli. *Poincaré's Philosophy: From Conventionalism to Phenomenology*. Chicago: Open Court, 2001. Traces the development of Poincaré's philosophy, discussing his thoughts about general science, geometry, mathematics, relativity, and other subjects.

Pythagoras

Greek philosopher and mathematician

Pythagoras set an inspiring example with his energetic search for knowledge of universal order. His specific discoveries and accomplishments in philosophy, mathematics, astronomy, and music theory make him an important figure in Western intellectual history.

Born: c. 580 B.C.E.; Samos, Ionia, Greece
Died: c. 500 B.C.E.; Metapontum, Lucania (now in Italy)

Early Life
Pythagoras (pih-THAG-oh-ruhs), son of Mnesarchus, probably was born about 580 B.C.E. (various sources offer dates ranging from 597 to 560). His birthplace was the Greek island of Samos in the Mediterranean Sea. Aside from these details, information about his early life—most of it from the third and fourth centuries B.C.E., up to one hundred years after he died—is extremely sketchy. On the other hand, sources roughly contemporary with him tend to contradict one another, possibly because those who had been his students developed in many different directions after his death.

Aristotle's *Metaphysica* (335-323 B.C.E.; *Metaphysics*, 1801), one source of information about Pythagorean philosophy, never refers to Pythagoras himself but always to "the Pythagoreans." Furthermore, it is known that many ideas attributed to Pythagoras have been filtered through Platonism. Nevertheless, certain doctrines and biographical events can be traced with reasonable certainty to Pythagoras himself. His teachers in Greece are said to have included Creophilus and Pherecydes of Syros; the latter (who is identified as history's first prose writer) probably encouraged Pythagoras's belief in the transmigration of souls, which became a major tenet of Pythagorean philosophy. A less certain but more detailed tradition has him also studying under Thales of Miletus, who built a philosophy on rational, positive integers. In fact, these integers were to prove a stumbling block to Pythagoras but would lead to his discovery of irrational numbers such as the square root of two.

Following his studies in Greece, Pythagoras traveled extensively in Egypt, Babylonia, and other Mediterranean lands, learning the rules of thumb that, collectively, passed for geometry at that time. He was to raise geometry to the level of a true science through his pioneering work on geometric proofs and the axioms, or postulates, from which these are derived.

A bust now housed at Rome's Capitoline Museum (the sculptor is not known) portrays the philosopher as having close-cropped, wavy Greek hair and beard, his features expressing the relentlessly inquiring Ionian mind—a mind that insisted on knowing for metaphysical reasons the *exact* ratio of the side of a square to its diagonal. Pythagoras's eyes suggest an inward focus even as they gaze intently at the viewer. The furrowed forehead conveys solemnity and powerful concentration, yet deeply etched lines around the mouth and the hint of a crinkle about the eyes reveal that this great man was fully capable of laughter.

LIFE'S WORK
When Pythagoras returned to Samos from his studies abroad, he found his native land in the grip of the tyrant Polycrates, who had come to power about 538 B.C.E. In the meantime, the Greek mainland had been partially overrun by the Persians. Probably because of these developments, in 529 Pythagoras migrated to Croton, a Dorian colony in southern Italy, and entered into what became the historically important period of his life.

At Croton he founded a school of philosophy that in some ways resembled a monastic order. Its members were pledged to a pure and devout life, close friendship, and political harmony. In the immediately preceding years, southern Italy had been nearly destroyed by the strife of political factions. Modern historians speculate that Pythagoras thought that political power would give

Pythagoras. (Library of Congress)

his organization an opportunity to lead others to salvation through the disciplines of nonviolence, vegetarianism, personal alignment with the mathematical laws that govern the universe, and the practice of ethics in order to earn a superior reincarnation. (Pythagoras believed in metempsychosis, the transmigration of souls from one body to another, possibly from humans to animals. Indeed, Pythagoras claimed that he could remember four previous human lifetimes in detail.)

His adherents he divided into two hierarchical groups. The first was the *akousmatikoi*, or listeners, who were enjoined to remain silent, listen to and absorb Pythagoras's spoken precepts, and practice the special way of life taught by him. The second group was the *mathematikoi* (students of theoretical subjects, or simply "those who know"), who pursued the subjects of arithmetic, the theory of music, astronomy, and cosmology. (Though *mathematikoi* later came to mean "scientists" or "mathematicians," originally it meant those who had attained advanced knowledge in a broader sense.) The *mathematikoi*, after a long period of training, could ask questions and express opinions of their own.

Despite the later divergences among his students—fostered perhaps by his having divided them into two classes—Pythagoras himself drew a close connection between his metaphysical and scientific teachings. In his time, hardly anyone conceived of a split between science and religion or metaphysics. Nevertheless, some modern historians deny any real relation between the scientific doctrines of the Pythagorean society and its spiritualism and personal disciplines. In the twentieth century, Pythagoras's findings in astronomy, mathematics, and music theory are much more widely appreciated than the metaphysical philosophy that, to him, was the logical outcome of those findings.

Pythagoras developed a philosophy of number to account for the essence of all things. This concept rested on three basic observations: the mathematical relationships of musical harmonies, the fact that any triangle whose sides are in a ratio of 3:4:5 is always a right triangle, and the fixed numerical relations among the movement of stars and planets. It was the consistency of ratios among musical harmonies and geometrical shapes in different sizes and materials that impressed Pythagoras.

His first perception (which some historians consider his greatest) was that musical intervals depend on arithmetical ratios among lengths of string on the lyre (the most widely played instrument of Pythagoras's time), provided that these strings are at the same tension. For example, a ratio of 2:1 produces an octave; that is, a string twice as long as another string, at the same tension, produces the same note an octave below the shorter string. Similarly, 3:2 produces a fifth and 4:3 produces a fourth. Using these ratios, one could assign numbers to the four fixed strings of the lyre: 6, 8, 9, and 12. Moreover, if these ratios are transferred to another instrument—such as the flute, also highly popular in that era—the same harmonies will result. Hippasus of Metapontum, a *mathematikos* living a generation after Pythagoras, extended this music theory through experiments to produce the same harmonies with empty and partly filled glass containers and metal disks of varying thicknesses.

Pythagoras determined that the most important musical intervals can be expressed in ratios among the numbers 1, 2, 3, and 4, and he concluded that the number 10—the sum of these first four integers—comprehends the entire nature of number. Tradition has it that the later Pythagoreans, rather than swear by the gods as most other people did, swore by the "Tetrachtys of the Decad" (the sum of 1, 2, 3, and 4). The Pythagoreans also sought the special character of each number. The tetrachtys was called a "triangular number" because its components can readily be arranged as a triangle.

By extension, the number 1 is reason because it never changes; 2 is opinion; 4 is justice (a concept surviving in the term "a square deal"). Odd numbers are masculine and even numbers are feminine; therefore, 5, the first number representing the sum of an odd and an even number (1, "unity," not being considered for this purpose), symbolizes marriage. Seven is *parthenos*, or virgin, because among the first ten integers it has neither factors nor products. Other surviving Pythagorean concepts include unlucky 13 and "the seventh son of a seventh son."

To some people in the twentieth century, these number concepts seem merely superstitious. Nevertheless, Pythagoras and his followers did important work in several branches of mathematics and exerted a lasting influence on the field. The best-known example is the Pythagorean theorem, the statement that the square of the hypotenuse of a right triangle is equal to the sum of the squares of the other two sides. Special applications of the theorem were known in Mesopotamia as early as the eighteenth century B.C.E., but Pythagoras sought to generalize it for a characteristically Greek reason: This theorem measures the ratio of the side of a square to its diagonal, and he was determined to know the *precise* ratio. It cannot be expressed as a whole number, however, so Pythagoras found a common denominator by showing a relationship among the *squares* of the sides of a right triangle. The Pythagorean theorem is set forth in book 1 of Euclid's *Stoicheia* (*Elements*), Euclid being one of several later Greek thinkers whom Pythagoras strongly influenced and who transmitted his ideas in much-modified form to posterity.

Pythagoras also is said to have discovered the theory of proportion and the arithmetic, geometric, and harmonic means. The terms of certain arithmetic and harmonic means yield the three musical intervals. In addition, the ancient historian Proclus credited Pythagoras with discovering the construction of the five regular geometrical solids, though modern scholars think it more likely that he discovered three—the pyramid, the tetrahedron, and the dodecahedron—and that Theaetetus (after whom a Platonic dialogue is named) later discovered the construction of the remaining two, the octahedron and the icosahedron.

The field of astronomy, too, is indebted to Pythagoras. He was among the first to contend that the earth and the universe are spherical. He understood that the sun, the moon, and the planets rotate on their own axes and also orbit a central point outside themselves, though he believed that this central point was the earth. Later Pythagoreans deposed the earth as the center of the universe and substituted a "central fire," which, however, they did not identify as the sun—this they saw as another planet. Nearest the central fire was the "counter-earth," which always accompanied the earth in its orbit. The Pythagoreans assumed that the earth's rotation and its revolution around the central fire took the same amount of time—twenty-four hours. According to Aristotle, the idea of a counter-earth—besides bringing the number of revolving bodies up to the mystical number of ten—helped to explain lunar eclipses, which were thought to be caused by the counter-earth's interposition between sun and moon. Two thousand years later, Nicolaus Copernicus saw the Pythagorean system as anticipating his own; he had in mind both the Pythagoreans' concept of the day-and-night cycle and their explanation of eclipses.

Like Copernicus in his time, Pythagoras and his followers in their time were highly controversial. For many years, the Pythagoreans did exert a strong political and philosophical influence throughout southern Italy. The closing years of the sixth century B.C.E., however, saw the rise of democratic sentiments, and a reaction set in against the Pythagoreans, whom the democrats regarded as elitist.

Indeed, this political reaction led either to Pythagoras's exile or to his death—there are two traditions surrounding it. One is that a democrat named Cylon led a revolt against the power of the Pythagorean brotherhood and forced Pythagoras to retire to Metapontum, where he died peacefully about the end of the sixth century B.C.E. According to the other tradition, Pythagoras perished when his adversaries set fire to his school in Croton in 504 B.C.E. The story is that of his vast library of scrolls, only one was brought out of the fire; it contained his most esoteric secrets, which were passed on to succeeding generations of Pythagoreans.

Whichever account is true, Pythagoras's followers continued to be powerful throughout Magna Graecia until at least the middle of the fifth century B.C.E., when another reaction set in against them, and their meetinghouses were sacked and burned. The survivors scattered in exile and did not return to Italy until the end of the fifth century. During the ensuing decades, the leading Pythagorean was Philolaus, who wrote the first systematic exposition of Pythagorean philosophy. Philolaus's influence can be traced to Plato through their mutual friend Archytas, who ruled Taras (Tarentum) in Italy for many years. The Platonic dialogue *Timaeus* (360-347 B.C.E.), named for its main character, a young Pythagorean astronomer, describes Pythagorean ideas in detail.

SIGNIFICANCE

"Of all men," said Heraclitus, "Pythagoras, the son of Mnesarchus, was the most assiduous inquirer." Pythagoras is said to have been the first person to call himself a philosopher, or lover of wisdom. He believed that the universe is a logical, symmetrical whole, which can be understood in simple terms. For Pythagoras and his students, there was no gap between the scientific or mathematical ideal and the aesthetic. The beauty of his concepts and of the universe they described lies in their simplicity and consistency.

Quite aside from any of Pythagoras's specific intellectual accomplishments, his belief in universal order, and the energy he displayed in seeking it out, provided a galvanizing example for others. Sketchy as are the details of his personal life, his ideals left their mark on later poets, artists, scientists, and philosophers from Plato and Aristotle through the Renaissance and down to the twentieth century. Indirectly, through Pythagoras's disciple Philolaus, his ideas were transmitted to Plato and Aristotle, and, through these better-known thinkers, to the entire Western world.

Among Pythagoras's specific accomplishments, his systematic exposition of mathematical principles alone would have been enough to make him an important figure in Western intellectual history, but the spiritual beliefs he espoused make him also one of the great religious teachers of ancient Greek times. Even those ideas of his that are seen as intellectually disreputable have inspired generations of poets and artists. For example, the Pythagorean concept of the harmony of the spheres, suggested by the analogy between musical ratios and those of planetary orbits, became a central metaphor of Renaissance literature.

Thomas Rankin

FURTHER READING

Guthrie, Kenneth Sylvan, ed. *The Pythagorean Sourcebook and Library: An Anthology of Ancient Writings Which Relate to Pythagoras and Pythagorean Philosophy*. Grand Rapids, Mich.: Phanes Press, 1988. This anthology of Pythagorean writings contains the four ancient biographies of Pythagoras

as well as later Pythagorean and Neopythagorean writings.
Kahn, Charles H. *Pythagoras and the Pythagoreans*. Indianapolis: Hackett, 2001. Surveys Pythagorean tradition from Pythagoras's time to early modern times, including his influence on early modern math, music, and astronomy. Indexed by ancient and early modern name and by modern name.
Kirk, Geoffrey S., and John E. Raven. *The Presocratic Philosophers*. New York: Cambridge University Press, 1983. Provides a good account of Pythagoras and his followers, in their historical context, from a philosopher's point of view.
Muir, Jane. *Of Men and Numbers*. New York: Dover, 1996. Written for lay readers. Contains a chapter on Pythagoras's mathematical work and its influence on later scientists, especially Euclid.
Philip, J. A. *Pythagoras and Early Pythagoreanism*. Toronto: University of Toronto Press, 1968. Attempts to separate the valid information from the legends surrounding Pythagoras and his teachings. Includes notes and a selected bibliography.
Strohmeier, John, and Peter Westbrook. *Divine Harmony: The Life and Teachings of Pythagoras*. Berkeley, Calif.: Berkeley Hills Books, 2003. Describes Pythagoras's travels in Egypt, Phoenicia, Babylonia, and Greece and examines Pythagorean ideas as taught at his scholarly community in southern Italy. Includes illustrations, map, introduction, and bibliography.

BERTRAND RUSSELL
Welsh philosopher and mathematician

Russell's original work in the areas of logic, mathematics, and the theory of knowledge was complemented by several important volumes of philosophical popularization. In his later years Russell emerged as a major figure in the peace movement.

Born: May 18, 1872; Trelleck, Monmouthshire, Wales
Died: February 2, 1970; Plas Penrhyn, near Penrhyndeudraeth, Wales
Also known as: Bertrand Arthur William Russell (full name)

EARLY LIFE
Bertrand Russell was born in Trelleck, Monmouthshire, Wales. His mother, née Kate Stanley, was the daughter of the second Baron Stanley of Alderley and a leader in the fight for votes for women; his father, Lord Amberley, was the eldest son of the first Earl Russell and a freethinker who lost his seat in Parliament because of his advocacy of birth control. Both parents were considered extremely eccentric, and both died before Russell reached the age of four. Russell and his older brother were brought up by their rigidly conventional paternal grandmother, and they spent a rather solemn childhood being educated at home by a succession of governesses and tutors.

At the age of eighteen, Russell entered Trinity College, Cambridge, where it did not take him long to make a positive impression. He was taken under the wing of the philosopher Alfred North Whitehead, with whom he would later collaborate on *Principia Mathematica* (1910-1913), and was much influenced by fellow student G. E. Moore (1873-1958), who helped him to develop his early ideas on the independent existence of what is perceived by the senses. In 1894, Russell married Alys Pearsall Smith, an American Quaker five years older than he was, and in 1895 he was elected a Fellow of Trinity College for his dissertation "An Essay on the Foundations of Geometry." In the following year, he and his wife spent three months in the United States, thus beginning a lifelong interest in and involvement with American affairs.

In the late 1890's, Russell achieved wide recognition as a professional philosopher of promise, as he subjected the dominant Idealist thought of the period to an increasingly rigorous critique. His personal life revolved around the strains of a deteriorating marriage, which in 1902 reached a crisis when Russell told his wife that he no longer loved her. Although they continued to live together until 1911, the pressures of conflict at home and a demanding professional career made this the most difficult period of Russell's life. It was also, however, a very productive time for him, highlighted by the publication of perhaps his greatest single work: *The Principles of Mathematics* (1903), which took the groundbreaking step of removing metaphysical notions from the concept of numbers and arguing that logic alone could serve as the basis for a true science of mathematics. After the publication of this volume, even those who took issue with Russell's views had to acknowledge his status as a major contributor to contemporary philosophical and mathematical thinking.

Russell's striking personal appearance became part of the folklore of Cambridge. His tall, thin frame and sharply chiseled, almost hawkish facial lines were seldom observed at rest, as his penchant for vigorous intellectual disputation was matched by a passion for strenuous walking. Russell kept his distinctive looks to the end of his life, with the only significant change being a whitening of his full head of hair, which added a mature dignity to his craggy features. The heavy media coverage of his public appearances on behalf of the peace movement in the 1960's reflected the charismatic appeal of his majestically leonine figure, which seemed to many observers to possess an almost biblical air of wisdom and authority.

LIFE'S WORK

The decade preceding the outbreak of World War II found Russell achieving success as a professional philosopher and undertaking what would be the first in a tempestuous string of love affairs and marriages. His collaboration with Whitehead on the three volumes of *Principia Mathematica* developed the ideas touched on in *The Principles of Mathematics* into a coherent and influential formal system, and he was fruitfully stimulated by his pupil Ludwig Wittgenstein, who helped him to clarify his thoughts about the proper conduct of philosophical analysis. Russell's growing interest in the theory of knowledge resulted in his *The Problems of Philosophy* (1912), the first in what would be a series of books concerned with such perennial philosophical issues as the nature of reality and the operations of the mind. In 1911, he began an intense love affair with Lady Ottoline Morrell, which lasted until 1916 and put an end to his first marriage.

Russell was deeply affected by the horrors of World War I and found himself compelled to become active in the pacifist movement. His *Principles of Social Reconstruction* (1916) signaled a deepening involvement with questions of human relations, and his antiwar efforts led to his being fined in 1916, imprisoned for six months in 1918, and as a result deprived of his lectureship at Trinity College. In 1916 he met and in 1921 married his second wife, Dora Black, a fellow freethinker with whom he had two children and briefly ran an experimental school at Beacon Hill. A visit to postrevolutionary Russia produced *The Practice and Theory of Bolshevism* (1920), in which he recorded his disillusionment with the gap between the Soviet Union's promise and performance, and he became somewhat notorious for his advocacy of free love—a doctrine he practiced in a string of extramarital affairs—in the book *Marriage and Morals* (1929).

Russell continued to do serious work in philosophy during the interwar period, although his refusal to accept Trinity College's offer of reinstatement meant that he had to spend more time writing financially profitable books and journalistic articles. Thus, popularly oriented titles such as *The Conquest of Happiness* (1930) and *Education and the Modern World* (1932) were interspersed with the more technical philosophical works *Our Knowledge of the External World* (1929) and *An Inquiry into Meaning and Truth* (1940), and Russell also became a popular lecturer on topics such as divorce, sexual relations, and pacifism. His personal life continued to be turbulent, as he divorced his second wife in 1935 and married his third, Patricia Spence, in 1936, to whom a son was born in the following year.

In 1938, Russell and his new family moved to the United States, where he held several university posts and made a number of extensive lecture tours. He was in continual difficulty as right-wing pressure groups

Bertrand Russell. (Library of Congress)

attacked his liberal views, and in 1943 he even had to go to court to collect his salary from the outraged head of a charitable foundation. The main positive result of his American sojourn was *History of Western Philosophy* (1945), a popular success whose royalties would comfortably support him for the remainder of his life. The Russells returned to England in 1944, and he decided to accept a five-year lectureship at Trinity College and endeavor to settle down into a less hectic pattern of existence.

The award of the Order of Merit in 1949 and the Nobel Prize in Literature in 1950 indicated that Russell's achievements now commanded general respect. This did not, however, seem to have a stabilizing effect on his domestic life: He divorced his third wife in 1949 and married his fourth, Edith Finch, in 1952. Disappointed by the indifference of his academic colleagues to his last serious philosophical work, *Human Knowledge: Its Scope and Limits* (1948), Russell devoted more and more time to antiwar activities. He helped to found the Campaign for Nuclear Disarmament in 1958, organized its militant wing, the Committee of 100, in 1960, and was jailed for seven days for participating in a 1961 Whitehall sit-in. Although Russell's physical powers now began to weaken, he remained an effective propagandist for the pacifist movement and in the late 1960's was a prominent member of the international opposition to the U.S. presence in Vietnam. His final days were spent resting quietly at his home in northern Wales, where he died on February 2, 1970.

SIGNIFICANCE

Russell brought keen intellectual perception to every task that fired his imagination. As a philosopher, he was instrumental in the development of modern analytical techniques; as a mathematician, he helped to anchor speculative hypotheses on the firm ground of formal logic; and as a political activist, he cut through the verbiage of politicians with a clarion call to abandon nuclear weapons before they produced a global holocaust. Although the content of his ideas has in some cases been rejected by subsequent commentators, there is an almost universal acknowledgment of his methodological contributions to his areas of academic specialization.

However, Russell was not merely a man with a great mind. His charismatic public persona and his

---

> ### "Why I Am Not a Christian"
>
> *In a lecture delivered at Battersea Town Hall in London on March 6, 1927, Bertrand Russell discussed his ideas about religion, including his belief that fear was religion's foundation.*
>
> Religion is based, I think, primarily and mainly upon fear. It is partly the terror of the unknown and partly, as I have said, the wish to feel that you have a kind of elder brother who will stand by you in all your troubles and disputes.... Fear is the parent of cruelty, and therefore it is no wonder if cruelty and religion have gone hand-in-hand.... In this world we can now begin a little to understand things, and a little to master them by the help of science, which has forced its way step by step against the Christian religion, against the churches, and against the opposition of all the old precepts. Science can help us to get over this craven fear in which mankind has lived for so many generations. Science can teach us, and I think our own hearts can teach us, no longer to look around for imaginary supports, no longer to invent allies in the sky, but rather to look to our own efforts here below to make this world a fit place to live in, instead of the sort of place that the churches in all these centuries have made it.

---

ability to write for a general readership extended his influence into regions usually closed to the professional academic, and the range of his published work is extraordinarily impressive. His volatile emotional life also reflected the immense energy and appetite for new experiences that, in combination with his outstanding intellectual prowess, made Bertrand Russell one of the seminal figures of his time.

Paul Stuewe

FURTHER READING

Banfield, Ann. *The Phantom Table: Woolf, Fry, Russell, and the Epistemology of Modernism*. New York: Cambridge University Press, 2000. Examines Virginia Woolf's involvement with Russell and the other members of the Bloomsbury Group to describe how they developed their concepts of modernism.

Feinberg, Barry, and Ronald Kasrils. *Bertrand Russell's America*. Vol 1. New York: Viking Press, 1973. Vol. 2. Boston: South End Press, 1983. Traces Russell's attitudes toward and experiences in the United States. The authors make extensive use of previously unpublished letters and essays and succeed in presenting a comprehensive account of what are often very dramatic and at times even amusing episodes in Russell's career.

Griffin, Nicholas. *The Cambridge Companion to Bertrand Russell*. New York: Cambridge University Press, 2003. Collection of essays analyzing various aspects of Russell's work, including his mathematics, philosophy, and ideas about morality.

Jager, Ronald. *The Development of Bertrand Russell's Philosophy*. New York: Humanities Press, 1973. An excellent study of Russell's growth as a philosopher, aimed at general readers as well as specialists. Pays particular attention to the historical influences, ranging from Plato to Wittgenstein, that affected a thinker always conscious of the rich history of his discipline.

Nakhnikian, George, ed. *Bertrand Russell's Philosophy*. New York: Harper & Row, 1974. Contains fourteen essays on various aspects of Russell's thought, from logical issues to the theory of knowledge to political philosophy. Some essays assume a background in philosophy, but the majority should be accessible to nonspecialist readers. Includes an excellent bibliography.

Potter, Michael K. *Bertrand Russell's Ethics*. New York: Continuum, 2006. Focuses on Russell's ideas on ethics and how his ethics guided his social activism.

Roberts, George W., ed. *Bertrand Russell Memorial Volume*. New York: Humanities Press, 1979. Twenty-six essays assessing Russell's achievements in philosophy, mathematics, ethics, and politics, from a distinguished group of contemporary scholars. The book is strongest on Russell's philosophical accomplishments, although its exhaustive index makes it a good source of information on almost every facet of his career.

Russell, Bertrand. *The Autobiography of Bertrand Russell*. 3 vols. Boston: Little, Brown, 1967-1969. The judicious mixture of chronological narrative and extensive quotations from letters contributes to the impact made by this superb intellectual autobiography. On personal matters, the story needs to be fleshed out by the books by Clark and Tait, but for a fascinating account of the development of Russell's thought there is no better source than these eminently readable volumes.

Tait, Katharine. *My Father Bertrand Russell*. 1975. New ed. Bristol, England: Thoemmes, 1996. Tait, Russell's daughter by his second wife, concentrates on his failures as a father and demonstrates how his self-centeredness resulted in great pain for his family. A sensitive memoir that provides an unfamiliar angle on Russell as well as some implicit criticism of his educational and social theories. Includes a new introduction by Ray Monk.

Vellacott, Jo. *Bertrand Russell and the Pacifists in the First World War*. New York: St. Martin's Press, 1981. A welcome portrait of Russell at the beginning of his career as a political activist. The book provides much information about the origins of the pacifist movement in Great Britain, particularly its socialist and Quaker roots, and is very well researched and written.

Charlotte Angas Scott

English-born American mathematician

As the first professor of mathematics at Bryn Mawr College in Pennsylvania, Scott created an environment that encouraged a large number of women to work for higher degrees in mathematics. Her research and stature in the field of mathematics blazed a trail that was not matched by other women for many years.

Born: June 8, 1858; Lincoln, England
Died: November 10, 1931; Cambridge, England

Early Life

Charlotte Angas Scott was born in England, the second of seven children of Caleb and Eliza Ann Exley Scott. Her father was a pastor at a nonconformist (non-Anglican Protestant) church; when Charlotte was seven years old, he was elevated to the headship of Lancashire College. Charlotte was educated primarily at home, and her family could afford to provide her with a sequence of excellent tutors. In addition, the family often played mathematical games with her to provide an additional level of encouragement.

Charlotte's mathematical talents were sufficiently evident that she received a scholarship at Girton College, one of the newly created women's colleges at Cambridge University. At that time, women were not allowed to receive degrees from either Cambridge or

Oxford University, the two most prestigious institutions of higher education in England. However, that limitation had not prevented the creation of women's colleges, in which female students could study, even if they could not receive degrees. The idea of offering higher education to women was then largely a subject of mockery. An example of public attitudes can be seen in W. S. Gilbert and Arthur Sullivan's musical play *Princess Ida* (1884). It took great resolution for women students to put up with antifeminist attitudes in education, but Charlotte Scott was a model of resolution.

Scott received her first—and largest—dose of public attention when she took the Tripos examination at Cambridge. This test took its name from a three-legged stool on which students had originally sat, but it had evolved into the most challenging mathematical exam in England. It was taken by students working for undergraduate degrees in mathematics—a highly select group. Women were allowed to take the exam, but their results were not reported along with those of the male students. When Scott took the exam, she earned a high score equivalent to "eighth wrangler," but her name was omitted from the list of the results, according to university policy.

The Times of London, the most influential newspaper in Great Britain, took up Scott's case and transformed the question of reporting women's exam results into a national issue. The humorous weekly *Punch* pleaded her case, even though it was not known for holding progressive views. A campaign among the alumni of Cambridge University encouraged the university to change its policy by including women's exam results along with those of men. This change took place the year after Scott's success and was one of the first steps in the direction of allowing women to take degrees from the senior universities in England.

Life's Work

After her years as a student at Girton, Scott stayed on at Cambridge as a lecturer at the same college. Since the University of London did not have the same prohibition on women taking degrees that prevailed at Oxford and Cambridge, Scott earned an undergraduate degree there in 1882 and a doctorate in 1885, following in the footsteps of the relatively small number of women who had received advanced degrees in mathematics elsewhere in Europe. Her work was in the area of algebraic geometry, a subject that was at the forefront of mathematical research at the time. While she was at Cambridge, she had the chance to work with Arthur Cayley, one of the outstanding mathematicians in Europe and a founder of both modern algebra and its application to geometry.

The idea behind algebraic geometry is to use algebraic arguments to prove results about geometric objects. This process involves finding algebraic methods of describing curves and surfaces—which was usually done with equations. If one could work with the equations, one could apply the techniques of algebra rather than having to fall back on the principles of geometry, which sometimes were difficult to make precise. Many different types of geometry were being introduced into academic curricula during the late nineteenth century, and each came with a different range of algebraic techniques. Scott herself became especially well known for her applications of algebra to projective geometry, the branch in which curves are treated as identical when one can be projected onto the other.

During the year 1885, Bryn Mawr College opened in Bryn Mawr, Pennsylvania, to provide education of the highest quality, including graduate study, to women. The new college hired Scott to head its mathematics department, in part because she was one of the few women in the world with a doctorate in the subject. It is a tribute to her teaching skills that Bryn Mawr rapidly became a center for women pursuing doctorates in mathematics. During her tenure there, Bryn Mawr trailed only the University of Chicago and Cornell University in numbers of doctoral degrees in mathematics awarded to women. During that same period, of all U.S. doctorates awarded in mathematics, 14 percent went to women. By comparison, during the 1950's, only 5 percent of doctorates in mathematics went to women. Only in later years have the percentages of doctorates awarded to women exceeded the levels achieved by Scott at Bryn Mawr.

While she was teaching and supervising students at Bryn Mawr, Scott continued to do research. In 1894, she published a textbook with the ponderous title *An Introductory Account of Certain Modern Ideas and Methods in Plane Analytical Geometry* and later saw another edition through the press. The book was well received, and her gift for exposition was described as lucid.

During Scott's years at Bryn Mawr, she made annual pilgrimages back to England to be in a more active mathematical environment. However, she also promoted the professionalization of mathematics in the United States and was active in the newly formed New York Mathematical Society, the ancestor of the American Mathematical Society. She served on the council of the society on many occasions and served as its vice president in 1906. She was the first woman to hold that position, and she did not have a female successor for many decades. In recognition of her status, the first edition of *American Men of Science* (1906) included an entry for her.

Scott's views on education did not always make her life easy at Bryn Mawr. She disliked student behavior that she regarded as immoral, such as smoking and wearing makeup. On one occasion she took the president of the college to task for diluting the quality of women's education. However, she also made signal contributions to campus life, such as bringing the distinguished British mathematician and philosopher Bertrand Russell there to speak. Her teaching helped to put Bryn Mawr on the intellectual map in both England and the United States.

Scott taught at Bryn Mawr for forty years. By her last years there, ill health was interfering with her teaching as she became deaf and suffered from arthritis. When she retired to Cambridge, England, in 1925, she took up gardening and also spent some of her time and money betting on horse racing. On November 10, 1931, she died at her Cambridge home, at the age of seventy-three.

Significance

Charlotte Scott achieved her greatest prominence in the world at large while she was still a student. Her success on the Tripos exam indicated that women were at Cambridge for reasons other than to find husbands or make polemical statements. Her subsequent distinction as a mathematician made Cambridge University proud to have been her alma mater, even if it would not grant her a degree.

As a professional mathematician, Scott did not revolutionize the subjects in which she worked, but her solid accomplishments paved the way for women to achieve success in graduate work and beyond. She guaranteed that Bryn Mawr College would provide serious mathematical training for women, even if her insistence on high standards did not always make her an easy colleague. The fact that during the 1930's, Emmy Noether, the world's most eminent woman mathematician, left Germany because of the rise of Adolf Hitler and went to Bryn Mawr is a tribute to the mathematical environment that Charlotte Scott had created there.

Thomas Drucker

Further Reading

Farquhar, Diane, and Lynn Mary-Rose. *Women Sum It Up*. Christchurch, New Zealand: Hazard Press, 1989. Study of female mathematicians that covers aspects of Scott's family background and personal life, rather than her mathematical career.

Gray, J. J. "Charlotte Angas Scott." In *Oxford Dictionary of National Biography*, edited by H. C. G. Matthew and Brian Harrison. Oxford, England: Oxford University Press, 2004. Mathematically well-informed summary of Scott's life against the background of the times.

Green, Judy, and Jeanne Laduke. "Contributors to American Mathematics." In *Women of Science: Righting the Record*, edited by G. Kass-Simon and Patricia Farnes. Bloomington: Indiana University Press, 1990. Recognizes Scott's work in encouraging women to higher education in mathematics.

Kenschaft, Patricia Clark. "Charlotte Angas Scott." In *Complexities: Women in Mathematics*, edited by Bettye Anne Case and Anne M. Leggett. Princeton, N.J.: Princeton University Press, 2005. One of several articles on Scott by Kenschaft; useful for filling in some cultural background.

Macaulay, F. S. "Dr. Charlotte Angas Scott." *Journal of the London Mathematical Society* 7 (1932): 230-240. The source on which most subsequent biographers draw, paying full attention to Scott's mathematical work.

Seki Kōwa

Japanese mathematician and government official

Seki Kōwa was a key figure in the development of premodern Japanese mathematics. The expansion of commerce under Tokugawa rule created a new need for practical mathematical accounting techniques. Seki and his student Takebe Katahiro were leaders in developing such techniques, and they independently duplicated some major discoveries of leading Western mathematicians, such as Gottfried Wilhelm Leibniz and Sir Isaac Newton.

Born: March, 1642; Fujioka, Kozuke province, Japan
Died: October 24, 1708; Edo (now Tokyo), Japan
Also known as: Seki Takakazu

Early Life

Seki Kōwa (sehk-ee koh-wo) was born in 1642, the second son of Uchiyama Shichibei. Uchiyama had been an adviser to a once-powerful domain lord who fell out of favor with the shogun ten years before Seki's birth, and that lord was forced to commit suicide. As a result, Uchiyama's own future prospects were in doubt. Consequently, it may have significantly benefited Seki when he was adopted at an early age by a well-placed samurai named Seki Gorozaemon.

There were a number of retainers in attendance at Gorozaemon's mansion, and they took care of young Seki and kept him company. One day, one of these retainers was reading a book on abacus calculation that happened to attract Seki's attention. The retainer taught Seki the basics of abacus use, and Seki then obtained an abacus and worked through all the exercises in the book on his own. From that point on, he spent most of his waking hours learning as much about mathematics as he could, from every book he could obtain. Seki is known to have studied Yoshida Mitsuyoshi's *Jinkōki* (1627; *Book of Large and Small Numbers*, 2000), working through it page by page on his own.

It was the custom at that time for Japanese mathematicians to write books containing some mathematical problems without solutions, as a challenge for readers. One book Seki read, containing one hundred challenging problems by another brilliant young mathematics prodigy, Isomura Yoshinori, stimulated Seki to work out solutions to all one hundred problems. Throughout his career, Seki worked out original solutions to mathematics problems by Isomura and others, though he seems not to have created many such problems himself.

Seki obtained a modest official position as an accountant at the Edo mansion of Tokugawa Tsunashige, the domain lord of Kōfu, in modern Yamanashi prefecture. When Tsunashige was made the heir to the reigning shogun, Tokugawa Ietsuna, Seki rose in status along with the rest of Tsunashige's household. In addition to his accounting work, Seki served as the mathematics tutor of the offspring of Tsunashige, his retainers, and other Edo aristocrats.

Life's Work

The chances of Seki Kōwa's patron Tokugawa Tsunashige becoming shogun were reduced over time as the result of court intrigues by his rivals, and this in turn lowered Seki's own prospects for advancement in Edo. Tsunashige died in 1678, two years before Ietsuna, but this in fact turned out to enhance Seki's prospects.

Tokugawa Ienobu, Tsunashige's son, to whom Seki had given lessons and for whom he had subsequently worked as accountant, was in a much better position at the shogun's court than Tsunashige had been. Ienobu succeeded to the position of domain lord of Kōfu in 1678, and he was later adopted as the heir of Shogun Ietsuna's successor, Tokugawa Tsunayoshi, in 1704. Ienobu did not actually become shogun until 1709, a year after Seki's death, but Seki's close relationship with Ienobu assured him a good position as an accountant and tutor at the shogun's court. He enjoyed this position in Edo from 1678 to 1708, providing Seki with both the leisure and the resources he needed to conduct his research.

The early Tokugawa era was an important developmental period for mathematics in Japan. The Tokugawa shogunate had restored order to the country after more than 150 years of civil war. Commerce had once again begun to develop and flourish, and this created a pressing national need for accounting, inventory, and other commercial and financial forms of record-keeping. Practical mathematics had fallen into relative disuse in Japan, so the latest techniques were imported from China. A number of new abacus schools *(soroban-juku)*, using Chinese methods, soon began to appear in Edo, the Kyōto-Ōsaka area, and elsewhere.

One of these abacus schools was run by Mōri Shigeyoshi, a former samurai *(rōnin)* said to have fought on the losing side against the Tokugawas. Mōri was the author of the earliest Japanese mathematics book still extant, the *Warizansho* (1622; writings on division), though it was actually a general abacus calculation manual. Yoshida Mitsuyoshi, who had learned mathematics at Mōri's academy, became the mathematics tutor of the feudal domain lord of Kumamoto and wrote a derivative work of his own, the *Jinkōki*. Because Chinese methods still largely focused on basic abacus calculation methods, however, aspiring Japanese mathematicians began to see a need to move beyond this stage. Building on the familiar Chinese base, these mathematicians created a new set of uniquely Japanese mathematical techniques, known as *wasan*.

Yoshida Mitsuyoshi had established his reputation as a mathematical scholar and tutor to the aristocracy, so it was relatively easier for Seki to achieve the same sort of status a generation later. Most of Seki's posthumous fame, in fact, resulted from accounts of his life and work by his aristocratic students. Unlike most of his contemporaries, Seki did not actively seek to publicize his own mathematical accomplishments. In fact, though he wrote many manuscripts on mathematics, Seki published only one book during his lifetime, *Hatsubi sanpō* (1674), a book of solutions to mathematical problems.

Among Seki's aristocratic students were Takebe Katahiro and Katahiro's older brother Katakira, the sons of an important official at the shogun court in Edo. Together with Seki, they worked on an encyclopedic study of all current Japanese mathematical knowledge, the *Taisei sankei* (1710; collection of classic mathematical texts). It took twenty-eight years to complete this twenty-volume work, which was finally published two years after Seki's death.

Among Seki's other pioneering accomplishments in Japanese mathematics, he developed a workable system of algebraic notation, a theory of determinants, and a formula for calculating the circumference of a circle. Seki also approximated roots of higher-order equations and is believed to have discovered the concept eventually named "Bernoulli numbers" even before Jakob I Bernoulli did. Seki also developed a form of calculus and anticipated important discoveries by Gottfried Wilhelm Leibniz and Sir Isaac Newton.

Takebe Katahiro is considered by historians to be Seki's true intellectual heir, who refined and further developed his teacher's mathematical ideas and methods. Katahiro carefully went through Seki's many manuscripts, systematizing his ideas and preserving his writings for future generations. Katahiro also wrote his own mathematics books and made additional independent discoveries, succeeding Seki as the court mathematician for several more shoguns until his own death in 1739. The period of more than six decades during which Seki and his students dominated mathematics in Edo assured the continued predominance of Seki's methods and concepts, until Western mathematics replaced traditional Japanese mathematics completely.

Significance

From the seventeenth century to the second half of the nineteenth century, the field of Japanese mathematics was dominated by *wasan*, as developed by Seki Kōwa, Takebe Katahiro, and others. *Wasan* was also the form of mathematics initially chosen by Japanese modernizers for use in the national school system, until it was officially replaced by European-style mathematics in 1873. A preexisting knowledge of *wasan* methods actually functioned as a bridge to the future, serving as the foundation upon which Meiji-era students were able to learn the new Western style of mathematics. Like Seki and his followers, many nineteenth century *wasan* experts were also from the samurai class, well-placed and well-educated people who often made the transition to become teachers of the new Western-style mathematics.

Kikuchi Dairoku (1855-1917), a pioneer in developing the study of modern mathematics in Japan, began learning Western mathematics at an early age, but later, as a leading mathematics teacher in Japan, he took advantage of his students' basic *wasan* knowledge, using it as a basis on which to teach them equivalent new Western methods. Fujisawa Rikitarō (1861-1933), one of Kikuchi's earliest students who had learned *wasan* as a child, subsequently became a leading figure in modern Japanese mathematics. Far from dismissing the importance of the *wasan* methods developed by Seki Kōwa, Fujisawa gave lectures on those methods in Japan and Europe. When the Tokyo Sugaku Kaisha, the first modern mathematical society in Japan, was established in 1877, more than half of its members had originally been *wasan* mathematicians.

Michael McCaskey

Further Reading

Horiuchi, Annick. *Les mathematiques japonaises a l'epoque d'Edo, 1600-1868: Une étude des travaux de Seki Takakazu, ?-1708, et de Takebe Katahiro, 1664-1739*. Paris: Mathesis, 1994. A four-hundred-page study of the work of Seki and Takebe, the only book on this subject so far in a Western language.

Morris-Suzuki, Tessa. *The Technological Transformation of Japan: From the Seventeenth to the Twenty-first Century*. New York: Cambridge University Press, 1994. Contains an essay on Tokugawa technological development.

Nakayama, Shigeru. *Academic and Scientific Traditions in China, Japan, and the West*. Tokyo: University of Tokyo Press, 1984. A standard comparative history of scientific thought and research by a leading Japanese authority.

Smith, David, and Mikami Yoshio. *A History of Japanese Mathematics*. Mansfield Center, Conn.: Martino, 2002. The definitive history of Japanese mathematics up to the early twentieth century in English.

Jakob Steiner

Swiss mathematician

One of the greatest geometers of the first half of the nineteenth century, Steiner wrote books and dozens of articles on geometry that established him as a chief authority on isoperimetric geometry and as the founder of modern synthetic geometry in Germany.

Born: March 18, 1796; Utzentorf, Canton of Bern, Switzerland
Died: April 1, 1863; Bern, Switzerland

Early Life

Jakob Steiner (SHTI-ner) was born into a family of thrifty, humble, and hardworking Swiss farmers. Though the youngest of five children, he contributed from a very early age to the family income, the family expecting nothing more than the most modest intellectual development. Consequently, he remained illiterate until he was fourteen and continued farm work until he was nineteen. According to his later recollections, before he had any formal education he developed an astounding capacity for spatial conceptualization.

Contrary to the desires of his father, Jakob entered the school of the Swiss educational reformer Johann Pestalozzi at Yverdon. Out of conformity with Swiss educational precepts, Pestalozzi continued stressing the pedagogical importance of individual training and direct experience for his students. Before Pestalozzi's institution failed, Steiner had become a teaching assistant. Thereafter, Steiner entered Heidelberg University, where he pursued numerical perceptions in connection with imaginative spatial concepts. From 1818 until 1821, while earning a living as a teacher, Steiner worked with one of the institution's leading geometers, whose lectures and ideas he profoundly disdained. Notwithstanding, Steiner obtained his doctorate from Heidelberg, thereafter accepting a teaching position as a tutor at a private school.

The eldest son of the famed German statesman and philologist Wilhelm von Humboldt was one of Steiner's pupils. Steiner's acquaintance with the distinguished Humboldt family altered his fortunes. The Humboldts introduced him to Berlin's premier mathematicians, and Steiner was encouraged to accept a teaching post at a Berlin vocational institution during the next decade. Eventually the University of Berlin created an endowed chair, which Steiner was to fill—indeed, he had, since 1834, been a member of the Berlin Academy on the basis of his previous mathematical, or geometrical, writings.

Life's Work

Steiner's mathematical publications commenced in 1826, while he still tutored at his vocational school. This creative production coincided with the founding by August Leopold Crelle of what became one of the nineteenth century's most famous mathematical publications, *Journal für die reine und angewandte Mathematik* (the journal for pure and applied mathematics). Professionally, Steiner expanded his reputation in 1832 with his *Systematische Entwicklung der Abhängigkeit Geometrischer Gestalten* (systematic evolution of the mutual dependence of geometrical forms), a planned introduction to a five-part series never to be completed.

Steiner's work does not readily reduce to layman's terms. It is projective geometry, built upon synthetic constructions. Geometry's basic forms are based on planes. Projective geometry moves from the fundamental plane to lines, planar pencils of lines to pencils of planes, bundles of lines, bundles of planes—and then into space itself, steadily generating higher geometric forms. For Steiner, one form in this projective hierarchy related with the others.

It was not the originality of Steiner's work that was dominant, although the questions he raised were then novel considering geometers' principal preoccupations. Steiner's own view was that "the writings of the present day have tried to reveal the organism by which the sundry phenomena of the external world are bound to one another." What he sought to determine was how "order enters into chaos," how all parts of the external world fit naturally into one another, and how related parts join to form well-defined groups. Specifically, it was the brilliantly stated and systematic treatment Steiner lent to his inquiries that gained for him his reputation.

The unique and justly famed French École Polytechnique, with its unparalleled training of France's intellectual elite and special concentration of intensive mathematical training, had long before Steiner's day divided geometry into two branches: the analytical and the synthetic, or projective. During the early seventeenth century, René Descartes had explained how numbers could be utilized to describe points in a plane or in space algebraically. Steiner, however, concentrated on the other branch: projective geometry, which did not usually resort to the measurements or lengths of angles.

Steiner learned something from Johann Pestalozzi and his eccentric preoccupation with right triangles, and as a pedagogue Steiner, like Pestalozzi, encouraged his students' independent and rigorously logical search for learning. As might be expected, Steiner avoided figures to illustrate his lectures. His own intuitions were so much a part of his character, he sought both in teaching and writing to use them. He did not neglect his own disciplined scholarship. He read exhaustively the works of his European counterparts, staying on the cutting edge of his investigations.

Mathematical authorities agree that in midcareer Steiner still fell short of his goals by rejecting the achievements of some of his predecessors and contemporaries. For example, he lost the chance to employ signs drawn from Karl August Möbius's synthetic geometry and therefore the opportunity for the full deployment of his imagination. It is small wonder that Steiner sometimes wrote of "the shadow land of geometry."

Steiner's practical ambitions, related to, but lying near or on the margin of his geometrical scholarship, were not as shadowy. Perhaps this was understandable, for in class-conscious Berlin and the German academic world, his social origins were not advantageous. His special professorship or chair created for him at Berlin University was partly an effort to avoid this implicit embarrassment. Moreover, the timing of his publications was partly calculated to advance him toward the directorship of Berlin's planned Polytechnic Institute. Hence, in 1833 he published a short work, *Die geometrischen Konstruktionen, ausgeführt mittelst der geraden Linie und eines festen Kreises* (*Geometrical Constructions with a Ruler, Given a Fixed Circle with Its Center*, 1950), which was intended for high schools and for practical purposes. Indeed, following his appointment to his Berlin chair, he never completed what he promised would be a comprehensive work.

Steiner apparently was not surprised when analytical geometricians discovered that his own results could often be verified analytically. It was not so much that Steiner disdained others' analyses. Rather, he was headed in a different direction of inquiry, and he believed that analysis prevented geometricians from seeing things as they actually are. Like other projective geometers, he thought that because projective geometry could advance so swiftly from a few fundamental concepts to significant statements, he, like them, eschewed the formidable axiomatic studies that were the hallmark of Euclidean geometry. Most mathematicians argued against him, however, that despite Steiner's disclaimers, there was no royal road to a new geometry. No matter how logical, clear, and intuitive Steiner's projective geometry was, most geometricians actually wanted to see—metrically and analytically—what the projectivists were describing. Geometry was, for its nineteenth century scholars, simply too full of irrationals to make its results completely tenable.

Significance

Jakob Steiner was not the originator of projective or synthetic geometry. Nevertheless, his contributions were substantial and significant in the revival and advancement of synthetic geometry. This was the result of his clearly presented intuitions and his marvelous systematization of his projections. Before the close of his career, moreover, he had both trained others through the clarity of his lectures and writings and encouraged other geometricians such as Julius Plücker, Karl Weierstrass, and Karl von Staudt to resolve problems that had eluded or defeated him. These geometricians, through their own citations and references to his work, spread his name further throughout the European mathematical community. In addition, Steiner left a substantial body of published works.

By the 1850's, his health declined and the eccentricities of an always contentious character increased. He journeyed from spa to spa seeking the rejuvenation of his health. He died on April 1, 1863, at such a spa in Bern, Switzerland. However, his repute and the respect of geometricians for his revitalization of synthetic geometry and its conundrums outlasted him.

Clifton K. Yearley

Further Reading

Courant, Richard, and Herbert Robbins. *What Is Mathematics? An Elementary Approach to Ideas and Methods*. Revised by Ian Stewart. 2d ed. New York: Oxford University Press, 1996. This overview of mathematical history, aimed at readers with some knowledge of the subject, contains information about Steiner's geometric constructions and "Steiner's problem."

Klein, Felix. *Development of Mathematics in the Nineteenth Century*. Translated by M. Ackerman. Brookline, Mass.: Math-Sci Press, 1979. Klein's work is indispensable, as little of Steiner's writing has been translated into English. Filled with technical mathematical signs, symbols, and equations, it nevertheless contains much that is understandable to lay readers. His expositions include biographical material on all mathematicians treated, with Steiner prominently among them, as well as good contextual explanations of their objectives, problems, and results. Contains ample illustrations.

Kline, Morris. *Mathematical Thought from Ancient to Modern Times.* New York: Oxford University Press, 1972. A layperson's survey, which largely ignores Steiner but places his work in a broad comprehensible framework. Contains illustrations, a good select bibliography, and an index.

Newman, James R., ed. *The World of Mathematics: A Small Library of the Literature of Mathematics from A'h-mosé,* the Scribe, to Albert Einstein. 4 vols. New York: Simon & Schuster, 1956. Volume 2 of this work is pertinent to Steiner's context and to defining aspects of his work. Illustrations help nonspecialists appreciate the nature of some synthetic, isoperimetric geometrical problems and their attempted solutions. A fine explication of certain projective geometrical investigations. There are bibliographical citations scattered throughout and a select bibliography and usable index at the end of the second volume.

Porter, Thomas Isaac. "A History of the Classical Isoperimetric Problem." In *Contributions to the Calculus of Variations, 1931-1932: Theses Submitted to the Department of Mathematics of the University of Chicago.* Chicago: University of Chicago Press, 1933. Rather than a raw thesis, this essay is an excellent survey of the synthetic geometrical problems Steiner, among others, tackled. Illustrated and readily understandable for those lacking special math training. Includes a substantial, if somewhat dated, bibliography.

Torretti, Roberto. *Philosophy of Geometry from Riemann to Poincaré.* Boston: Reidel, 1978. This important study is critical for a sound understanding by specialists as well as nonspecialists of a creative period in the development of both German and French mathematics, once again placing Steiner in a somewhat different historical context from that of the works cited above. It has some illustrations, a select bibliography, and an index.

SIMON STEVIN

Flemish mathematician, scientist, and engineer

Stevin is best known for his advocacy of the decimal system, his discoveries in hydrostatics, his work on the inclined plane, and his musical theory of consonance. His many and varied contributions have merited him a place in histories of mathematics, accounting, science, engineering, and music.

Born: 1548; Brugge, Flanders (now in Belgium)
Died: February, 1620; The Hague, Holland, United Provinces (now in the Netherlands)
Also known as: Simon Stevinus

EARLY LIFE

Simon Stevin (steh-VIHN) lived through an age of turbulent change that would eventually divide his homeland. He was born in Brugge, a former port in the southern Netherlands that, because of silting, had lost its access to the sea. Although Antheunis Stevin and Cathelijne van de Poort, Simon's parents, were wealthy, Simon was born out of wedlock (his mother had earlier borne two illegitimate daughters with Brugge's burgomaster).

Because of his later expertise in languages, including Latin and Greek, and his knowledge of ancient mathematics and science, Stevin must have received an excellent education, but not much is known about the details. Some scholars think that he learned bookkeeping at a private school, whereas others believe that he may have attended the Catholic university at Louvain.

His first position was in bookkeeping and tax collecting for Brugge's city administration. He later moved to the commercial city of Antwerp, where he worked as an accountant for a rich merchant. During this time, revolution erupted in the Netherlands, destroying the union of the Catholic south and Protestant north. Stevin left the Netherlands and traveled to Prussia, Poland, and Norway. When he returned to his divided country in 1581, he settled in Leiden. He attended the University of Leiden, which was recently founded to train jurists, physicians, and Protestant theologians for the new Dutch Republic. Unlike Louvain, which emphasized traditional Humanistic studies, Leiden became a center for the study of new scientific ideas. Stevin made excellent use of both his classical and modern education in his career.

LIFE'S WORK

Stevin published his first book in 1582. Its subject was simple and compound interest, and he computed tables for the rapid calculation of annuities. This and his later work were based on the writings of Luca Pacioli, the Italian "father of accounting," who took a pragmatic approach to bookkeeping. Stevin understood the long history of accounting, but he built on this understanding to help create several modern accounting practices. For example, accounting historians consider him to be the inventor of the income statement, a summary of revenues and expenses for a given period.

After his work in business mathematics, Stevin turned to geometry, decimal fractions, and algebra. In 1583, he published a book on geometrical problems, which grew out of his fascination with the works of the ancient mathematician Archimedes. Although he was not the first to use decimal fractions, his book on this subject was the stimulus for their widespread use by bookkeepers. He also encouraged the use of decimals in coinage and in weights and measures. In algebra, he invented a powerful and widely used exponential notation that conveniently designated ordinary, fractional, and negative powers.

While working on these books, and as a citizen of the Low Countries, he became interested in the control of water. In 1588, the States-General granted him a patent for a high-capacity drainage mill that was able to lift four times as much water as old mills, and several of these wind-driven mills were built. He also invented a system of sluices, the floodgates of which could be opened to inundate lands before an enemy could occupy them.

The year 1586 is considered Stevin's *annus mirabilis*, or wonderful year, because of the amount of significant work he produced. One of Stevin's most important books was on hydrostatics, the science that deals with liquids at rest and under pressure, and was published that year. For some, this work warranted him the title of "founder of modern hydrostatics." In it, he clearly stated the hydrostatic law that a liquid's pressure depends only on its vertical height, not on the shape of its container.

Also published in 1586 was an influential book on the art of weighing. As in his work on mathematics and hydrostatics, this Flemish book built on the earlier studies of Archimedes, in particular on the problem of the equilibrium of a stationary object under the influence of a vertical force such as gravity. In fact, he is most famous for the *clootcrans* theorem, his clever derivation of the law of the inclined plane through a device called the wreath of spheres. Using a triangle with unequal sides and a string of evenly weighted beads, he proved the law that "two bodies on two different, inclined planes are in balance if their weights are proportional to the lengths of the two planes."

A further discovery in 1586 was Stevin's experimental demonstration that the velocity of two freely falling lead spheres, one ten times heavier than the other, was independent of this weight difference. Galileo is often given credit for refuting Aristotle's claim that heavy objects fall faster than light ones, but he most likely never did the actual experiment, and even if he did, it would have taken place after Stevin's well-evidenced demonstration.

During the late sixteenth and early seventeenth centuries, the religious and political turmoil in the Netherlands intensified, and the Dutch prince Maurice of Nassau played an important role in helping to establish the Dutch Republic. Prince Maurice had met Stevin when they were both students at Leiden, and in 1604, Stevin became quartermaster in the Dutch army under the prince. Stevin's knowledge of practical mechanics made him extremely useful. For example, he wrote a treatise on the art of fortification that Prince Maurice used during his military campaigns. Stevin also studied how to best train, equip, and make use of troops in war, and how to effectively finance these tasks. He was one of the first to write a book on governmental accounting, and "Bookkeeping for War and Other Extraordinary Finances" was an important part of this treatise. For the Dutch navy, he figured how to use magnetic declinations and a Mercator map to accurately guide ships.

In addition to his official work for the Dutch military, Stevin also tutored Prince Maurice in science and mathematics. He often wrote out the prince's lessons in great detail, and the prince, deeply impressed, had them published between 1605 and 1608. This massive work of fifteen hundred pages contained a comprehensive account of Stevin's accomplishments in mathematics, accounting, mechanics, and astronomy. His chief book on astronomy was also published during this time (1608), and it contained his analysis of the Copernican heliocentric system, which he strongly supported.

While he was engaged in his military work, Stevin amazingly found time for the scientific study of music. In particular, he devised an influential theory of consonance to explain which combinations of musical sounds are pleasant. He was unique in rejecting the ancient Pythagorean idea that pleasing sounds coincided with simple integral ratios. He argued that geometric, not arithmetic, division of the octave yielded genuine consonantal ratios. Though some have criticized Stevin's theory for its insensitivity to the practices of actual musicians, others have seen his work as anticipating the later development of a scale of equal temperament.

Despite marrying late in life (he was sixty-two), he and his wife, Catherine Gray, had two boys and two girls. During his final years, he was deeply admired for his many contributions to the Dutch Republic. He lectured in Dutch at the University of Leiden and helped organize its engineering school. His belief that the Dutch language was particularly suitable for mathematical and scientific works probably contributed to their lack of influence in foreign countries, where Latin was still the language preferred by most scholars.

Significance

Stevin was very much a Renaissance man, accomplished in such Humanistic fields and political and musical theory, such technical fields as military science and engineering, and such scientific fields as mechanics and astronomy. Because of his deep understanding of both ancient and modern science, he helped bridge the gap between the two approaches. Because he was both a theoretical scientist and a practical engineer, he was able to use his theories in mechanics to develop improved windmills, lifting devices, and military fortifications.

His many published books exhibit his versatility and curiosity, his ability to combine theory and practice, and his skill in crafting clear and creative arguments to support his many original ideas. A major figure in the scientific revolution, he exhibited the movement's confidence in reason to solve the many puzzles of a natural world, a world he believed was governed by beautiful mathematical laws.

Robert J. Paradowski

Further Reading

Boyer, Carl B. *A History of Mathematics*. Revised by Uta C. Merzbach. 2d ed. New York: John Wiley & Sons, 1991. This classic textbook, expanded and updated by Merzbach, contains an analysis of Stevin's contributions to mathematics. Includes bibliographies with each chapter, a general bibliography, and an index.

Chatfield, Michael, and Richard Vangermeersch, eds. *The History of Accounting: An International Encyclopedia*. New York: Garland, 1996. This pioneering book contains a biographical article on Stevin that analyzes his contributions to such topics as balance sheets and compounds entries. Bibliographies at the ends of articles. Comprehensive index.

Dijksterhuis, E. J. *Simon Stevin: Science in the Netherlands Around 1600*. The Hague, the Netherlands: Martinus Nijhoff, 1970. This English translation of a work originally published in Dutch in 1943 has been abridged and edited by R. Hooykaas and M. G. J. Minnaert. It remains the best English account of Stevin's life and work.

Grout, Donald J., Claude V. Palisca, and Peter J. Burkholder. *A History of Western Music*. 7th ed. New York: W. W. Norton, 2004. This textbook contains an analysis of Stevin's contributions to music theory. Bibliographies at the ends of each chapter. Comprehensive index.

Alan Mathison Turing

British mathematician

Through his research on computable functions and artificial intelligence, Turing prepared the foundation for modern computer science. His work during World War II breaking German codes for the British government was of major value to the Allied effort in Europe.

Born: June 23, 1912; London, England
Died: June 7, 1954; Wilmslow, Cheshire, England

Early Life

Alan Mathison Turing (MATH-ih-sehn TEWR-ihng) was born in London, England. His mother, née Ethel Sara Stoney, and his father, Julius Mathison Turing, were both of middle-class, Protestant British families. Both Turing and his brother John were educated in British boarding schools while their father served in the British civil service in India. Turing matriculated at the Sherbourne School in Dorset, rather than at one of the more fashionable public schools because his family believed that the school would offer a more conducive environment for their socially awkward child. At Sherbourne, Turing distinguished himself in mathematics and science but was an indifferent student of other subjects. Already, his lifelong interest in scientific matters and passionate determination to learn by self-discovery rather than through reading were well established.

LIFE'S WORK

Turing's most enduring scientific accomplishment was conceived during his undergraduate years at Cambridge University. He entered King's College, Cambridge, in 1931 and remained there until 1936, during which time he completed his baccalaureate and master's degrees with honors. Turing was tutored by mathematician M. H. A. Newman, who introduced him to the work of David Hilbert, Kurt Gödel, and others, on the logical foundations of mathematics—a subject of considerable mathematical interest and activity in the 1920's and early 1930's. Turing was attracted to one of the open questions in this subject: determining which mathematical functions are computable, that is, which functions can be calculated entirely in a machinelike way by a fixed set of deterministic rules. Following an independent and very clever line of reasoning, Turing found a solution. His answer involved defining an abstract machine, known today as the universal Turing machine, which could calculate all those functions that are computable. His machine embodied the concept of the general-purpose, stored-program computer; thus the paper describing his result is regarded as a classic of both mathematical logic and computer science.

In 1936, Turing accepted a year's fellowship to visit Princeton University, where he could work with Alonzo Church and other mathematicians interested in the subject of computability. While at Princeton, Turing was persuaded to stay a second year and complete a Ph.D. in mathematics. His dissertation on ordinal logics made yet another important contribution to mathematics and computer science. Turing might have settled into a productive academic career as a mathematical logician, had World War II not intervened.

Turing returned to war-troubled Great Britain in 1938. He reassumed his fellowship at Cambridge, but soon he volunteered his services to the Government Code and Cypher School in Bletchley, a small town between Oxford and London, where a massive effort was under way to develop techniques and construct calculating machines capable of decoding machine-encrypted German diplomatic and military messages. Turing played a crucial role in the design of this equipment and in the development of procedures for breaking the codes: The work at Bletchley was critical to the Allied successes in Europe, and Turing was highly decorated for his contributions.

This wartime experience gave Turing a familiarity with electronics that complemented his mathematical training. In 1945, he moved to the National Physical Laboratory in Teddington, where he was given the responsibility for designing an electronic, stored-program computer for use in government work. What Turing designed, in effect, was a physical embodiment of his universal Turing machine that used electronic circuitry. The plans he drew up called for an ambitious modern computer using vacuum tubes for logical switching and arithmetic calculations, and mercury delay lines (a radar technology) for storing information. A scaled-down version, known as Pilot Automatic Computing Engine (Pilot ACE), was completed in 1950—one of the first modern computers placed in operation—and used for important government and industrial research, including aircraft design.

Frustrated at the slow and somewhat political process of building the computer, Turing resigned from the National Physical Laboratory in 1947. After a short time at Cambridge, Turing accepted a position at Manchester University as the chief programmer for a powerful new computer, the Mark I, being built there. Joining this project, he was reunited with Newman, his college teacher and Bletchley colleague. His duties allowed him ample time and opportunity to pursue his main interest, what is now called "artificial intelligence." Turing became the champion of artificial intelligence research. He experimented with programming the Mark I computer to execute what he regarded as intelligent actions: proving mathematical theorems, translating from one language into another, breaking codes, and playing games such as chess. He also published a widely read paper setting out his counterarguments to the most popular objections to the possibility of artificial intelligence.

This paper introduced his "Imitation Game," better known today as the Turing Test. The test is the most frequently cited method for determining whether a computer has achieved intelligence and is based on Turing's belief in behaviorist philosophy: If a computer exhibits intelligent behavior, it is intelligent. In the test, an interrogator is connected by terminal to either a human or a computer at a remote location. The interrogator is allowed to pose any questions he wishes. Based on the answers, the interrogator must decide whether a human or a computer is at the other end of the line. If the interrogator cannot distinguish between the two in a statistically significant number of cases, then Turing's test claims artificial intelligence to have been achieved.

Turing's productive and extraordinarily creative scientific career came to a sudden end in 1954 as the result of a fatal dose of cyanide poisoning. A few months earlier, Turing had been convicted of homosexual activity, at that time a felony in Great Britain. He had been sentenced to mandatory estrogen treatments, which caused strong physiological and psychological changes in him. These changes severely depressed Turing, and many people believe they caused him to take his life. At the time of his death, Turing had begun an ambitious investigation of the chemical basis of morphogenesis—that is, the information process in living organisms that determines why and how single cells grow into differentiated organs with specific functions.

Significance

Turing made many contributions to mathematics, logic, and statistics, and was a marathon runner of world-class distinction. He is best remembered, however, for his contributions to computability, machine design, and artificial intelligence. His work on computability, especially the universal Turing machine concept, was the first modern work on the theory of computation and is a central idea in recursive function theory (an active area of research in mathematical logic) and in automata theory (an important theoretical discipline within computer science). The value of Turing's efforts in the design of code-breaking equipment to the war effort should not be underestimated. His postwar work on computer design is much harder to evaluate. His work at the National Physical Laboratory resulted in one of the first operating modern computers, which was used for important scientific and engineering applications in the 1950's. Turing's work also influenced the design of computers built by two companies, English Electric and Bendix, but his design ideas did not fall within the mainstream of computer design developments. Turing was the foremost champion of artificial intelligence research in the first decade of modern computing. He introduced the distinction between robotics and artificial intelligence research, arguing in opposition to researchers such as Grey Walter and W. Ross Ashby that the future of artificial intelligence research lay in the use of the stored-program computer, not in the construction of special-purpose robots that could mimic vision or other human attributes. The Turing Test has endured as the principal test of success in artificial intelligence research.

William Aspray

Further Reading

Carpenter, B. S., and R. W. Doran, eds. *A. M. Turing's ACE Report of 1946 and Other Papers*. Cambridge, Mass.: MIT Press, 1986. The volume reprints Turing's design report for the ACE computer, a 1947 lecture by Turing on computing, and a paper by M. Woodger on the history of computing at the National Physical Laboratory. The editors provide a historical introduction to computers that provides the context for these works.

Copeland, B. Jack, ed. *Alan Turing's Automatic Computing Engine: The Master Codebreaker's Struggle to Build the Modern Computer*. New York: Oxford University Press, 2005. A detailed history of Turing's contributions to computer science, describing how he designed and built some of the first computers. Includes diagrams and illustrations explaining the hardware, software, and other features of Turing's computers.

Hodges, Andrew. *Alan Turing: The Enigma*. New York: Simon & Schuster, 1983. Hodges provides the definitive biography of Turing. It supplies detailed descriptions of his scientific accomplishments in computing equipment design, artificial intelligence, and cryptology. The effect, if any, of Turing's homosexuality and his social and personality traits on his career are explored in detail.

Leavitt, David. *The Man Who Knew Too Much*. New York: W. W. Norton, 2006. A comprehensive biography that explains Turing's complex scientific achievements in an accessible style.

Teuscher, Christof, ed. *Alan Turing: Life and Legacy of a Great Thinker*. New York: Springer, 2004. Collection of essays assessing Turing's life, career, and contributions to artificial intelligence, computing, mathematics, physics, cryptology, and other fields.

Turing, Alan. "Computing Machinery and Intelligence." *Mind*. n.s. 59 (1950): 433-460. The paper gives Turing's counterarguments to common objections against artificial intelligence and introduces the Turing Test, which decides when a computer has achieved intelligence.

_____. "On Computable Numbers, with an Application to the *Entscheidungsproblem*." In *Proceedings of the London Mathematical Society*. 2d ser. 42 (1937): 230-265. This famous article sets forth the concept of the Turing machine and its application to problems in mathematical logic.

Turing, Sara. *Alan M. Turing*. Cambridge, England: Heffers, 1959. A memoir of Turing written by his mother when she was seventy-five years old. It provides useful information about his family life and a rather different interpretation of the facts surrounding his death, but it is weak on scientific matters.

John Wallis

English mathematician

Wallis made advances in mathematical notation and created new methods for making mathematical discoveries. He paved the way for the work of Sir Isaac Newton and consequently for the invention of the calculus.

Born: December 3, 1616; Ashford, Kent, England
Died: November 8, 1703; Oxford, England

Early Life

John Wallis (WAHL-uhs) was born in the village of Ashford, Kent, the son of the village rector, John Wallis, senior, and his second wife, Joanna Wallis, née Chapman. Wallis's father died in 1622. He was left with sufficient resources to obtain a good elementary education and enrolled at an Essex school run by Martin Holbech in 1630. There, he was told by the master that he was the best prepared student in mathematics he had encountered. Wallis was, in addition to his mathematical learning, something of a calculating prodigy. From Holbech's school, he proceeded to Emanuel College, Cambridge, where he studied medicine and took courses in physics as well as moral philosophy.

After receiving a bachelor's degree in 1637 and a master's degree in 1640, Wallis became a fellow at Queen's College. As was typical for the period, such fellowships were limited to unmarried scholars, so when he married Susanna Glyde in 1645, he was deprived of his fellowship. The marriage lasted forty-two years, until his wife's death in 1687. They had one son and two daughters. In the early years of his marriage, Wallis was able to make a living as a private chaplain in Yorkshire and Essex, where he had connections.

The English Civil War soon intervened in Wallis's life, however. Wallis was active on behalf of the Parliamentary side, and, in particular, he put his code-breaking talents to use to help with deciphering intercepted Royalist messages. When the Parliamentary forces triumphed, this put Wallis in an enviable position, and he became Savilian Professor of Geometry at Oxford, thanks to the previous occupant's having had the bad fortune to support the Royalist cause. Since Wallis's mathematical skills at the time (1649) had scarcely even been tested, the political nature of the appointment is evident.

There were many political appointments made at the time that were reversed with the Restoration of the monarchy in 1660. Wallis, however, was not evicted from his chair under Charles II, and the explanation for his retention is usually found in Wallis's having signed

John Wallis. (Library of Congress)

the remonstrance against the execution of Charles I. Charles II, Charles I's son, remembered those who had not acquiesced in the death of his father. It was also true that by the time of the Restoration, Wallis had made a name for himself in mathematics worthy of the position he held.

LIFE'S WORK

The first work of Wallis to achieve notice was his *Arithmetica Infinitorum* (1655; the arithmetic of infinitesimals). In this book, he sought to combine the ideas of the Italian mathematician Bonaventura Cavalieri with the style of argument made famous by the French mathematician and philosopher René Descartes. Cavalieri had tried to break down geometrical objects in a given dimension into slices of one dimension lower. Thus, Cavalieri's principle argued that two solids would have the same volume if the areas of all the corresponding cross-sections were equal. Descartes's use of analytical methods helped to make the ideas of Cavalieri more palatable.

In 1659, Wallis published a treatise on conic sections, a subject familiar since the days of the Greeks. What was distinctive about Wallis's approach, though, was that he provided an alternative to the traditional approach of deriving the properties of the conic sections (like the circle and the ellipse) from their geometric origin. Wallis, instead, showed that certain quadratic equations would generate the points that made up the conic section and that the properties of the curves could be derived from the equations without having to go back to cones.

One general subject upon which Wallis was employed was trying to find the areas under certain curves. In order to go beyond the work done previously in the area by mathematicians like Pierre de Fermat, he worked out some of the values for the areas for curves that were already known and then showed how other curves could be obtained as the complements of those curves. In general, Wallis made progress by the use of two assumptions, induction and interpolation. Wallis would establish a result for the first few positive whole number values of the variable and would then claim that it must hold for the rest of the whole numbers. Having done that, he would then claim that some principle of continuity would also guarantee that the result would hold for the rest of the real numbers as well.

Wallis's, then, was scarcely a rigorous procedure, but it had a great influence on Sir Isaac Newton, whose early calculations of areas were quite similar to those of Wallis. If Newton made any progress beyond Wallis in these early years, it was by virtue of recognizing the advantages of taking variable limits for the regions whose area he was trying to find. This enabled him to recognize patterns where Wallis had only obtained numerical values.

One of Wallis's most celebrated results was an infinite product for the number $4/\pi$. Again, this was not the first arithmetic expression that involved π, but Wallis's approach served as the basis for further work in the subject. In addition to his work on areas, Wallis also looked at the volumes of three-dimensional solids of rotation. He was the first to use the symbol ∞ for infinity and provided the notation subsequently used for logarithms as well.

Typical of Wallis's later years was his volume on algebra published in 1685. It included the use of notation to represent complex numbers well before what became the standard representation was devised. In addition, the book also included a good deal of questionable history, in which Wallis took potshots at the work of Descartes and held up the work of Thomas Harriot, an English mathematician, as superior. Subsequent historians have seen Wallis as going well beyond the evidence both in his praise of Harriot and in his disregard for Descartes. On the other hand, it was indicative of Wallis's nationalism and his willingness to ignore generally accepted valuations.

Among Wallis's most notorious quarrels were those with Thomas Hobbes and the defenders of the German mathematician Gottfried Wilhelm Leibniz as an inventor of the calculus. Hobbes was nowhere near the mathematical level that Wallis achieved, but he was a careful reader and pointed out some of the shortcomings in Wallis's attempts to explain the foundations of calculus, shortcomings he shared with the other mathematicians of the time. Wallis kept the argument with Hobbes going for many years, as he found repugnant Hobbes's attempt to found his philosophical materialism on mathematics. With regard to the creation of the calculus, Wallis was determined to keep Newton and Leibniz at odds for the sake of the glory of English mathematics. He also opposed the Gregorian calendar.

After the Glorious Revolution of 1688, Wallis was once more confirmed in his academic positions, possibly because of his continued willingness to put his code-breaking talents to use for the state. During much of his later years, he devoted time to the preparation of editions of other authors' works, as well as his own collected works. While his quarrels with contemporaries were many, it is worth mentioning that Samuel Pepys, the diarist, ordered a portrait of Wallis painted by the Sir Godfrey Kneller. Since Pepys did not get along with everyone, this is something of a contrast to Wallis's feuds.

Significance
Wallis was arguably the most important precursor of Newton in England. His willingness to go out on a limb in pursuit of a mathematical discovery served as a model for Newton's efforts and led to the latter's discovery of the proof for the binomial theorem for general exponents. Wallis was one of the first to make a serious effort to provide an arithmetic form for geometry, especially for the heavily geometric parts of Euclid's *Stoicheia* (c. 300 B.C.E.; *The Elements of Geometrie of the Most Auncient Philosopher Euclide of Megara*, 1570, commonly known as the *Elements*), such as Books II and V. His innovations in notation helped to make mathematics easier to communicate, even across international boundaries.

Thomas Drucker

Further Reading
Fauvel, John, et al., eds. *Oxford Figures: 800 Years of the Mathematical Sciences*. New York: Oxford University Press, 2000. The sixth chapter is devoted to Wallis and traces his contributions to the teaching of mathematics at Oxford and the growth of the university's mathematical reputation.

Mahoney, Michael Sean. *The Mathematical Career of Pierre de Fermat, 1601-1665*. Princeton, N.J.: Princeton University Press, 1973. The last chapter of the book takes up Fermat's arguments and relations with British mathematicians toward the end of his life and points to Wallis's disappointing reaction to questions proposed by Fermat.

Merton, Robert K. *On the Shoulders of Giants: A Shandean Postscript*. Chicago: University of Chicago Press, 1993. Looks at the impression Wallis created among his contemporaries of being greedy for glory and taking credit for the work of others.

Pears, Iain. *An Instance of the Fingerpost*. London: Jonathan Cape, 1997. This historical novel is a mystery in four parts, the third of which is narrated by Wallis, who has the reader's sympathy by the end of the narrative.

Scott, J. F. *The Mathematical Work of John Wallis, D.D., F.R.S., 1616-1703*. 1938. Reprint. New York: Chelsea, 1981. Seeks to restore Wallis's reputation among historians of science. While defending Wallis's mathematics, much of his history is treated as falsification.

Scriba, Christoph J. "John Wallis." In *Dictionary of Scientific Biography*, edited by Charles C. Gillispie. Vol. 14. New York: Charles Scribner's Sons, 1971. Much briefer than Scott's survey of Wallis's mathematics but clearer to the modern reader.

Stillwell, John. *Mathematics and Its History*. New York: Springer-Verlag, 1989. Looks at Wallis primarily as a precursor to Newton.

Westfall, Richard S. *Never at Rest: A Biography of Isaac Newton*. New York: Cambridge University Press, 1980. The best explanation of how Newton imitated Wallis's manipulations.

Alfred North Whitehead
British philosopher

Striving for a more comprehensive and unified system of human knowledge, Whitehead made major contributions to mathematical logic and produced a wholly original and modern metaphysics.

Born: February 15, 1861; Ramsgate, Isle of Thanet, Kent, England
Died: December 30, 1947; Cambridge, Massachusetts

Early Life
Alfred North Whitehead was the last of four children born to Alfred Whitehead, a schoolmaster and clergyman, and Maria Sarah Buckmaster. Whitehead's father was a typical Victorian country vicar who tirelessly tended to the needs of the people of the Isle of Thanet and was well loved by them. His grandfather, Thomas Whitehead, was more remarkable intellectually. The son of a prosperous farmer, he had single-handedly created a successful boys' school at Ramsgate, unusual for its time in its emphasis on mathematics and science.

Ramsgate was a small, close-knit community in which history was a physical presence in the form of many ancient ruins, including Norman and medieval churches and Richborough Castle, built by the Romans when they occupied Britain. The surrounding waters were notoriously treacherous, and Whitehead remembered as a child hearing at night the booming of cannon and seeing rockets rise in the night sky, signaling a ship in distress. He believed that over the generations this environment instilled in the people an obstinacy and a tendency toward lonely thought.

Because he was small for his age and appeared frail, young Whitehead was not allowed to attend school or participate in children's games. Instead, his father tutored him in Latin, Greek, and mathematics. Whitehead learned his lessons quickly and had free time for periods of solitary thought and rambles through the wild coastal countryside with its mysterious ruins.

In 1875, Whitehead left home and entered Sherborne in Dorsetshire, a well-regarded public school from which both of his brothers had been graduated. He had grown to love mathematics, and he excelled at it enough to be excused from some of the standard courses in classical languages and literature to study it more deeply. Ignoring his "frailty" he took up Rugby, developing his athletic skills with seriousness and tenacity. As captain of the team he compensated for his size with intelligence and leadership and became one of the best forwards in the history of the school. Later in life he said that being tackled in a Rugby game was an excellent paradigm for the "Real" as he meant the term philosophically.

Before his last year at Sherborne, he chose to take the grueling six-day scholarship examination for Trinity College, Cambridge, an examination that would determine not only entry and the needed financial assistance but, more important, eligibility for a fellowship and, therefore, his hopes for a career in mathematics. Whitehead took the examination a year earlier than he needed to, and passed.

Whitehead entered Trinity College in the autumn of 1880 as a participant in a special honors program that allowed him to study in his area of specialty, mathematics, exclusively for the full three years of undergraduate work. In the Cambridge of that time, however, perhaps more than today, important education also took place outside the classroom in lively, spontaneous discussions with other students, an experience that Whitehead described as being like "a daily Platonic Dialogue," and that sometimes ran late into the night and into the early morning, ranging over politics, history, philosophy, science, and the arts. For a time, Whitehead became intensely interested in Immanuel Kant's *Critique of Pure Reason* (1781), in which one of

Alfred North Whitehead.

Whitehead's high scores on final examinations and his dissertation on Maxwell's *Treatise on Electricity and Magnetism* (1873) won for him a six-year fellowship, allowing him free room and board at Cambridge and unlimited freedom for mathematical research.

Unlike most research fellows, however, Whitehead did not become immediately productive. By character he was not a piecemeal problem solver who worked in ever narrower and more refined areas of a subject, but rather an explorer seeking a wider and more unified perspective. He discovered and was impressed with the works of Hermann Günther Grassman, an all but forgotten German mathematician who had developed a new kind of algebra. Grassman's *Die lineale Ausdehnungslehre, ein neuer Zweig der Mathematik* (1844), along with George Boole's *The Mathematical Analysis of Logic* (1847) and William Hamilton's *The Elements of Quaternions* (1866), seemed to Whitehead to portend a whole new field of algebras of logic not limited to number and quantity and with exciting unexplored applications. Whitehead envisioned a work in which all these ideas would be brought together in a general theory that would include giving a spatial interpretation to logic,

161

which would provide a more powerful general theory of geometry.

On a visit to his parents in Broadstairs in June of 1890, at a time when his great work did not seem to be going anywhere and he was contemplating conversion to Roman Catholicism, Whitehead was introduced to Evelyn Wade and fell in love. She was twenty-three years old, with black eyes and auburn hair and a vibrant personality. Though English, she spoke French as her native language, having been reared in a convent at Angers. Whitehead wasted no time and proposed to her romantically in a cave under the garden in his father's vicarage. They were married in December of 1891. Their marriage was to produce three children. The youngest would be tragically killed in aerial combat in 1918. Evelyn loved and cared for Alfred and always made a place where he could work without interruption wherever they lived, but she had no interest in science or mathematics. However, she perfectly complemented his analytic temperament with a deep interest in people and a wonderful aesthetic sense. Her example, according to Whitehead, taught him that "beauty, moral and aesthetic, is the aim of existence: and that kindness, and love, and artistic satisfaction are among its modes of attainment"—logic and science being important because they are useful in providing the conditions for periods of great art and literature. Whitehead's later worldview incorporated these ideas.

Marriage was good for Whitehead. He soon began work in earnest on his project, the first volume of which was published in 1898 as *A Treatise on Universal Algebra*. The second volume was never written, partly because of another momentous event that occurred in the same year as Whitehead's marriage: In 1890, Bertrand Russell entered Cambridge.

Whitehead, who happened to be reading examinations at the time, recognized Russell's potential and, even though Russell's scores were disappointingly low, saw to it—some say by burning the scores—that Russell received a scholarship. Their future collaboration on *Principia Mathematica* (1910-1913) was to be one of the most fruitful in the history of mathematics.

By 1903, Russell and Whitehead were colleagues. Russell had won a fellowship to Trinity College with a dissertation on the foundations of geometry. The two men found their individual interests and aims increasingly converging. With their wives, they attended the First International Congress of Philosophy in Paris in 1900 and met the great mathematician Giuseppe Peano, who had devised a symbolic notation that he was applying to the clarification of the foundations of mathematics. Russell, in his own work, had extended some of Peano's ideas and in 1903 published *Principles of Mathematics*, in which he attempted to demonstrate that all mathematics, including geometry, could be deductively derived from a few concepts of logic. At the same time, Whitehead was working along similar lines in his projected second volume to his *A Treatise on Universal Algebra*. Their friendship and their intellectual excitement with what seemed to be a revolution in mathematics led inevitably toward collaboration, and they spent more and more time together.

At one point, the Russells even moved in with the Whiteheads. This, however, did not work out. Russell fell secretly and unrequitedly in love with Evelyn Whitehead—he was by then very unhappy with his own wife—and the Russells, without anything coming into the open, moved out amiably. The collaboration, however, was unaffected by this turn of events. The life of the mind was the greater passion for both men, and their work was proceeding well.

In 1903, they decided, rather than publish two separate additions to their own works, that they would concentrate on one joint work. This would become the monumental *Principia Mathematica*. They originally believed that it would take only about a year to complete. It took approximately nine. The manuscript ran to more than four thousand pages and required a four-wheeler to transport it to Cambridge Press. Whitehead, almost fifty, was now stooped from leaning over his desk for long periods. During the work on *Principia Mathematica*, Whitehead found time to write a short work that he considered to be his most original, "On Mathematical Concepts of the Material World." Published in 1906, it is concerned with "the possible relations to space of the ultimate entities which (in ordinary language) constitute the 'stuff' of space." Anticipating Albert Einstein's general theory of relativity, he criticized the classical concept of an absolute space occupied by pointlike atoms, defending instead a relativistic view that space is not independent of the things in it but is rather dependent for its structure on objects within it. Further, he proposed that the fundamental constituents of matter are not pointlike particles but are more complex entities such as lines of force, with particles being the result of the interactions of these lines. Neither this work nor *Principia Mathematica* met with great success when first published, being considered too philosophical by specialists.

In 1910, Whitehead rather mysteriously left his assured position at Trinity College and moved to London, where he had no position. Whitehead may have simply found himself in a rut; the reasons, however, are probably more complex. For one thing, he had become politically active in his last few years at Cambridge, speaking out for women's rights and adopting a liberal position that outraged the university elite with their bias toward the rich and titled—he and his wife were once pelted with oranges and rotten eggs while sitting behind a Labour Party speaker. A clue to his leaving Cambridge may also be found elsewhere. Years before, in 1887, as a member of the prestigious Cambridge Apostles discussion group, Whitehead had responded to the topic question, which asked which was more important in life, "Study or Marketplace?" with the terse comment, "Study with windows." Whitehead probably had begun to sense that he could not commit himself to the ivory tower life of a don without trying to make some connection between his work and the wider world.

Indeed, his subsequent actions bear out this inference. Soon after settling in London, he wrote a popular exposition of mathematics for the layperson, *An Introduction to Mathematics*, which was published in 1911. From 1911 to 1914, he taught at University College, London, and from 1914 to 1924 he held a professorship at the Imperial College of Science and Technology in Kensington, all the while serving on various committees and councils setting the policies for London education. The aim of these institutions was to bring education to the masses, rationally adapting it to their needs and circumstances. Whitehead believed that the old elitist structure of the university must give way to new forms and that no less than the salvation of civilization was at stake. He compared this enterprise of mass education to the activities of the monasteries of the Middle Ages. His ideas on this subject appeared later in his books *The Aims of Education, and Other Essays*, published in 1929, and *Essays in Science and Philosophy*, published in 1947.

In 1923, another major shift occurred in Whitehead's life. While still at the Imperial College of Science and Technology, Whitehead received an invitation to come to Harvard University to join the philosophy department. This invitation was no doubt based on his reputation as a philosopher of science. Whitehead was then sixty-two years old, an age when most people are thinking of retirement. Whitehead, however, had never taught philosophy and liked the idea, and his wife wholeheartedly supported the move. It turned out to be the beginning of a massive creative surge in a new direction, reaching far beyond the specialized boundaries of the philosophy of science.

Beginning with *Science and the Modern World*, based on the Lowell lectures he gave in 1925 and published that year, Whitehead called into question the view that values have nothing to do with the basic constituents of nature, which, according to the prevailing view that he called "scientific materialism," consisted really only of matter in motion. In 1929, continuing this trend toward an all-inclusive worldview, there followed *Process and Reality*, which is considered to be one of the greatest works of metaphysics of all time, as well as one of the most forbidding. It propounds what Whitehead called "the philosophy of organism," in a novel terminology. *Adventures of Ideas*, published in 1933, was his last major work and probably his most accessible book on philosophy, with extensive explorations of sociology, cosmology, philosophy, and civilization as revealed from the perspective of his new metaphysical views.

Whitehead's reception in the United States was warm, and it provided him with an audience for the products of his far-ranging and independent intellect that he perhaps could not have had in his native England. He is remembered fondly by students and faculty as rosy-cheeked and cherubic, giving freely of his time, meeting with students often on Sunday evenings, and lecturing widely at eastern and midwestern universities. His last work, a small, lucid, and accessible work titled *Modes of Thought* (1938), was based on lectures he gave at Wellesley College and the University of Chicago. Whitehead died at Cambridge, Massachusetts, on December 30, 1947.

Significance

For more than fifty years, Whitehead applied his unique intellectual gifts successively to mathematics, education, and speculative philosophy and cosmology, always striving for an understanding of the nature of reality that could be applied to the betterment of humankind. Where he saw value in the ideas of others, he unselfishly helped those ideas to reach fruition. As a teacher, he brought out the best in his students, always with kindness and respect for their distinctive gifts. His later metaphysics is unique in the field for its scientific sophistication, yet it is free of the dogmatic rejection of the preeminence of human value in the world often found in the natural sciences.

Scott Bouvier

Further Reading

Emmet, Dorothy M. "Whitehead, Alfred North." Vol. 7 in *The Encyclopedia of Philosophy*, edited by Paul Edwards. New York: Macmillan, 1967. An excellent topical overview of Whitehead's philosophy.

Epperson, Michael. *Quantum Mechanics and the Philosophy of Alfred North Whitehead.* New York: Fordham University Press, 2004. Examines the relationship between relativity theory, quantum mechanics, and Whitehead's view of the cosmos, describing how his ideas incorporate science and philosophy.

Lowe, Victor. *Alfred North Whitehead: The Man and His Work.* 2 vols. Baltimore: Johns Hopkins University Press, 1985-1990. The fullest available account of Whitehead's life through his tenure at Cambridge. Volume 2 covers Whitehead's life in London and at Harvard. Lowe was a student of Whitehead and is an eminent authority on his work.

Malone-France, Derek. *Deep Empiricism: Kant, Whitehead, and the Necessity of Philosophical Theism.* Lanham, Md.: Lexington Books, 2007. Malone-France finds critical comparisons between Immanuel Kant's transcendental realism and Whitehead's organic realism.

Rose, Philip. *On Whitehead.* Belmont, Calif.: Wadsworth/Thomson Learning, 2002. A concise overview of Whitehead's ideas, one in a series of books designed to provide students and other readers an introduction to the thoughts of significant philosophers.

Russell, Bertrand. *The Autobiography of Bertrand Russell: 1872-1914.* Boston: Little, Brown, 1951. Valuable for its account of the writing of *Principia Mathematica*, even though it is related from Russell's somewhat biased point of view.

Schilpp, Paul Arthur, ed. *The Philosophy of Alfred North Whitehead.* 2d ed. New York: Tudor, 1951. A collection of critical essays on Whitehead that includes Whitehead's autobiographical sketch and Lowe's insightful essay "The Development of Whitehead's Philosophy." Contains complete bibliography of Whitehead's works.

Whitehead, Alfred North. *Alfred North Whitehead: An Anthology.* Compiled by F. S. C. Northrop and Mason W. Gross. New York: Macmillan, 1953. Contains selections from all Whitehead's major works and is thus an excellent starting point for anyone wishing to become familiar with his ideas.

_____. *An Introduction to Mathematics.* London: Williams and Norgate, 1911. Rev. ed. London: Oxford University Press, 1948. For anyone interested in Whitehead or in an exposition of the power of mathematics.

Norbert Wiener

American mathematician

Wiener was a distinguished American mathematician credited with a founding of cybernetics, a science that facilitates comparison of biological and electronic systems by focusing on communication, feedback, and control.

Born: November 26, 1894; Columbus, Missouri
Died: March 18, 1964; Stockholm, Sweden

Early Life

In *Ex-Prodigy* (1964), the first volume of his two-volume autobiography, Norbert Wiener (WEE-nuhr) describes in detail his precocious youth and his relationship with his father, a brilliant and very forceful personality. Wiener's father, Leo Wiener, was born in the ghetto area of czarist Russia. At the age of thirteen, Leo became self-supporting. In spite of anti-Semitic laws and customs, he managed to graduate from a Warsaw gymnasium. After emigrating to the United States, he worked in factories and on farms. He never received a university education, but he became a professor of Slavic languages, first at the University of Missouri and then at Harvard. Leo's wife and Norbert's mother, Bertha Kahn Wiener, was born in Missouri of German Jewish parents. Her family was in the process of being assimilated; each of her brothers had married a non-Jew. Norbert was not reared with an awareness of his own Jewish heritage and later bitterly resented what he came to believe was his mother's anti-Semitism.

Leo Wiener believed in beginning the education of children at a very early age and in expecting substantial intellectual progress during childhood. Coached by his father, Norbert Wiener learned to read by the time that he was three. His father educated him at home until he was nine, when he entered Ayer High School. At eleven, he entered Tufts College, where he became especially interested in physics, chemistry, and biology. After graduating from Tufts, he enrolled before his fifteenth year in zoology at Harvard Graduate School. Since Norbert lacked the coordination and eyesight essential for laboratory work in the life sciences, at his father's suggestion, he shifted his program of study to philosophy. Wiener later resented his father's involvement in this decision.

Wiener's doctoral thesis involved a comparison of Ernst Schroeder's algebra of relatives with that of Alfred North Whitehead and Bertrand Russell. Of his thesis, he later commented, "When I came to study under Bertrand Russell in England, I learned that I had missed almost every issue of true philosophical significance."

During his last year at Harvard, Wiener was awarded a traveling scholarship, which he used to visit England and Germany. At Cambridge, his main course work was with Russell. Wiener enjoyed his time in Cambridge, finding it a more sympathetic environment than Harvard because eccentricity and individuality were not only tolerated but even highly valued in this English university town. Since Russell was to be away from Cambridge during the spring term, Wiener visited Göttingen, where he met some of the greatest of the mathematicians and physicists of his day. His experiences in Göttingen also contributed to his social maturity, because he found that he could get along with many different types of people.

In 1919, after an instructorship at the University of Maine and service on the ballistics staff at Aberdeen Proving Ground, Wiener joined the faculty of the Massachusetts Institute of Technology (MIT). In 1924, he was promoted to an assistant professor, and in 1926, he married Margaret Engemann; his parents strongly approved of this marriage, which they even promoted. The Wieners' two daughters were named Barbara and Peggy. The first volume of Wiener's autobiography concludes with his marriage at the age of thirty-one.

Like his father, Wiener was a vegetarian. To his contemporaries he seemed highly eccentric in appearance. Stephen Toulmin, director of the Nuffield Foundation Unit for the History of Ideas in London, described him in 1964 in the *New York Review* as

> the most *peculiar* American in my experience, and even in England I can liken him only to the late Sir Thomas Beecham. . . . Both of them were short, myopic, tubby. . . . With it, there went a rotundity of expression in public conversation—I nearly said monologue—which was too puckish to be called pompous, and an assumed air of prejudice and self-importance so extreme it became a joy to observe.

Norbert Wiener. (Library of Congress)

Wiener described his father, Leo, as "brilliant," "absentminded," and "hot-tempered," adjectives his contemporaries thought equally apt as a description of him. Wiener would go to sleep and even snore during discussions and classroom seminars. Since he seemed able to process information while sleeping, much to the surprise and even chagrin of his colleagues and students, he would sometimes awaken and make very perceptive comments concerning the topic being discussed.

LIFE'S WORK

In 1929, Wiener was given the rank of associate professor, and after the appearance of important papers on generalized harmonic analysis (1930) and Tauberian theorems (1932), he was promoted to a professorship in 1932. Although he remained based at MIT, he welcomed opportunities to travel abroad, spending a year in Beijing, China, during the academic year 1935-1936.

Throughout his career, Wiener experienced doubts concerning his reputation among other mathematicians. In 1933, however, he was awarded the Bôcher Prize, which is awarded only every five years by the American Mathematical Society. Since his work on generalized harmonic analysis and Tauberian theorems had appeared in 1930 and 1932, it is clear that the society recognized his achievements as soon as it was possible to do so.

During the war years, Wiener received a small grant to work on a design for an apparatus that would direct antiaircraft guns effectively. Solving this problem led Wiener to develop a theory of prediction and pointed the way toward cybernetics, the major contribution of his adult years. Wiener's prediction theory could have been synthesized from his previous work, but it was his solution to a concrete problem that prompted this important synthesis. He had to determine the position and direction of flight of airplanes and then extrapolate over the flight time of the projectile to be sure that the projectile would reach the airplane.

From a mathematical perspective, this theory was Wiener's principal contribution to cybernetics. When his book *Cybernetics: Or, Control and Communication in the Animal and the Machine* was published in 1948, Wiener became a public figure. Sometimes called the philosopher of automation, Wiener contributed to cybernetics as an organizer, popularizer, and enthusiastic interpreter.

The word "cybernetics" derives from the Greek for "helmsman." Wiener proposed the term as a replacement for the title Conference for Circular Causal and Feedback Mechanisms in Biological Systems. Cybernetics concerns the science of communication and control theory, especially in regard to the comparative analysis of automatic control systems, the brain and nervous system as compared with mechanical-electrical communication systems. According to Wiener, the intellectual concerns of cybernetics distinguish modern civilization from that of previous centuries: "The thought of every age is reflected in its technique. . . . If the seventeenth and early eighteenth centuries are the age of the clocks, and the later eighteenth and the nineteenth centuries constitute the age of steam engines, the present time is the age of communication and control." As early as 1948, Wiener viewed the new concepts of message, information, feedback, and control as a supplement to physics that would facilitate a fully scientific description of an organism. These concepts now pervade neurobiology, biochemistry, genetics, psychology, and other disciplines concerned with organisms.

In 1964, in recognition of Wiener's contribution to theoretical mathematics and the sciences, U.S. president Lyndon B. Johnson awarded him the National Medal of Science. Shortly thereafter, while traveling in Sweden, Wiener died, on March 18, 1964.

Significance

In the years preceding World War II, Wiener was plagued by continual doubts about his productivity. His colleagues and students had to assure and reassure him that his current work was indeed excellent. This insecurity, which Wiener himself attributed to the pressures he experienced in early childhood, was never entirely overcome but did not keep Wiener from taking a very independent position on military contracts. Advocating noncooperation as a policy for scientists, Wiener commented, "It is perfectly clear also that to disseminate information about a weapon *in the present state of our civilization* is to make it practically certain that that weapon will be used." Wiener's position on noncooperation was supported by Albert Einstein, but as a consequence of his refusal to accept military grants and contracts, he had to discover other means of generating funds to maintain his research. His books on cybernetics and technology were addressed to the general public and were financially successful.

Wiener is acknowledged as the founder of cybernetics, although his contributions represented a synthesis of many concepts already implicit in the methodology of the social sciences. The basis for treating humans and machines with the same theory derives not from analysis of physical constituents but from patterns of communication and control. Wiener was skeptical about the efficacy of quantitative description in these fields, but he did show that patterns of communication can be described mathematically by using statistics.

At the same time, Wiener opposed any simplistic application of the human/machine model. He opposed the use of game theory as a means of determining military or political strategy. In spite of the great impact that cybernetics had on the social sciences, Wiener remained skeptical about sociological and economic predictions, arguing that the statistical runs were too short and that the observations were conditioned by interaction between the social scientist and his subject.

Author of a number of books addressed to the educated general public, such as *The Human Use of Human Beings* (1950), Wiener set out to inform the public about both the potential and the pitfalls of communications and computation technology. He was concerned about the impact of automation on the employment of laborers but recognized that machines might also help to improve working conditions. Advocating the independent and unbiased study of the relationship between humans and machines, Wiener tried to assess what the relationship should be between a human and a mechanical translator, between a computerized diagnosis and a physician's diagnosis. He pointed out as well that the paradigm of "man as master" and "machine as slave" dangerously ignores the way in which machines may influence decisions and shape the course of events, if only because of their much greater speed of action.

Jeanie R. Brink

FURTHER READING

Conway, Flo, and Jim Siegelman. *Dark Hero of the Information Age: In Search of Norbert Wiener, the Father of Cybernetics*. New York: Basic Books, 2005. The authors reassess Wiener's legacy, arguing that he has "fallen between the cracks of the Information Age." They trace his life, his development of cybernetics, and how he later realized the implications of the science he had created.

Grattan-Guiness, I. "Wiener on the Logics of Russell and Schroeder: An Account of His Doctor's Thesis, and of His Discussion of It with Russell." *Annals of Science* 32 (1975): 103-132. Discussion of Wiener's intellectual relation to Bertrand Russell.

Levinson, Norman. "Wiener's Life." *Bulletin of the American Mathematical Society* 72 (1966): 1-32. Biographical description of Wiener and discussion of his contributions to mathematics. Levinson, Wiener's student and colleague, regards most of Wiener's work in cybernetics as not mathematical.

Struik, Dirk J. "Norbert Wiener: Colleague and Friend." *American Dialog* 3 (March, April, 1966): 34-37. Essay by a close personal friend of Wiener. As a result of his Marxist views, Struik was indicted during the anticommunist crusades of the 1950's. Wiener strongly supported his friend, threatening to resign from MIT if the institution failed to support Struik.

Wiener, Norbert. *Collected Works with Commentaries*. Edited by P. Masani. 4 vols. Cambridge, Mass.: MIT Press, 1976-1985. Comprises all of Wiener's scholarly publications other than books and some previously unpublished material. Organized to show the author's intellectual evolution and supplemented with commentaries by important scholars. Best for advanced students. Includes a complete bibliography.

_____. *Ex-Prodigy: My Childhood and Youth*. Cambridge, Mass.: MIT Press, 1964. First volume of Wiener's autobiography, covering the years from 1894 to 1926.

_____. *The Human Use of Human Beings*. Boston: Houghton Mifflin, 1950. Very popular nonmathematical treatment of cybernetics. Newer editions available.

_____. *I Am a Mathematician*. Cambridge, Mass.: MIT Press, 1964. Second volume of Wiener's autobiography, covering the years from 1926 to 1964.

_____. "A Scientist Rebels." *The Atlantic Monthly*. 179 (1947): 46. Proposes that scientists should refuse to engage in weapons research.

Resource Guide

Books

Aaboe, Asger. *Episodes From the Early History of Mathematics.* Washington, DC: Mathematical Association of America, 1975.

Adrian, Yeo. *The Pleasures of Pi and Other Interesting Numbers.* Singapore: World Scientific Publishing, 2006.

Agresti, A. *Categorical Data Analysis.* Hoboken, NJ: Wiley, 2002.

Aho, A. V., J. E. Hopcrotf, and J. D. Ullman. *The Design and Analysis of Computer Algorithms.* Reading, MA: Addison-Wesley, 1976.

Albert, Jim, and Jay Bennett. *Curve Ball: Baseball, Statistics, and the Role of Chance in the Game.* New York: Springer-Verlag, 2001.

Ascher, Marcia. *Mathematics Is Everywhere: An Exploration of Ideas Across Cultures.* Princeton, NJ: Princeton University Press, 2002.

Ball, W. W. Rouse. *A Short Account of the History of Mathematics.* New York: Sterling Publishing Company, 2001.

Barnett, Raymond, Michael Ziegler, and Karl Byleen. *Calculus for Business, Economics, Life Science, and Social Science.* Upper Saddle River, NJ: Prentice-Hall, 2005.

Baumohl, Bernard. *The Secrets of Economic Indicators: Hidden Clues to Future Economic Trends and Investment Opportunities.* 2nd ed. Upper Saddle River, NJ: Pearson Education, 2008.

Beckmann, Petr. *A History of π (Pi).* New York: Barnes & Noble, 1971.

Behrends, Ehrhard. *Five-Minute Mathematics.* Providence, RI: American Mathematical Society, 2008.

Bell, Eric Temple. *Men of Mathematics.* New York: Simon & Schuster, 1937.

Bennett, Jay, and James Cochran. *Anthology of Statistics in Sports.* Philadelphia, PA: Society for Industrial and Applied Mathematics, 2005.

Berggren, Lennart, Jon Borwein, and Peter Borwein. *Pi: A Source Book.* New York: Springer-Verlag, 1997.

Berlekamp, Elwyn R., John H. Conway, and Richard K. Guy. *Winning Ways for Your Mathematical Plays.* Natick, MA: AK Peters, 2001.

Blackwell, William. *Geometry in Architecture.* Hoboken, NJ: Wiley, 1984.

Blatner, David. *The Joy of π.* New York: Walker & Co., 1997.

Blue, Ron, and Jeremy White. *The New Master Your Money: A Step-by-Step Plan for Gaining and Enjoying Financial Freedom.* Chicago: Moody, 2004.

Blum, Raymond. *Mathemagic.* New York: Sterling Publishing, 1992.

Bodie, Zvi, Alex Kane, and Alan Marcus. *Investments.* Chicago, IL: McGraw-Hill/Irwin, 2008.

Borwein, Jonathan, and Peter Borwein. *A Dictionary of Real Numbers.* Pacific Grove, CA: Brooks/Cole Publishing Co., 1990.

Boyer, C. B. *A History of Mathematics.* Hoboken, NJ: Wiley, 1968.

Boyer, C. B. *The History of the Calculus and Its Conceptual Development.* New York: Dover Publications, 1949.

Brealey, Richard A., Stewart C. Myers, and Franklin Allen. *Principles of Corporate Finance.* 9th ed. New York: McGraw-Hill, 2008.

Bressoud, David. *The Queen of the Sciences: A History of Mathematics.* Chantilly, VA: The Teaching Company, 2008.

Broverman, Samuel A. *Mathematics of Investment and Credit.* Winsted, CT: ACTEX Publications, 2008.

Burkett, Larry, and Brenda Armstrong. *Making Ends Meet: Budgeting Made Easy*. Gainesville, GA: Crown Financial Ministries, 2004.

Burton, David M. *The History of Mathematics: An Introduction*. New York: McGraw-Hill, 2005.

Calinger, Ronald. *A Contextual History of Mathematics*. Upper Saddle River, NJ: Prentice-Hall, 1999.

Clagett, Marshall. *Archimedes in the Middle Ages*. Madison: University of Wisconsin Press, 1964.

Closs, Michael. *A Survey of Mathematics Development in the New World*. Ottawa: University of Ottawa, 1977.

Closs, Michael, ed. *Native American Mathematics*. Austin: University of Texas Press, 1986.

Coe, Michael D. *Breaking the Maya Code*. New York: Thames and Hudson, 1992.

Copeland, Thomas E., J. Fred Weston, and Kuldeep Shastri. *Financial Theory and Corporate Policy*. 4th ed. Upper Saddle River, NJ: Pearson Education, 2005.

Cullen, Christopher. *Astronomy and Mathematics in Ancient China: The Zhou Bi Suan Jing*. Cambridge, England: Cambridge University Press, 1996.

Cuomo, Serafina. *Ancient Mathematics*. London: Routledge, 2001.

Davenport, Harold. *The Higher Arithmetic: An Introduction to the Theory of Numbers*. Cambridge, England: Cambridge University Press, 1999.

Davis, Morton D. *The Math of Money: Making Mathematical Sense of Your Personal Finances*. New York: Copernicus, 2001.

De Mestre, Neville. *The Mathematics of Projectiles in Sport*. Cambridge, England: Cambridge University Press, 1990.

Devlin, Keith. *The Math Gene: How Mathematical Thinking Evolved and Why Numbers Are Like Gossip*. New York: Basic Books, 2001.

———. *The Unfinished Game: Pascal, Fermat, and the Seventeenth-Century Letter That Made the World Modern*. New York: Basic Books, 2008.

Drobat, Stefan. *Real Numbers*. Upper Saddle River, NJ: Prentice-Hall, 1964.

Dudley, Underwood. *Numerology or What Pythagoras Wrought*. Washington, DC: Mathematical Association of America, 1997.

Eastway, Rob, and John Haigh. *Beating the Odds: The Hidden Mathematics of Sport*. London: Robson Books, 2007.

Eglash, Ron. *African Fractals: Modern Computing and Indigenous Design*. New Brunswick, NJ: Rutgers University Press, 1999.

Eves, Howard. *An Introduction to the History of Mathematics*. New York: Saunders College Publishing, 1990.

Flegg, G. *Numbers: Their History and Meaning*. New York: Schocken Books, 1983.

Friberg, Jöran. *Unexpected Links Between Egyptian and Babylonian Mathematics*. Singapore: World Scientific Publishing Co., 2005.

Friedman, Arthur. *World of Sports Statistics: How the Fans and Professionals Record, Compile, and Use Information*. New York: Athenaeum, 1978.

Fries, Christian. *Mathematical Finance: Theory, Modeling, Implementation*. Hoboken, NJ: Wiley, 2007.

Frumkin, Norman. *Guide to Economic Indicators*. Armonk, NY: M. E Sharpe, 2000.

Gamow, George. *One, Two, Three... Infinity*. New York: Viking Press, 1947.

Gardner, David, and Tom Gardner. *The Motley Fool Personal Finance Workbook: A Foolproof Guide to Organizing Your Cash and Building Wealth*. New York: Fireside Books, 2003.

Gardner, Martin. *Mathematics, Magic and Mystery*. New York: Dover, 1956.

Gay, Timothy. *The Physics of Football*. New York: HarperCollins, 2005.

Gerdes, Paulus. *Geometry From Africa: Mathematical and Educational Explorations*. Washington, DC: Mathematical Association of America, 1999.

Gillings, R. J. *Mathematics in the Time of the Pharaohs*. New York: Dover Publications, 1982.

Gutstein, Eric, and Bob Peterson, eds. *Rethinking Mathematics: Teaching Social Justice by the Numbers*. Milwaukee, WI: Rethinking Schools, 2005.

Hadamard, Jacques. *A Mathematician's Mind*. Princeton, NJ: Princeton University Press, 1996.

Hardy, G. H. *A Mathematician's Apology*. Cambridge, England: Cambridge University Press, 1941.

Henry, Granville C. *Logos: Mathematics and Christian Theology.* Lewisburg, PA: Bucknell University Press, 1976.

Hersh, Rueben. *What Is Mathematics, Really?* New York: Oxford University Press, 1997.

Hoyle, Joe Ben, Thomas F. Schaefer, and Timothy S. Doupnik. *Fundamentals of Advanced Accounting.* New York: McGraw-Hill, 2010.

Kalbfleisch, John D., and Ross L. Prentice. *The Statistical Analysis of Failure Time Data.* Hoboken, NJ: Wiley, 2002.

Katz, Victor J., ed. *Mathematics of Egypt, Mesopotamia, China, India, and Islam: A Sourcebook.* Princeton, NJ: Princeton University Press, 2007.

Kellison, Stephen G. *Theory of Interest.* New York: McGraw-Hill, 2009.

Kimmel, Paul D., Jerry J. Weygandt, and Donald E. Keiso. *Financial Accounting: Tools for Business Decision Making.* Hoboken, NJ: Wiley, 2009.

King, Jerry. *The Art of Mathematics.* New York: Plenum Press, 1992.

Klein, John P., and Melvin L. Moeschberger. *Survival Analysis: Techniques for Censored and Truncated Data.* New York: Springer-Verlag, 1997.

Kline, M., *Mathematical Thought From Ancient to Modern Times.* New York: Oxford University Press, 1972.

Koetsier, T., and L. Bergmans, eds. *Mathematics and the Divine: A Historical Study.* Amsterdam: Elsevier, 2005.

Longe, Bob. *The Magical Math Book.* New York: Sterling Publishing, 1997.

Martzloff, Jean-Claude. *A History of Chinese Mathematics.* New York: Springer-Verlag, 1987.

Moses, Robert P., and Charles E. Cobb, Jr. *Radical Equations: Civil Rights From Mississippi to the Algebra Project.* Boston: Beacon Press, 2001.

Mullis, Darrell, and Judith Handler Orloff. *The Accounting Game: Basic Accounting Fresh From the Lemonade Stand.* Naperville, IL: Sourcebooks, 2008.

Nahin, Paul J. *Dr. Euler's Fabulous Formula.* Princeton, NJ: Princeton University Press, 2006.

Nasar, Sylvia. *A Beautiful Mind: The Life of Mathematical Genius and Nobel Laureate John Nash.* New York: Simon & Schuster, 2001.

Oliver, Dean. *Basketball on Paper: Rules and Tools for Performance Analysis.* Washington, DC: Brassey's, 2004.

Pullan, J. M. *The History of the Abacus.* New York: F. A. Praeger, 1969.

Rafiquzzaman, M. *Fundamentals of Digital Logic and Microcomputer Design.* Hoboken, NJ: Wiley, 2005.

Rudin, W. *Principles of Mathematical Analysis.* New York: McGraw-Hill, 1953.

Salem, Lionel, Frédéric Testard, and Coralie Salem. *The Most Beautiful Mathematical Formulas.* Hoboken, NJ: Wiley, 1992.

Schwarz, Alan. *The Numbers Game: Baseball's Lifelong Fascination with Statistics.* New York: St. Martin's Press, 2004.

Smith, D. E. *History of Mathematics.* Vol. 2. New York: Dover Publications, 1958.

Solow, Daniel. *How to Read and Do Proofs: An Introduction to Mathematical Thought Process.* Hoboken, NJ: Wiley, 1982.

Steen, Lynn A. *On the Shoulders of Giants: New Approaches to Numeracy.* Washington, DC: National Academy Press, 1990.

Sterrett, Andrew. *101 Careers in Mathematics.* Washington, DC: The Mathematical Association of America, 1996.

Suzuki, Jeff. *A History of Mathematics.* Upper Saddle River, NJ: Prentice Hall, 2002.

Taylor, Alan D. *Mathematics and Politics: Strategy, Voting Power, and Proof.* New York: Springer-Verlag, 1995.

van der Waerden, B. L. *Geometry and Algebra in Ancient Civilizations.* Berlin: Springer, 1983.

Venema, G.A. *The Foundations of Geometry.* Upper Saddle River, NJ: Pearson Prentice Hall, 2006.

Weygandt, Jerry J., Paul D. Kimmel, and Donald E. Keiso. *Managerial Accounting: Tools for Business Decision Making.* Hoboken, NJ: Wiley, 2008.

Winkler, Peter. *Mathematical Puzzles: A Connoisseur's Collection.* Natick, MA: AK Peters, 2004.

Wright, Tommy, and Joyce Farmer. *A Bibliography of Selected Statistical Methods and Development Related to Census 2000.* Washington, DC: U.S. Bureau of the Census, 2000.

Yeldham, F. A. *The Teaching of Arithmetic Through Four Hundred Years (1535–1935)*. London: G. G. Harrap & Company, 1935.

Yong, L. L., and A. T. Se. *Fleeting Footsteps*. Singapore: Word Scientific Publications, 2004.

Zaslavsky, Claudia. *Africa Counts: Number and Pattern in African Culture*. Chicago: Lawrence Hill Books, 1999.

Zill, D. G. *Calculus with Analytic Geometry*. Boston: Prindle, Weber & Schmidt, 1985.

Journals and Magazines

The AMATYC Review
The American Mathematical Monthly
Association for Women in Mathematics Newsletter
Biometrics
Chance
The College Mathematics Journal
Experimental Mathematics
The Fibonacci Quarterly
Historia Mathematica
IMU-Net
Involve
Journal of Humanistic Mathematics
Journal of Integer Sequences
Journal of Recreational Mathematics
Journal of Statistics Education
Loci
MAA FOCUS
Math Horizons
Mathematics Magazine
Mathematics Teacher
NAM Newsletter
Notices of the American Mathematics Society
The Pentagon
Pi Mu Epsilon Journal
Plus Magazine
PRIMUS
Rose-Hulman Undergraduate Mathematics Journal
SIAM Review
Scholastic Math
Significance
Teaching Children Mathematics
Undergraduate Mathematics and Its Applications

Internet

American Institute of Mathematics
  www.aimath.org
The Algebra Project
  www.algebra.org
AMATYC
  www.amatyc.org
American Mathematical Society
  www.ams.org
American Statistical Association
  www.amstat.org
Association for Women in Mathematics
  www.awm-math.org
CryptoKids
  www.nsa.gov/kids
Datamath Calculator Museum
  www.datamath.org
Illuminations
  illuminations.nctm.org
MacTutor History of Mathematics
  www-history.mcs.st-and.ac.uk
Mathematical Fiction
  http://kasmana.people.cofc.edu/MATHFICT
Math for America
  www.mathforamerica.org
Math Forum
  www.mathforum.com
Math Fun Facts!
  www.math.hmc.edu/funfacts
MathDL
  mathdl.maa.org/mathDL
Mathematical Association of America
  www.maa.org
Mathematical Science Research Institute
  www.msri.org
The Museum of Mathematics
  www.momath.org
National Association of Mathematicians
  www.nam-math.org
National Council of Teachers of Mathematics
  www.nctm.org
RadicalMath
  www.radicalmath.org
Society for Industrial and Applied Mathematics
  www.siam.org
We Use Math
  www.weusemath.org
Wolfram MathWorld
  www.mathworld.wolfram.com

Index

Abel, Niels Henrik, 1, 66, 70
Academy of Sciences, France, 5, 108
Academy of Sciences, Italy, 5
Academy (Plato's), 51
Activism, peace
    world, 144, 145
Agnesi, Maria Gaetana, 4
    girls education, 4
    mathematics, 4
Alembert, Jean le Rond, 34, 35
Alexandria
    mathematics at, 10, 47
    science at, 14
Algebra, 46, 48, 55, 66, 69, 90, 114
    Italy, 28
    Netherlands, 154
Algebraic geometry, 147
Alhazen, 12
Alice's Adventures, 32
Alp Arslan, 126
Apollodorus (chronicler), 50
Apollonius of Perga, 10, 49, 85, 129
Arabic, 46
Archimedes, 13, 129
Aristarchus of Samos, 85
Aristotle, 7
Artificial intelligence, 156
Aryabhaṭa, 17
Āryabhaṭa the Elder, 25
Astrolabes, 84
Astronomy
    France, 93
    Germany, 20, 69
    India, 18, 25
    Iran, 126
    Iraq, 8
    mapping, 84
    mathematics and, 12
    Netherlands, 154
    Pythagoras and, 141

    Scotland, 76
    United States, 123
Atomic bomb
    development of, 120

Bacon, Francis, 41
Bayess theorem, 94
Benedict XIV, 5
Berkeley, George
    criticism of Isaac Newton, 107
Berlin Academy, 35, 54, 89
Berlin Society of Sciences, 100
Bernoulli, Jakob I, 150
Bessel, Friedrich Wilhelm, 20
Bhāskara II, 26
Bolyai, János, 104
Born-Haber cycle, 24
Born, Max, 23
Born-Oppenheimer approximation, 24
Boyle, Robert, 98
Brahmagupta, 19
Briggs, Henry, 118
Brinkley, John, 79
Bryn Mawr College, 147

Calculating machine
    France, 131
    Germany, 98
    Japan, 149
Calculus
    England, 159
    France, 56
    Germany, 98
    Japan, 150
    Scotland, 77
Calendar
    Persian solar, 126
Calvinism, 35
Cantor, Georg, 39
Cardano, Gerolamo, 28

Carroll, Lewis, 30
Catholic Church
    and the , 35
Cauchy, Augustin-Louis, 2, 66
Cavalieri, Bonaventura, 159
Cavendish, Henry, 113
Chaos theory, 111
Charles II (king of England)
    and James Gregory, 77
    and John Wallis, 158
Charles I (king of England)
    and John Wallis, 159
Chemistry
    Gaspard Monge, 113
Christianity
    France, 133
Christina (queen of Sweden)
    and Blaise Pascal, 131
    and René Descartes, 43
Chrysippus of Cnidus, 50
Clarke, Samuel, 99
Code breaking, 156
Cohen, Paul J., 75
Colburn, Zerah, 79
Computers
    development of, 156
    John von Neumann, 120
Conica, 11, 130
Conic sections, 10
Conon of Samos, 10, 14
Copernicus, Nicolaus, 12, 86
Crelle, August Leopold, 2
Cybernetics, 166

Dedekind, Richard, 37
Degen, Ferdinand, 1
Del Ferro, Scipione, 28
Delian problem, 51
Delos, Oracle at, 51
Demetrius of Phalerum, 48

173

Descartes, René, 5
    and Blaise Pascal, 131
    and Gottfried Wilhelm Leibniz, 97
    and John Wallis, 159
    and Pierre de Fermat, 56
Diderot, Denis, 35
Diophantus, 19, 26, 44
Diopter, invention of, 84
Dirichlet, Peter Gustav Lejeune, 38

Eclipses
    prediction of, 85
École Polytechnique, 60, 67, 72, 92, 96, 110, 113, 114, 138, 153
Economy
    United States, 124
Education
    women, 147
Egyptian Institute, 114
Einstein, Albert, 138
    uncertainty principle, 24
Engineering
    Gaspard Monge, 114
England, 32, 158
Enlightenment
    France, 115
Epicycles, 85
Equations, determinate and indeterminate, 44
Equinoxes, precession of the, 84
Eratosthenes of Cyrene, 14, 85
Euclid, 10, 47
Eudoxus of Cnidus, 15, 50
Euler, Leonhard, 2, 53
Ex Prodigy, 164

Fāṭimid caliph al-Ḥākim, 6
Fermat, Pierre de, 29, 46, 56, 159
Fibonacci. *See* Leonardo of Pisa
Fibonacci sequence, 102
Fiore, Antonio, 28
Fourier, Joseph, 59
Fractals, 110
France, 34, 40, 56, 66, 72, 89, 93, 112
Frederick II, 101
Frederick the Great (king of Prussia)
    Jean le Rond dAlembert, 35
Frege, Gottlob, 62
French Academy, 90
Fresnel, Augustin, 80

Function, 61
    automorphic functions, 138
    Bessels functions, 21

Galen, 7
Galileo, 17
    and René Descartes, 41
Galois, Évariste, 65, 70
Game theory, 122
Gauss, Carl Friedrich, 16, 68, 72, 103
Geocentrism, 12, 85
Geodesy, 70
Geographica, 84
Geography
    Alexandrian, 84
Geometry
    algebraic, 138
    England, 159
    Euclid and, 48
    France, 41, 57
    Greek, 10, 15, 128
    India, 18
    projective, 12, 151
    Pythagoras and, 140
    Russia, 104
    Scotland, 77
Geometry, study of, 82
Germain, Sophie, 71
Germany, 70, 87, 97
Gödel, Kurt, 73, 83
Greece, 19
    mathematics, 19
Gregory, James, 76

Haber, Fritz, 24
Halleys Comet, 20
Hamilton, Sir William Rowan, 78
Hansteen, Christopher, 1
Harriot, Thomas, 159
Heisenberg, Werner, 24
Heliocentrism, 12, 85
Hero of Alexandria, 25
Heyne, Christian Gottlob, 69
Hiero II, 14
Hilbert, David, 81
Hilbert space, 83
Hipparchus (astronomer), 84
Hobbes, Thomas, 159
Holmboe, Bernt Michael, 1
Humboldt, Alexander von, 70

Huygens, Christiaan, 77, 97
Hydrogen bomb
    development of, 121
Hypatia, 12
Hypsicles of Alexandria, 48

Iamblichus, 46
Incompleteness theorem, 74
Indeterminate analysis, 16
India
    astronomy, 18, 25
    geometry, 18
    mathematics, 18, 25
    poetry, 18, 25
Inquisition
    Galileo, 41
    Roman, 29
Inventions
    Carl Friedrich Gauss, 70
Iran
    astronomy, 126
    poetry, 126
Iraq
    astronomy, 8
    optics, 6
Irrational numbers, 38
Isomura Yoshinori, 149
Italy, 4, 28

Jacobi, Carl Gustav, 3, 70
Jacobite Rebellion (1745-1746), 108
Jansen, Cornelius Otto, 132
Jansenism, 131
Japan, 149
Jesuits, 132
Jordanus Nemorarius, 101

Kepler, Johannes, 12, 118
Khwamacrrizmimacr, al-Khwarizmi, al, 28
Klein, Christian Felix, 138
Kovalevskaya, Sofya, 86

Lagrange, Joseph-Louis, 2, 60, 89
Laplace, Pierre-Simon, 60, 93
Lavoisier, Antoine-Laurent, 113
Legendre, Adrien-Marie, 2, 72
Leibniz, Gottfried Wilhelm, 5, 96
    and John Wallis, 159
    and Pierre de Fermat, 56

and Seki Kōwa, 150
Leonardo of Pisa, 101
Levers, 16
Liddell, Alice, 31
Liouville, Joseph, 67
Literature
    England, 32
Lloyd, Humphrey, 80
Lobachevsky, Nikolay Ivanovich, 103
Locke, John, 99
logic, 63
Louis XIV
    and Gottfried Wilhelm Leibniz, 97
Lull, Raymond, 29

Maclaurin, Colin, 106
Mahāvīra, 26
Malebranche, Nicolas de, 97
Malik-Shāh, 126
Mandelbrot, Benoit B., 109
Manhattan Project, 120
Maps
    astronomical, 84
Marcellus, Marcus Claudius (268?-208 B.C.E.), 16
Mathematike syntaxis, 12, 84, 129
Maurice of Nassau, 154
Mechanics, 4, 34, 55, 90
Menelaus of Alexandria, 86
Metric system, 91
Mézières, France (military school), 112
Military
    Netherlands, 154
Monasticism
    Pythagorean brotherhood and, 140
Monge, Gaspard, 60, 112
Moore, G. E., 143
Morgenstern, Oskar, 122
Mōri Shigeyoshi, 149
Museum of Alexandria
    mathematics, 47
Music
    Netherlands, 154
Musical intervals, 141
Muslim, 101

Napier, John, 116
Napoleon I
    Gaspard Monge, 114
    Joseph Fourier, 60
Neoplatonism
    mathematics and, 129

Netherlands
    military, 154
Neumann, John von, 119
Newcomb, Simon, 123
Newton, Isaac, 16
Newton, Sir Isaac, 56, 77, 98, 106, 150, 159
Niẓām al-Mulk, 126
Nihilism, 87
Niẓāmī 'Arūzī, 126
Norway, 1
Numbers
    polygonal, 46
Number theory, 90
Numeral system, Hindu-Arabic, 101

Olbers, Wilhelm, 21
Omar Khayyám, 125
Oppenheimer, J. Robert, 24, 121
Optics
    France, 57
    Scotland, 76
Optics, Iraq, 6

Pappus, 11, 128
Pappus problem, 130
Pascal, Blaise, 29, 56, 97, 131
Pauli, Wolfgang, 24
Pepys, Samuel, 160
Peri ton mechanikon theorematon, 15
Philosophes, 35
Philosophy
    France, 34, 41, 133
    Germany, 97
Phlogiston theory, 113
Photography
    England, 31
Physics
    celestial mechanics, 137
    conical refraction, 80
    Gaspard Monge, 113
    heat diffusion, 60
    Leonhard Euler, 53
    mathematical, 83
    Netherlands, 154
    relativity theory, 138
    theory of heat, 113
Piazzi, Giuseppi, 69
Picard, Émile, 134
Picards theorem, 134
Pi, value of, 16
Planck, Max, 138

Planetary motion, theories of, 52
Plato, 51
Platonism
    Pythagoras and, 140
Poetry
    India, 18, 25
    Iran, 126
Poincaré, Henri, 136
Poisson, Siméon-Denis, 66, 72
Polycrates, 140
Poverty and the poor
    Maria Gaetana Agnesi, 5
Precession of the equinoxes, 84
Pretenders to the throne
    England, 108
Probability, 93
    Cardanos studies of, 29
Proclus, 141
Projective geometry, 12
Protestantism
    Scotland, 117
Ptolemy, 7, 25
Ptolemy I Soter, 47
Punic War, Second (218-202 B.C.E.), 16
Pythagoras, 139
Pythagorean theorem, 48, 141

Quadrature, 15
Quantum theory, 24, 120
Quaternions, 80
Qusṭā ibn Lūqā al-Ba'labakkī, 44

Relativity theory, 138
Religion
    science and, 141
Revolution of 1830, 66
Robespierre, 60
Royal Society, London, 54, 106
Royal Society of England, 98
Russell, Bertrand, 64, 143, 162
Russell Paradox, 64

Schrödinger, Erwin, 24
Science
    France, 41
    religion and, 141
Scotland, 77, 106, 117
Scott, Charlotte Angas, 146
Seki Kowa, 148
Seljuk Turks, 126
Set theory, 75
Spinoza, Baruch, 98

Star catalogs, 84
Steiner, Jakob, 151
Stetigkeit und Irrationale Zahlen, 38
Stevin, Simon, 153
St. Petersburg Academy of Sciences,
    Russia, 54
Strato of Lampsacus, 48
Strindberg, August, 88
Switzerland, 53, 151
Synagoge, 11, 129

Takebe Katahiro, 150
Takebe Katakira, 150
Tartaglia, Niccolò Fontana, 28
Telescopes, 76
Thales of Miletus, 140
Theaetetus, 51

Theorems and proofs, 50
Time, measurement of, 18
Toghrïl Beg, 125
Tokugawa Ienobu, 149
Tokugawa Ietsuna, 149
Tokugawa Tsunashige, 149
Tokugawa Tsunayoshi, 149
Transmigration of souls, 140
Trigonometry, 85
Turing, Alan Mathison, 83, 155
Turing machine, 156
Turing Test, 156

Uncertainty principle, 24

Vitruvius, 14
Voltaire, 35

Wallis, John, 158
War machines, engineering of, 16
Weierstrass, Karl, 87
Whitehead, Alfred North, 143, 160
Wiener, Norbert, 164
"Witch of Agnesi," 5
Women
    mathematics, 4

Yoshida Mitsuyoshi, 149

Zero, concept of, 102